Fifty Fascinating Molecules from A to Z

Fifty Fascinating Molecules from A to Z: Volume 1 follows on from the author's earlier books *Every Molecule Tells a Story* and *Every Molecule Matters* and relates the stories of a wide range of molecules.

They encompass familiar names – medicines including aspirin, cisplatin and morphine; flavourings such as menthol and vanillin; and substances of abuse such as cocaine and methamphetamine. But there are the unexpected, like the poisons that appear in the stories of James Bond and Dr Crippen; the molecules that give chillies and mustard their particular charm; the polymers that became a part of everyday life in the twentieth century – polythene, nylon and terylene; and other everyday molecules, such as hexenal, which plays a role in the smell of a summer's day; dimethyl sulphide, which can transport you to the seaside; and water, a substance that many people take for granted. It is not just about molecules. Chemicals, including the medicines we rely on, were brought into use by scientists. Read about these molecules and their creators here.

- Many important medicinal molecules are featured, including Artemisinin, Aspirin, Atropine, Cisplatin, Codeine, Ivermectin, Mefloquine, Morphine, Paracetamol, Penicillin, Quinine, Streptomycin, Sulfanilamide, Taxol, Tetracycline, and Thalidomide.
- Numerous abused drugs of importance are featured, including Cathinones and Khat, Cocaine, Desomorphine, Fentanyl, Heroin (Morphine), LSD, Methamphetamine, and Oxycodone, and 'Spice'.
- Several important polymers are discussed, including Nylon, Polythene, and Terylene.
- Several molecules important to the environment are included, such as Dichlorodifluoromethane (CF_2Cl_2), Chloromethane, Dimethylsulfide, Octane (and isooctane), Water.
- The text also covers a number of molecules with distinctive smells and/or flavours, such as 2-acetylpyrroline, Capsaicin, Chanel No. 5, Cinnamaldehyde, Limonene, Menthol, and Vanillin.
- The book is complemented by a comprehensive bibliography comprising nearly 1,000 references from the original literature, providing all the background information one could reasonably expect for each one of the molecules detailed throughout the chapters.

Fifty Fascinating Molecules from A to Z

Volume 1

Simon Cotton

CRC Press
Taylor & Francis Group
Boca Raton London New York

CRC Press is an imprint of the
Taylor & Francis Group, an **informa** business

Cover image credit: Shutterstock

First edition published 2027
by CRC Press
2385 NW Executive Center Drive, Suite 320, Boca Raton FL 33431

and by CRC Press

4 Park Square, Milton Park, Abingdon, Oxon, OX14 4RN

CRC Press is an imprint of Taylor & Francis Group, LLC

ISBN: 978-1-041-32603-8 (hbk)
ISBN: 978-1-041-32602-1 (pbk)
ISBN: 978-1-003-78356-5 (ebk)

DOI: 10.1201/9781003783565

Typeset in Times
by Apex CoVantage, LLC

Dedication

This book is offered for the repose of the souls of Mary Nisbet and María de los Ángeles Santiago, two brave ladies who fought the same fell disease with the help of the taxol molecule.

Contents

Preface

I've been writing general articles about molecules for 30-odd years, starting from the time when I realised that using unfamiliar molecules (the ones that weren't 'in the syllabus') could be a useful vehicle for teaching UK A-level (pre-university) chemistry students. Then I started writing 'Molecules of the Month' for Paul May's site at the University of Bristol. So why this book? In the first instance, it is aimed at students on both sides of the school-university divide and their teachers. It is about 50 'lead' molecules and their relatives.

Each chapter is about a molecule – or a group of related chemicals. I have endeavoured to tell a story that non-chemists can follow, while also providing structures, syntheses and reaction details for chemically inclined readers. At the end of the book, there is a comprehensive bibliography for those who want to track down particular scientific papers. I'd add that I do not have online access to a library of chemical journals these days, but I have tracked down most of the papers I cite via the internet.

To give a few examples – Chapter 1 is concerned mainly with a molecule called 2-acetyl-1-pyrroline, which is a key odorant in popcorn, including its structure and how it is formed. Some other molecules that also contribute to the overall popcorn aroma are included. Similarly, Chapter 5 concentrates on aspirin, whilst explaining how it was developed and how it works, as well as telling the story of the so-called superaspirins. Chapter 7 is primarily about capsaicin, the molecule found in chilli peppers, associated with 'Indian' cuisine in the UK. In addition to explaining how chilli peppers, originating in central America, found their way to the rest of the world, this chapter explains the origin of the 'hot' sensation, and goes on to discuss their possible effects on health.

Chapter 15 begins with codeine, a molecule produced by the opium poppy and used as a painkiller for hundreds of years, right up to today. But codeine can be converted into other painkilling molecules. Oxycodone was first made in 1916, but it was not until it was marketed as OxyContin® in 1995 that the stage was set for a painkiller 'crisis' in the USA. And codeine can be used to make desomorphine, which has become another addictive drug, notably in Russia, where it is known as 'Krokodil'.

Chapter 27 begins with muscone, which is produced by the male musk deer. As the names of the molecule and the deer suggest, it has a 'musk' smell. At the end of the nineteenth century, it was a key ingredient in perfumes, but only available in small quantities. This chapter tells the story of how chemists used their ingenuity, as well as some luck, to create other 'musky' molecules. I hope that you get the idea.

The 50 molecules certainly don't claim to be the 50 most important molecules, although a few would definitely appear in many people's selections. In some cases, two or three chapters are linked to one another. Thus, there is a chapter on VX and the 'traditional' nerve agents, and there is another one on the Novichok agents, which have become well-known over the past decade. I thought that readers would not expect to find one without the other.

I've used three chapters to tell the story of anti-malarial medicines – one on the traditional 'natural treatment', quinine; another on the synthetic antimalarials such as mefloquine and chloroquine; and a third on artemisinin, again of natural origin, which forms the basis of many anti-malarial medications used today. Again, the chapters are complementary.

Chemistry does not just happen in test-tubes or in larger reaction vessels. Living matter – which includes us – is made of chemicals. Nature makes millions of different substances, and so do humans. People are intimately involved in the stories of diverse substances, whether the dangerous petrol additive tetraethyllead and the deadly nerve agent A-234 or antibacterial agents including the penicillin and the tetracycline family. This book tells you about some of the people involved.

Simon Cotton

Acknowledgements

I am very grateful to all those people who have discussed chemistry for many years, and to those who have provided copies of papers to which I did not otherwise have access, especially Liam Cox and John Snaith. I would also like to thank Hilary LaFoe for believing in this project and Adam Mlodzinski for his skilled editorial help.

About the Author

Simon Cotton obtained his BSc and PhD from the Department of Chemistry at Imperial College London, followed by research and teaching appointments at Queen Mary College, London, and the University of East Anglia. He subsequently taught chemistry in both state and independent schools for over 30 years. In 2011, he became an honorary senior lecturer in chemistry at the University of Birmingham, where he taught inorganic and organic chemistry for five years. He has published research on the chemistry of iron, cobalt, scandium, yttrium and the lanthanide elements in particular.

His 'Soundbite Molecules' feature ran as a regular column in the magazine *Education in Chemistry* from 1996 to 2012, reaching every secondary school in the UK. He has written over 100 'Molecules of the Month' articles, which are featured online (http://www.chm.bris.ac.uk/motm/motm.htm) and recognised worldwide. Additionally, he has delivered over 40 'Chemistry in Its Element' podcasts for the Royal Society of Chemistry's *Chemistry World* website (http://www.rsc.org/chemistry world/).

In 2005, he shared the Royal Society of Chemistry Schools Education Award, and in 2014, he was awarded the British Empire Medal for his contributions to chemistry and education.

He was the editor of the section on 'Lanthanide and Actinide Compounds' for the *Dictionary of Organometallic Compounds* and the *Dictionary of Inorganic Compounds* (Chapman and Hall) between 1984 and 1997. Moreover, he wrote the account of lanthanide coordination chemistry for the second edition of *Comprehensive Coordination Chemistry,* as well as the accounts of lanthanide inorganic and coordination chemistry for the first and second editions of the *Encyclopedia of Inorganic Chemistry*.

This is his tenth book. His previous books are:

D. J. Cardin, S. A. Cotton, M. Green, and J. A. Labinger, *Organometallic Compounds of the Lanthanides, Actinides and Early Transition Metals*, 1985.

S. A. Cotton, *Building the Late Mediaeval Suffolk Parish Church*, 2019.

S. A. Cotton, *Chemistry of Precious Metals*, 1997.

S. A. Cotton, *Every Molecule Tells a Story*, 2012: second edition 2026.

S. A. Cotton, *Every Molecule Matters*, 2026.

S. A. Cotton, *Lanthanide and Actinide Chemistry*, 2006; second edition 2024.

S. A. Cotton, *Lanthanides and Actinides*, 1991.

S. A. Cotton and F. A. Hart, *The Heavy Transition Elements*, 1975.

P. W. May and S. A. Cotton, *Molecules That Amaze Us*, 2015.

1 2-Acetylpyrroline

Popcorn is perhaps most often associated with mid-twentieth century cinema-going, but it originated some four thousand years ago in areas including Peru and New Mexico. Native people attempted to sell it to Columbus's crew in 1492, while the Aztecs were using it both as food and as a fashion accessory in 1519, as Cortes witnessed. They cooked it in pots over hot fires or sand, or in hot oil. Microwaves came on the scene after World War II. The best source of popcorn is the corn variety *Zea mays*, which gives the best popping. On cooking, the moisture within the kernel turns to steam; expansion of the endosperm takes place, with the expansion of the steam resisted by the pericarp until about 180 °C, when the gelatinised starch expands into bubbles.

Scientists have been identifying odorants in popcorn since 1970, when researchers named 2-acetylpyrazine (1.1) as the molecule with the most popcorn-like aroma of any of the compounds that they identified.

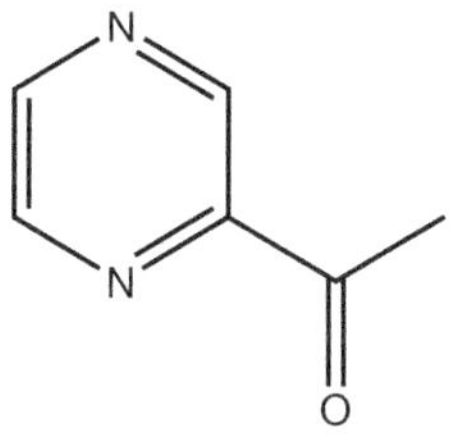

(1.1)

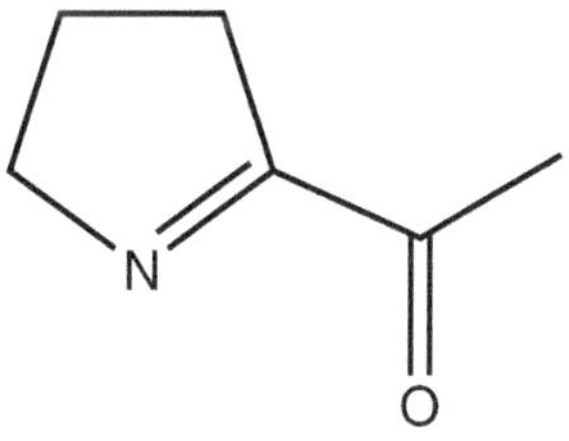

(1.2)

Then in 1991 the German expert Peter Schieberle identified some 23 odorant in popcorn, with 2-acetylpyrroline (1.2) having a 'roasty, popcorn-like' odour. He confirmed the presence of 2-acetylpyrazine and also found that 2-propionylpyrroline (1.3) and 2-acetyltetrahydropyridine (1.4) possessed the smell of popcorn.

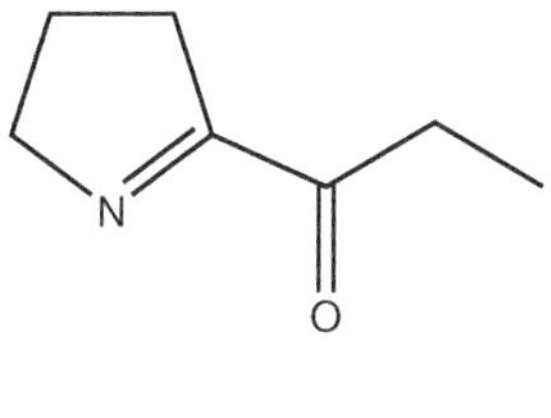

(1.3)

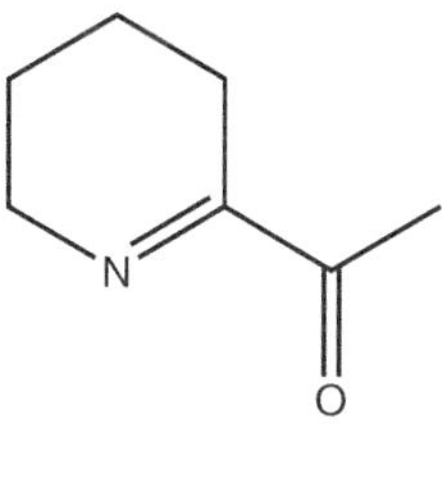

(1.4)

DOI: 10.1201/9781003783565-1

These four compounds do not contribute equally to the smell of popcorn, as they do not appear in equal concentrations and also have different odour thresholds, that is, the concentration at which they can first be detected. 2-Acetylpyrroline and 2-propionylpyrroline are the two compounds that can most easily be detected. However, taking everything into consideration, 2-acetylpyrroline and 2-acetyltetrahydropyridine are the main contributors to the smell of popcorn.

2-Acetylpyrroline had already been identified by two groups as a 'character impact compound' for the roasty, popcorn-like odour note of cooked rice (*Oryza sativa* L.) and wheat bread crust. It also crops up in other foods, such as sweet corn and roasted sesame. It should be noted, however, that a more recent study of popcorn (2006), which identified 51 volatile compounds, did not detect 2-acetylpyrroline in the varieties that were studied.

In its biosynthesis (Figure 1.1), 2-acetylpyrroline originates from the amino acid L-proline. Enzymes convert it into 1-pyrroline, which then reacts with glucose-derived methylglyoxal (2-oxopropanal), forming the 2-acetylpyrroline. Other amino acids, including ornithine, glutamate and putrescine, can also act as precursors.

However, these foodstuffs are only part of the 2-acetylpyrroline story, as the compound is also found in various animals. The bearcat (*Arctictis binturong*) uses 2-acetylpyrroline as a scent marker associated with mating. Its tail and feet get soaked in the urine – and, of course, 2-acetylpyrroline – allowing it to spread its that invitation to mate as it moves around. 2-Acetylpyrroline is similarly used as a scent marker by the Indian leopard, as well as the Bengal and Siberian tigers.

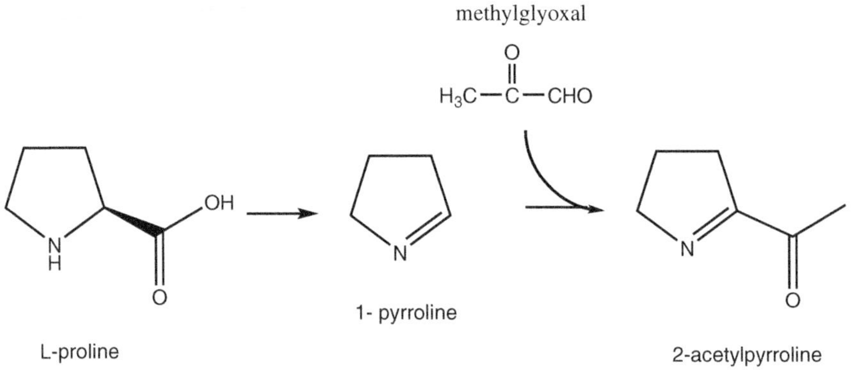

FIGURE 1.1 Synthesis of 2-acetylpyrroline from L-proline.

2 A-234 and the Novichok Agents

After the introduction and testing of VX (pages 214–217), the Americans prepared a considerable number of weapons containing VX, particularly artillery shells, bombs and rockets. However, in 1969, President Richard Nixon's administration renounced the first use of chemical weapons, resulting in the production of nerve agents such as VX being mothballed.

When the treaty banning chemical weapons materialised, the United States signed it in 1993 and ratified it in 1997, as did Russia. It transpired that, during the interval of just over 20 years between Nixon's initiative and the Chemical Weapons Convention, the USSR had developed a new class of nerve agents. This occurred during the 1970s and 1980s under a programme called 'Foliant'. These new agents were meant to be considerably more potent than existing agents such as VX and were known as новцчок (novichok = newcomer).

News of them first leaked out in 1991–1992 through Russian newspaper articles by a Russian scientist Vil Mirzayanov, formerly an employee at Russia's State Scientific Research Institute for Organic Chemistry and Technology. Mirzayanov was subsequently arrested and imprisoned. Upon his release, he moved to the USA, in 2008 publishing the book *State Secrets: An Insider's Chronicle of the Russian Chemical Weapons Program*, which brought these nerve agents to the attention of a wider audience.

The structures of these nerve agents remain controversial, the ones used here are given by Mirzayanov ((2.1)–(2.3)), although one alternative structure is shown in (2.4). Their properties differ significantly. A-230 reportedly has a very low melting point, allowing it to be used in freezing conditions. A-234 has a significantly higher melting point and is also decomposed by water only slowly.

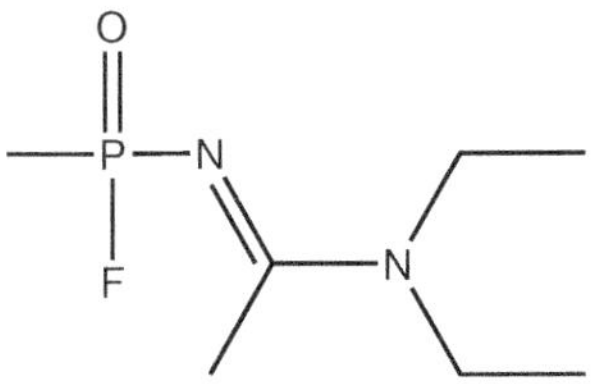

(2.1) A-230

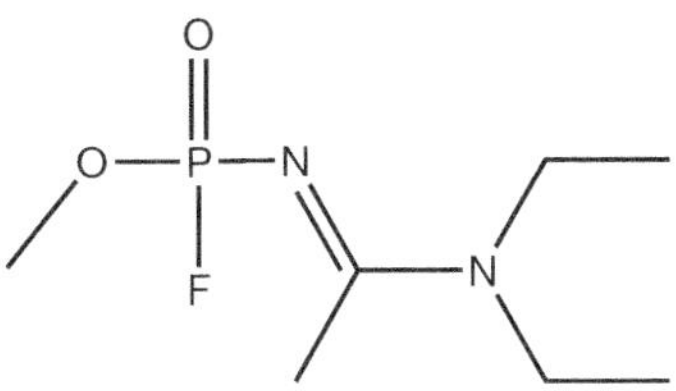

(2.2) A-232

DOI: 10.1201/9781003783565-2

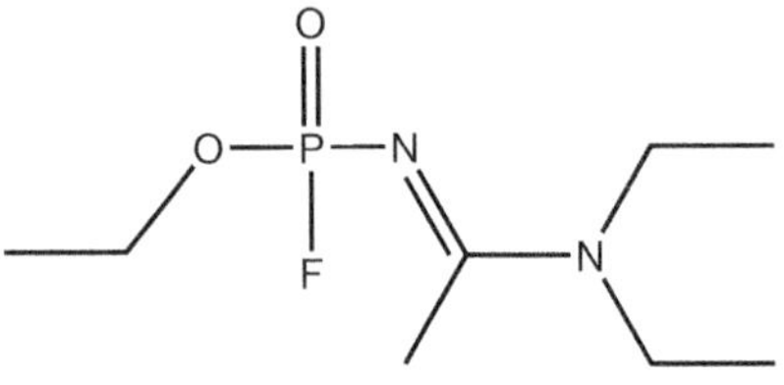

(2.3) A-234

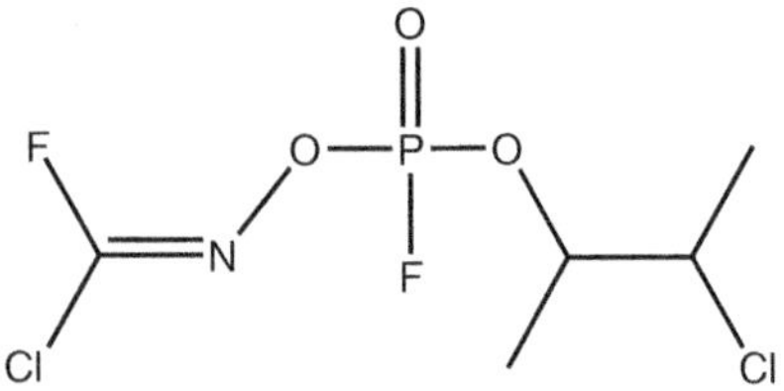

(2.4) 'alternative' A-234

Sergei Skripal (b. 1951) was an officer in the Soviet Union's military intelligence, the GRU, who was recruited as a MI6 agent in 1995. After a decade, the GRU discovered this, and he was imprisoned in 2006 with a 13-year sentence. However, he was released in 2010 as part of a 'spy swap' with the USA and subsequently moved to a house at 47 Christie Miller Road in the city of Salisbury, UK.

On Friday, 2 March 2018, a flight from Moscow landed at Gatwick Airport, near London, from which two Russians disembarked. Their names were given as Alexander Petrov and Ruslan Bushirov, although it subsequently emerged that these were aliases. From Gatwick, they travelled to London, booking into a hostel in Bow, in East London. The following day, Saturday, they travelled by train some 70 miles to the city of Salisbury, where they were captured on CCTV before returning to London. They repeated the journey the following day, Sunday, 4 April 2018 (their departure from Salisbury railway station was again recorded on CCTV). At some point, they are believed to have applied a smear of A-234 to the front door handle of the house at 47 Christie Miller Road. They then returned to London and flew back to Moscow.

Also on Saturday, 3 March 2018, Yulia Skripal arrived in London from Moscow on an Aeroflot flight. On arrival, she was collected by her father and two English friends before they drove to Salisbury.

The next day, Sunday, 4 March 2018, Sergei and Yulia left their house early in the afternoon, arriving at an Italian restaurant in Salisbury at 2.20 p.m. for a lunch that lasted about an hour. On leaving the restaurant, they went to sit on a park bench nearby, where they became ill, sweating heavily, shortly after 4.00 p.m. Yulia then collapsed.

While paramedics were being summoned, a doctor and a nurse – who happened to be passing – examined them and noted their laboured breathing, low pulse rates and low blood pressure. Moreover, the pupils of their eyes had constricted to pinpoints. The medics therefore administered atropine.

Meanwhile, police officers had arrived at the scene, checked the Skripals' pockets for identification, and passed the details to the Police National Computer. This promptly revealed that Sergei was a former Russian spy. An ambulance arrived within minutes and rapidly transported the two patients to the nearby hospital, where treatment was continued. Tests showed very low levels of acetylcholinesterase (AChE), confirming that they were victims of a nerve agent.

Eight days later, Theresa May, the then UK Prime Minister, informed Parliament that the Skripals had been victims of a Novichok attack. Subsequent tests showed that A-234 was responsible and that it was present at the Skripals' home, apparently having been smeared on the door handle. It was also detected on the door handle and steering wheel of Sergei's BMW, which he had used when they went out. Small amounts of A-234 were additionally detected at the restaurant and pub they had visited before falling ill.

There was soon another casualty – a police officer. The first officer to search the Skripals' house on the Sunday, Detective Sergeant Nick Bailey, tried the front door. Finding it locked, he went to the rear of the property and gained entry. After completing his search, he returned to his police station, removed his gloves and typed up a report. He became ill the following evening and was admitted to hospital on Tuesday. Although unwell, he was not in the same life-threatening condition as the Skripals and was soon discharged.

Evidently, when he touched the door handle at 47 Christie Miller Road, some nerve agent was transferred to his gloves. When he later removed the gloves at the police station, a small amount was transferred to his hands. In all three cases, the body's response to the nerve agent demonstrated that absorption through the skin is a much slower route of exposure than inhalation, which typically produces a far more rapid onset of symptoms.

The Skripals recovered slowly, with Yulia improving about a week ahead of her father. Sergei remained on the critical list for a month and was eventually discharged from the hospital on 18 May 2018.

However, there were two more victims, including one fatality. In June 2018, Charlie Burgess, a local resident, discovered a perfume bottle in a rubbish bin and gave it to his partner, Dawn Sturgess. The 'perfume' had no odour, so she tested it on her wrists. They were both hospitalised, with symptoms similar to those of the Skripals, and Dawn Sturgess died on 8 July. Analysis of the contents of the bottle revealed the presence of A–234.

The real identities of the Russians known as 'Alexander Petrov' and 'Ruslan Bushirov' were later revealed to be Anatoli Chepiga and Alexander Mishkin. They were operatives in the GRU, the Russian military intelligence service.

Two years later, an unspecified Novichok agent was used in an attempt to kill Alexei Navalny, a 44-year-old Russian opposition politician, during a domestic Russian flight on 20 August 2020. After he collapsed, the aircraft made an emergency landing, allowing him to be admitted to the toxicology unit of a hospital in Omsk. Two days later, he was transferred by air ambulance to a hospital in Berlin.

He recovered and was discharged from the hospital after a month. Upon returning to Russia, Navalny was imprisoned. It was reported in February 2024 that he had died in prison, and further reports in February 2026 alleged that the 'poison dart' toxin compound epibatidine had been used to kill him.

3 Allyl Isothiocyanate

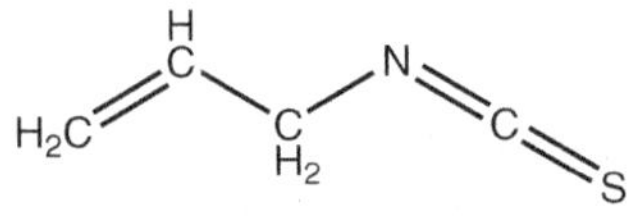

(3.1)

Allyl isothiocyanate (3.1) is a rather strange looking molecule, which plays an important role in mustard. It is not present in mustard seeds in its free form. Rather, it is generated from a substance called sinigrin, a glucosinolate chemically known as 2-propenyl glucosinolate (depicted as its potassium salt in Figure 3.1).

Sinigrin is classed as a plant 'secondary metabolite', a compound produced by a plant that is not essential for its life and growth but serves a secondary purpose. In this case, sinigrin 'stores' allyl isothiocyanate by binding it to a glucose molecule until it is needed to protect the plant from being eaten by herbivores. By itself, sinigrin is harmless. However, when you – or an insect – bite into a mustard seed, cell walls are broken. In the seed, the cell walls separate the sinigrin from an enzyme called myrosinase, which also produced by the plant. Once the cell walls rupture, myrosinase catalyses the decomposition of sinigrin into its constituent components, including allyl isothiocyanate.

Allyl isothiocyanate is in famous company, along with caffeine, cocaine and nicotine, which are other defence molecules made by plants. As far as we are concerned, it is an oily substance with a rather pungent smell. It stops herbivores such as insects from eating the mustard plant, by irritating nerve fibres within the insect. In humans, allyl isothiocyanate activates a receptor, the TRPA1 channel.

Our bodies contain certain nerve cells, known as neurons, which are sensitive to various stimuli, including temperature, and which send messages to the brain. It is the Transient Receptor Potential (TRP) cation channels in cell membranes that allow the flow of ions across the membrane, thereby transmitting these messages. The best-known TRP channel is the TRPV1 channel, which is activated by temperatures above 42 °C and by capsaicin (p. 23), the molecule that causes the 'heat' associated with chilli peppers. The TRPM8 channel is activated by cold temperatures and menthol (pp. 100–101). The TRPA1 channel, which is also sensitive to 'cool' temperatures, is another member of this family.

It is believed that allyl isothiocyanate works by forming covalent bonds with certain cysteine residues in the ion channel. One suggestion is shown here (Figure 3.2).

Allyl isothiocyanate is believed to affect the TRPV1 channel too. Allyl isothiocyanate is also found in horseradish and wasabi, the so-called Japanese horseradish. Other isothiocyanates are found in plants of the *Brassica* family. Sinalbin (3.2) is a glucosinolate found in white mustard, which acts as a source of 4-hydroxybenzyl isothiocyanate (3.3).

DOI: 10.1201/9781003783565-3

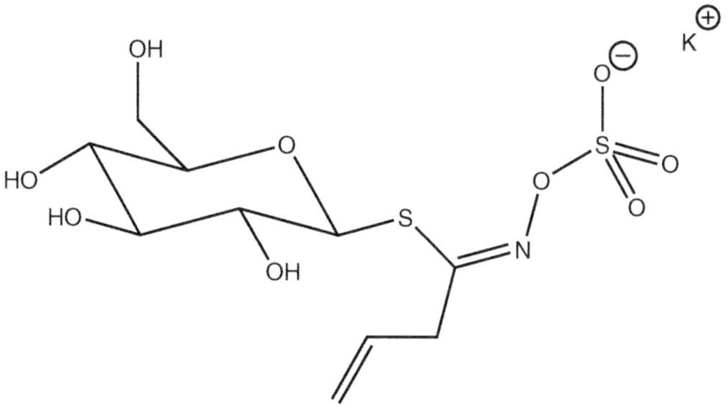

FIGURE 3.1 Structure of sinigrin.

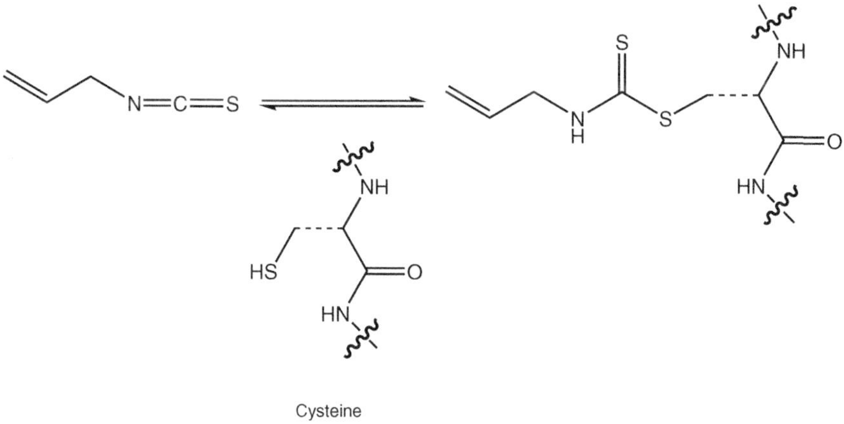

FIGURE 3.2 Allyl thiocyanate binds to cysteine residues.

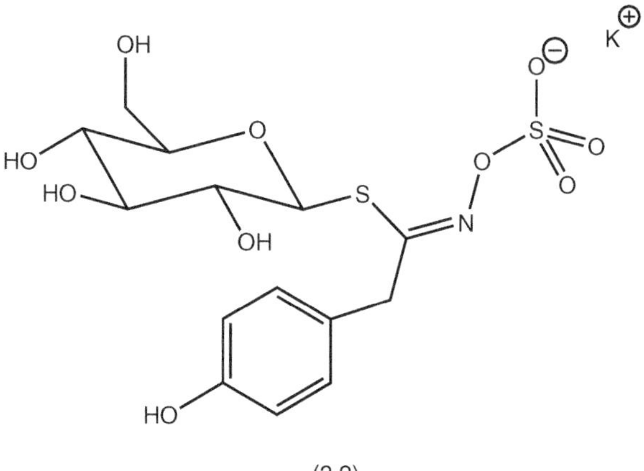

(3.2)

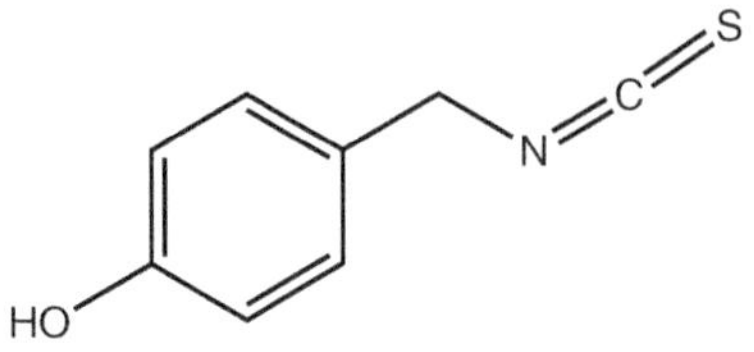

(3.3)

Benzyl isothiocyanate (3.4) is found in mustard plant Alliaria petiolate, native to the Eastern Hemisphere.

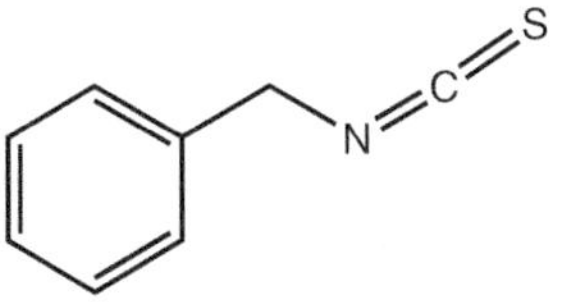

(3.4)

Mustard has been enjoyed for many centuries; it was used by the Romans and Greeks two thousand years ago. The Romans brought it to Gaul, now France, where it was widely produced in the Middle Ages, not least because it was cheaper than pepper. The word *moutarde* was introduced into the French language in the thirteenth century, by which time mustard production was established at Dijon. In Norwich, UK, Jeremiah Colman marketed Colman's Mustard from 1814.

It can be made in the laboratory, for example, by a (pseudo)halide-exchange reaction.

$$CH_2{=}CHCH_2Cl + KSCN \rightarrow CH_2{=}CHCH_2NCS + KCl$$

Alternatively, sulphur can be added to an isocyanide, using traces of a selenium catalyst.

$$CH_2{=}CHNC + S \rightarrow CH_2{=}CHCH_2NCS$$

So that is allyl isothiocyanate, a molecule which definitely 'cuts the mustard'.

4 Artemisinin

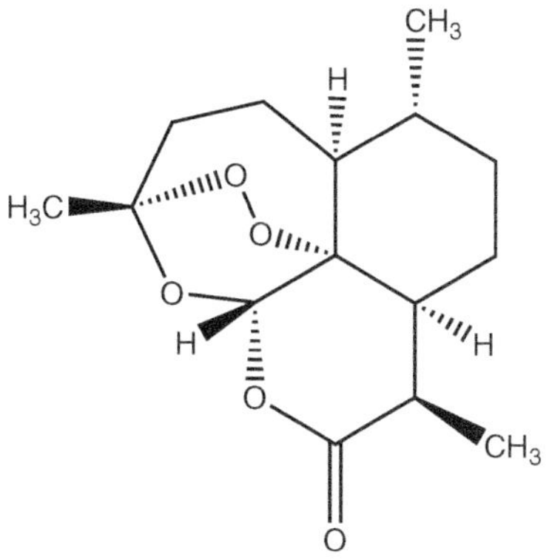

(4.1)

Nature is no respecter of nationality, so that during the Vietnamese war in the mid-1960s, malaria afflicted both protagonists: the North Vietnamese/Vietcong and the American forces. Whilst the Americans turned to synthetic antimalarials, notably mefloquine (pp. 95–96), the Vietnamese leader Ho Chi Minh asked Zhou en Lai, the Premier of the People's Republic of China, for assistance. Ho found a sympathetic ear because, so the story goes, Zhou en Lai himself had been afflicted by malaria some 40 years earlier.

In the short term, the Chinese supplied Vietnam with conventional antimalarial medications, such as pyrimethamine-dapsone and pyrimethamine-sulfadoxine combinations. In addition, the Chinese leadership established a secret military project, known as 'Project 523', named after the date on which the programme was launched: the 23rd day of the fifth month of 1967. Zhou en Lai directly supervised the project.

It should be pointed out that this was not the most propitious time for such a project, as the Cultural Revolution, launched in August 1966, was dominating Chinese life and was often inimical to research. Professor Youyou Tu (屠呦呦) was designated the leader of the '523 Research Team' in 1969 (and in 2015 she shared the Nobel Prize in Physiology or Medicine 'for her discoveries concerning a novel therapy against malaria').

At an early stage of the programme, researchers investigated more than 2,000 Chinese herbal preparations, selecting several hundred with possible antimalarial activity and carrying out tests in malaria-infected mice. An Artemisia annua L. extract was found to inhibit the growth of the malaria parasite. Close examination of the literature revealed instructions on the use of qinghao (the Chinese name for Artemisia annua L., 植物黄花蒿) to alleviate malaria symptoms in Ge Hong's *A Handbook of Prescriptions for Emergencies,* dating from around 317–420 AD: 'Another recipe:

DOI: 10.1201/9781003783565-4

qinghao, one bunch; take two sheng [2 x 0.2l] of water to soak it, wring it out, take the juice, and ingest it in its entirety'.

Apparently, peasants in rural areas had made a tea from Artemisia annua (using hot – but not boiling – water) or simply chewed fresh plants mixed with brown sugar as a means of treating fevers. The team carried out low-temperature extraction, using cold ether instead of water. They tested the compound they had extracted, first on themselves and then in clinical trials in Hainan Province on patients infected with both *Plasmodium vivax* and *Plasmodium falciparum*. This produced encouraging results, with rapid relief of symptoms (fever and malarial parasites in blood), in contrast to chloroquine, the standard treatment then. The compound responsible for this activity was identified in 1972 as colourless crystals melting at 156–157 °C, with the molecular formula [$C_{15}H_{22}O_5$]; it was given the name qinghaosu, or artemisinin.

The molecular structure of artemisinin was determined by Li Liang and Pengfei Li in 1975 (although this was not generally known in the West until a conference was held in 1981), revealing the molecule to be a hitherto unprecedented sesquiterpene lactone containing a bridging peroxo group (4.1). An important stage in the development of artemisinin was the study of its reactions, leading to a number of semisynthetic derivatives. Its reaction with sodium borohydride demonstrated the presence of a carbonyl group through its reduction to dihydroartemisinin (4.2), which contains a hydroxyl group (a secondary alcohol).

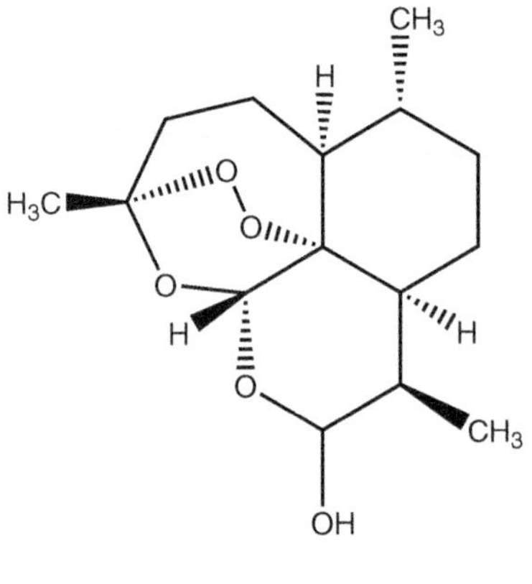

(4.2)

Unlike artemisinin, dihydroartemisinin is soluble in water and is a much more potent antimalarial agent than artemisinin itself; artemisinin therefore behaves as a prodrug. The presence of the hydroxyl group enables the formation of other derivatives using standard organic reactions (Figure 4.1), yielding compounds that are more useful clinically. Thus, if dihydroartemisinin is reacted with an alcohol and an acid catalyst (which may be a Lewis acid such as BF_3), the reaction forms ethers such as artemether and arteether, which are lipophilic and more rapidly absorbed. Reaction of dihydroartemisinin with succinic anhydride and the base trimethylamine

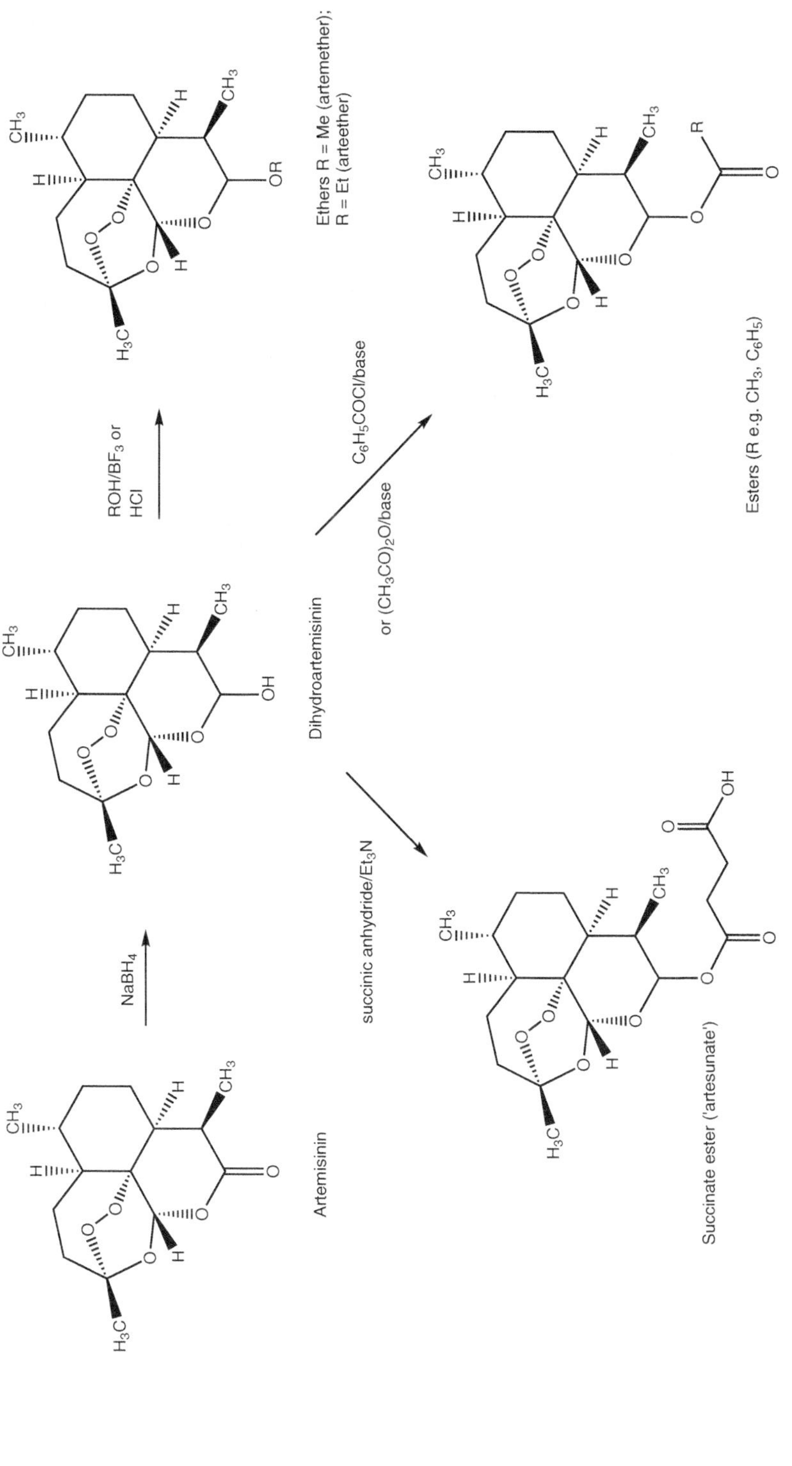

FIGURE 4.1 Syntheses of artemisinin derivatives.

generates artesunate (the monoester with succinic acid), which is the only artemisinin derivative that is sufficiently water-soluble to be administered intravenously and act rapidly (in vivo, it is hydrolysed to dihydroartemisinin, which rapidly clears the malaria infection). Other esters can be prepared using conventional acid derivatives such as ethanoic anhydride or benzoyl chloride.

More drastic reduction of artemisinin leads to the formation of deoxyartemisinin (4.3), which no longer contains the peroxide group, but instead contains only a single bridging oxygen atom. This molecule is totally ineffective against the malaria parasite, showing that the peroxide bridge is essential for activity.

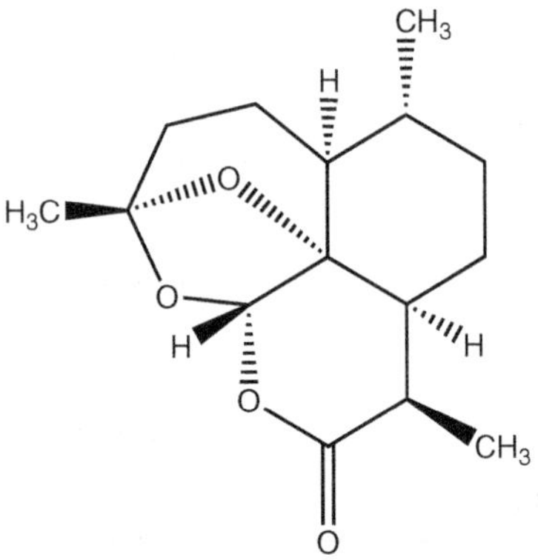

(4.3)

How artemisinin kills the malaria parasite is only partly understood. It is known that the bridging peroxide group is essential to its activity, and the presence of haem (iron bound to a porphyrin ring) is also required. Artemisinin kills the malaria parasite by binding proteins in the parasite. One possibility is that the Fe^{2+} ions in the haem group split the peroxide bridge, generating oxy-radicals. These can rearrange to form primary or secondary carbon centred radicals, which in turn attack the proteins, destroying the life cycle of the parasite. A factor in this is that artemisinin inhibits the *Plasmodium falciparium* ATP6 (PfATP6) enzyme, which is responsible for calcium transport.

Artemisinin derivatives have been a major component of the artemisinin combination therapy (ACT), favoured by the World Health Organisation (WHO) since 2005. Current chemotherapy uses artemisinin or an artemisinin derivative in combination with other antimalarial drugs, such as mefloquine, chloroquine, or prophylactic agents (for example, atovaquone and proguanil, marketed as Malarone), working on the principle that if one drug does not kill the parasite, the other will. Of course, other vector-control strategies, including insect repellents, insecticide-treated mosquito nets, indoor spraying and the elimination of stagnant water, are also an important part of the overall strategy.

Artemisinin derivatives also have potential as anticancer agents.

5 Aspirin

Molecules from willow have been used medicinally for over 3,000 years, the oldest record being a mention in the Ebers Papyrus (c. 1530 BC) of the use of willow to treat pain. Hippocrates (440–377 BC) said that the ancient Greeks used willow medicinally, and the Roman writer Pliny the Elder (who died in AD 79 following the eruption of Vesuvius) is also said to have recorded this. However, the first clear record of its use comes from eighteenth-century England.

In 1763, the Reverend Edward Stone of Chipping Norton (1702–1768) published an article entitled 'An Account of the Success of the Bark of the Willow in the Cure of Agues' in the Philosophical Transactions of the Royal Society. Chipping Norton had not yet attained its present status as a fashionable place to live, nor was it surprising that a clergyman of the Church of England should publish such an article. Clergymen were expected to be polymaths, and Stone's published interests extended to astronomy and algebra.

Health and safety standards were virtually non-existent back then. Suffering from the 'ague' (a fever, possibly malaria), Stone chewed a piece of willow bark, noting its bitterness. He linked it with the bitterness of 'Jesuit's bark', the cure for malaria, which we now know was bitter because of the quinine it contained. He speculated that willow bark, too, might be a cure for agues. Stone experimented with powdered willow bark on around 50 people, finding that it relieved ague fever from malaria and distempers, achieving a high degree of success.

Part of Stone's thinking stemmed from the so-called doctrine of signatories. As willow grew in damp places and, because agues and fevers were also associated with damp places, it was argued that willow might produce a medicine to treat them.

In the early nineteenth century, it was found that the glycoside salicin (5.1), extracted from willow bark, relieved pain. Salicin itself was first isolated in 1828 by Johann Andreas Buchner (not to be confused with Ernst Wilhelm Büchner, whose name is associated with Büchner flasks and funnels). Salicin and other willow-derived compounds take their name from the Latin word for willow tree, *salix*.

When salicin is ingested, it is hydrolysed in the stomach to produce glucose and salicyl alcohol (5.2). The salicyl alcohol is subsequently oxidised to form salicylic acid (5.3), a carboxylic acid whose systematic name is 2-hydroxybenzoic acid. Salicylic acid was first obtained from salicin in 1838 by the Italian chemist Raffaele Pirìa. Along with salicin, salicylic acid came into use as a treatment for fevers and pain, although gastrointestinal problems were an undesirable side effect. Another Italian chemist, Cesare Bertagnini, noted in 1855 that high doses of salicylate caused tinnitus.

DOI: 10.1201/9781003783565-5

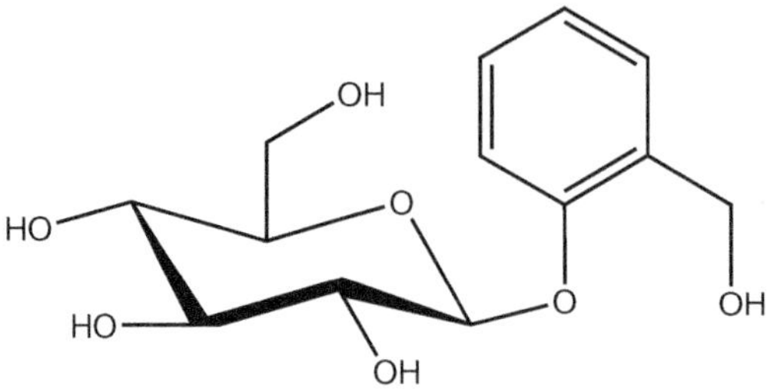

(5.1)

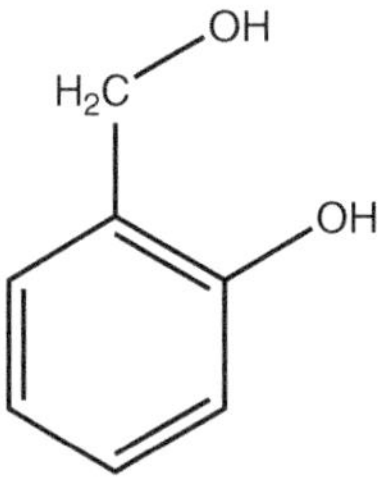

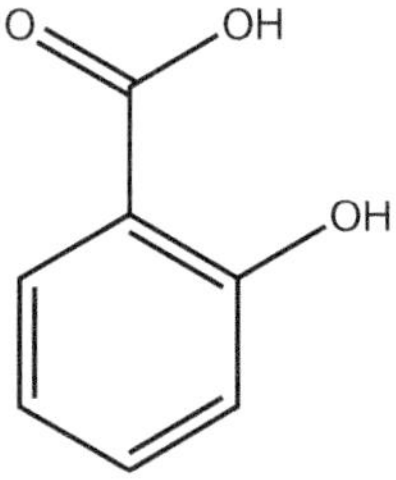

(5.2) (5.3)

At this point, acetylsalicylic acid enters the scene. It was the French chemist Charles Frédéric Gerhardt (1816–1856), a university professor at Montpellier, who, in 1853, reacted acetyl chloride with sodium salicylate and obtained impure acetylsalicylic acid (5.4). However, that was the limit of his contribution to the aspirin story; he did not take his discovery any further, possibly on account of his premature death.

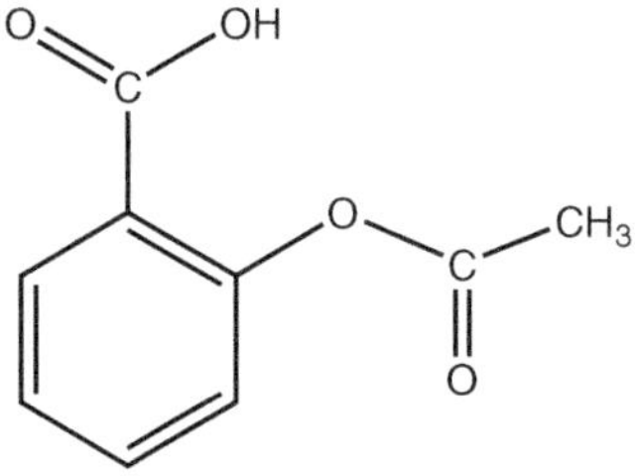

(5.4)

The story then shifts to Germany and the laboratories of the Bayer company in the town of Barmen (today part of Wuppertal), where, in 1897, Felix Hoffmann (1868–1946) reacted acetic anhydride with salicylic acid, obtaining acetylsalicylic acid. It is said that Hoffmann's father suffered from rheumatic pains and urged his son to produce a painkiller that was more palatable than salicylic acid or sodium salicylate.

This version of the story has been disputed by Hoffmann's former colleague Arthur Eichengrün, who claimed in 1949 that the synthesis of acetylsalicylic acid had been carried out under his supervision (a claim subsequently supported by an analysis of archival and published material by Walter Sneader, published in 2000). In addition, Heinrich Dreser, the head of the department in which Hoffman and Eichengrün worked, drew up an account of the development of the drug that omitted any mention of his two staff members. Dreser, moreover, became very rich through a deal concerning all drugs tested in his laboratory.

The manufacturing process was patented in the UK in 1898 and in the USA two years later. Bayer launched this new medicine in July 1899. The name 'Aspirin' reflected the meadowsweet plant *Spiraea ulmaria*, from which salicin, and thus salicylic acid (first known as *spiroylige Säure* and later as *Spirsäure*) were obtained. The name was prefixed by an 'A' in reference to its formation by acetylation and suffixed by 'in' to provide a characteristically medicinal-sounding ending.

Aspirin became a very popular medicine as an analgesic (painkiller), anti-inflammatory, and antipyretic (temperature-reducing agent). It was inexpensive, as Kolbe's synthesis (1856) of salicylic acid from sodium phenoxide and carbon dioxide ensured a ready supply of salicylic acid. Care has to be exercised when treating children, as aspirin can be fatal in those suffering from Reye's syndrome. Aspirin was widely used, often in very high doses, by patients during the 1918–1919 influenza pandemic, and may have contributed to the immense death total, believed to be in excess of 50 million worldwide.

It was another half century after the influenza pandemic before scientists started to understand how aspirin worked. By this time other painkillers had come into use, notably paracetamol and ibuprofen, which did not have the side effects of stomach irritation associated with aspirin, and whose availability contributed to a decline in aspirin's popularity.

The research and insights of John Vane (later Sir John) and his research group earned Vane a share in the Nobel Prize in Physiology or Medicine in 1982 by demonstrating that aspirin inhibited the production of prostaglandins. These are 'signalling molecules', hormones that the body's cells produce in response to trauma, such as infection. They cause inflammation because they promote import of white blood cells, which fight infection.

One of these molecules, prostaglandin E_2 (PGE_2), has an effect on temperature regulation in the hypothalamus, causing the body's temperature to rise, resulting in fever. What aspirin does is irreversibly inhibit two cyclo-oxygenase enzymes, COX-1 and COX-2. The mechanism for this involves the transfer of an acetyl group to a serine residue in the enzyme, which prevents the arachadonic acid substrate, used in prostaglandin synthesis, from accessing the enzyme's active site, where the prostaglandins are produced (Figure 5.1).

COX-1 has other roles. It synthesises prostaglandin PGI_2 (prostacyclin), which protects the stomach lining against irritation because it reduces the amount of gastric acid produced; it is also involved in the production of a protective mucus. Thus, inhibiting COX-1 has the effect of irritating the stomach and may damage the stomach lining, leading to peptic ulcers, with the acidity of acetylsalicylic acid also contributing to its effect. COX-1 stimulates platelet production, leading to blood clotting. Consequently, by inhibiting COX-1 activity in platelets, aspirin reduces the formation of thromboxane A2, which induces platelet aggregation and clotting. Hence, low

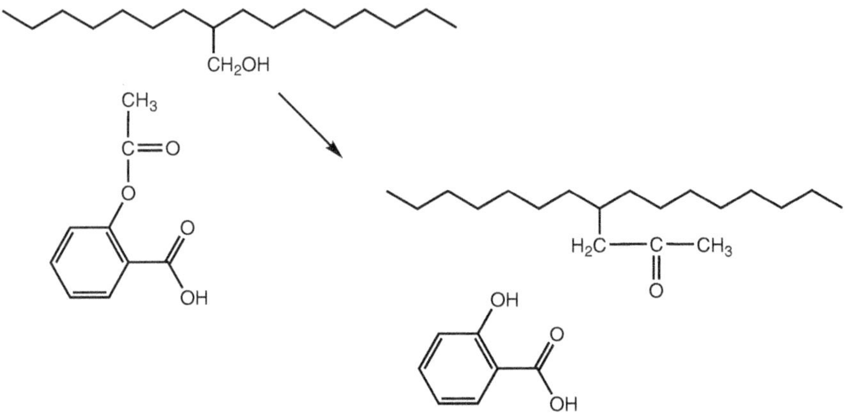

FIGURE 5.1 Acetylsalicylic acid transfers an acetyl group to a serine residue in the COX-1 enzyme, blocking prostaglandin synthesis.

doses of aspirin are prescribed as a cardioprotective agent, helping to prevent heart attacks and strokes.

COX-2 is inhibited by acetylsalicylic acid through a mechanism similar to that of COX-1. However, since the substrate access channel of COX-2 is both larger and more flexible than that of COX-1, acetylsalicylic acid has a lesser inhibitory effect on it. COX-2 is involved in the production of another prostaglandin, PGH_2 ; this is converted by other enzymes to several different prostaglandins that mediate pain, fever and inflammation, making inhibition of this enzyme therapeutically desirable. Thus, towards the end of the twentieth century, scientists sought to develop a 'superaspirin', which would selectively inhibit COX-2 but not COX-1. This would counteract inflammation and pain while, because it did not affect COX-1, it allowed the prostaglandins produced by that enzyme to continue protecting the stomach lining. The best-known of these drugs were celecoxib (Celebrex; (5.5) and rofecoxib (Vioxx; (5.6).

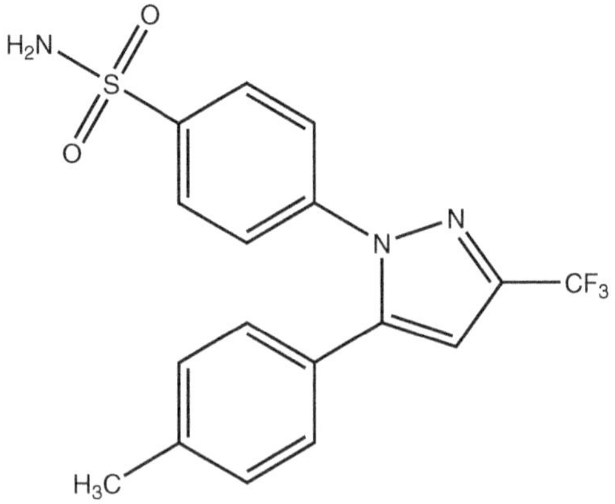

(5.5)

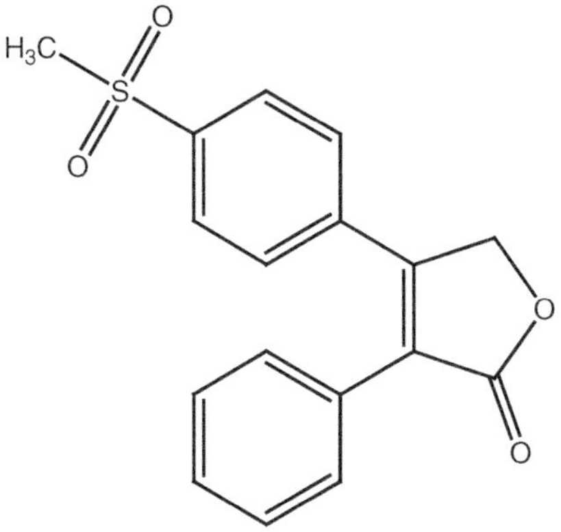

(5.6)

Rofecoxib (Vioxx) was approved by the FDA in May 1999 and quickly became a very popular drug for the treatment of arthritic pain. By 2003–2004, this drug generated annual sales of \$2.5 billion for Merck, its manufacturer. However, Vioxx was removed from the market the following year after it was found that people taking the drug over long periods were at an increased risk of heart attack.

Celecoxib (Celebrex) is currently used for pain and inflammation relief in patients suffering from osteoarthritis, rheumatoid arthritis, and various other painful conditions. It is believed that members of the coxib family of drugs may carry an increased risk of significant cardiovascular problems, though this is not believed to be the case for celecoxib.

In a different area, it has for many years been suggested that aspirin may play a role in the prevention of carcinogenesis, particularly colorectal cancer, an area that remains under investigation.

6 Atropine

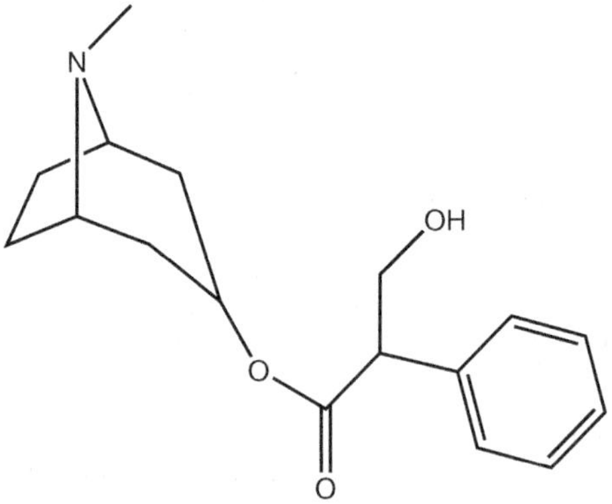

(6.1)

Atropine's structure features a heterocyclic ring that includes a nitrogen atom, making it basic and therefore qualifying it as an alkaloid. Atropine (6.1) is found in members of the nightshade family of plants (*Solanaceae*). The plants synthesise it from the amino acid L-ornithine via putrescine, $H_2N(CH_2)_4NH_2$. Its name reaches back to Greek mythology, for Atropos was one of the Three Fates – she cut a mortal's thread of life – reflecting atropine's use as the active agent in poisons such as deadly nightshade, dating back to mediaeval times and beyond. Atropine was first isolated in 1831 from the root of nightshade by a German apothecary named Mein.

It is not merely a poison. More than two thousand years ago, the Egyptian ruler Cleopatra used plant extracts to supply the atropine which dilated her pupils and added to her allure. During Renaissance, women in Venice and elsewhere followed the example of the Ptolemaic queen to make themselves more attractive, becoming beautiful women (Italian: bella donna). Thus, in naming the deadly nightshade shrub as *Atropa belladonna*, Linnaeus combined classical mythology with a more modern nod to Italy.

Today atropine has applications in medicine and optometry. In the latter, it is used in the treatment of myopia, as well as to relax the eye muscles and dilate the pupils. As a medicine, atropine can increase a low heart rate. Its lethal dose is not known but children have been killed by doses as low as 10 mg.

However, atropine tends to make headlines primarily as a poison. John Emsley has reminded us that in 44 AD, Livia Drusilla murdered her husband, the Roman Emperor Augustus, by adding atropine to one of his favourite figs. Two millennia later, the serial poisoner Graham Young experimented with atropine before turning

DOI: 10.1201/9781003783565-6

to thallium. In 1994, shortly after the crimes of Graham Young, an Edinburgh case once again brought atropine into the news.

Paul Agutter was a 49-year-old lecturer in Biology at Napier University, married to Alexandra, a lecturer in English at the same university. Agutter was successful in his job, but he was also having an affair with a younger woman, a mature student at the same university. Evidently, Agutter wanted to terminate his marriage without ending up in the divorce courts.

On the evening of Saturday, 28 August 1994, Agutter poured his wife a large gin and tonic, which contained enough atropine to kill her (he had obtained the atropine from a laboratory at the university). He reckoned that the quinine would conceal the bitterness of the atropine. Alexandra found the drink too bitter for her taste, and only drank part of it – not enough to kill her, but sufficient to cause dizziness, after which she collapsed. Her husband temporised by calling the family doctor and reaching the answering machine. Unfortunately for him, a locum was on duty, picked up the message, called for an ambulance, and headed for the Agutters' home. It was realised that Alexandra was seriously ill, possibly caused by something she had drunk. Hence, the ambulance man put the remains of the gin and tonic in a jar and took it away for analysis, while also seizing all the drink bottles in the house.

The seriously ill Mrs. Agutter spent some time in hospital but was eventually discharged, fully restored to health. She had probably consumed around 50 mg of atropine. By this time, her husband had adopted a cunning plan. He had spiked a dozen bottles of Indian tonic water on the shelves of a Safeway supermarket in Edinburgh with varying amounts of atropine, and purchased a couple for his own purposes. He then added extra atropine to Alexandra's gin and tonic. CCTV caught Paul Agutter visiting that supermarket, and a shelf-stacker remembered seeing him putting bottles back onto the shelves. Agutter had reasoned that other people would buy the contaminated bottles and become victims of atropine poisoning, allowing his wife's poisoning to be blamed on some unknown criminal. Well, several other people did indeed suffer atropine poisoning from these bottles. Among them were Elizabeth Sharwood-Smith and her 18-year-old son, Andrew Sharwood-Smith, who had similar symptoms to Alexandra Agutter and were likewise rushed to hospital.

As it happened, their husband and father, respectively, Geoffrey Sharwood-Smith, being an anaesthetist at the Royal Infirmary of Edinburgh, recognised the symptoms and informed the doctors at the hospital that he suspected atropine poisoning. Other victims of atropine poisoning were admitted to the same infirmary. All these victims, none of whom had received a fatal dose, were linked to the tonic water purchased from the same supermarket. Bottles from the Safeway's shelves were tested, and six were found to contain atropine. A national recall of the Safeway tonic water follow, and 50,000 bottles were destroyed as a precaution.

Crucially, forensic analysis of the remains of Alexandra's gin and tonic revealed that it contained a much higher concentration of atropine than in any of the spiked bottles, including those in the Agutter household. It was estimated that one-third of her gin and tonic contained a lethal dose. Traces of atropine were also found on a cassette case in Agutter's car. All these evidences proved crucial in the prosecution of Paul Agutter, who was tried for and found guilty of attempted murder in February 1995. He was sentenced to 12 years' imprisonment, of which he served seven. In 1998, Alexandra Agutter obtained a divorce. More recently, in 2022, a man was fatally poisoned by a 'friend' who emptied the contents of two bottles of atropine-based eye drops into his beer.

As a counterpoint, atropine has a legitimate use as part of an antidote, administered alongside an oxime such as pralidoxime, against some virulent poisons in the form of nerve agents, notably sarin (pp. 213–214) and VX (pp. 214–216). It counteracts the action of the neurotransmitter acetylcholine (ACh) by blocking ACh receptors, thereby preventing the signal from being transmitted. The oxime displaces the nerve agent from acetylcholine esterase (AChE), allowing the enzyme to resume its normal function of breaking down the acetylcholine after it has fulfilled its role in transmitting the signal.

7 Capsaicin

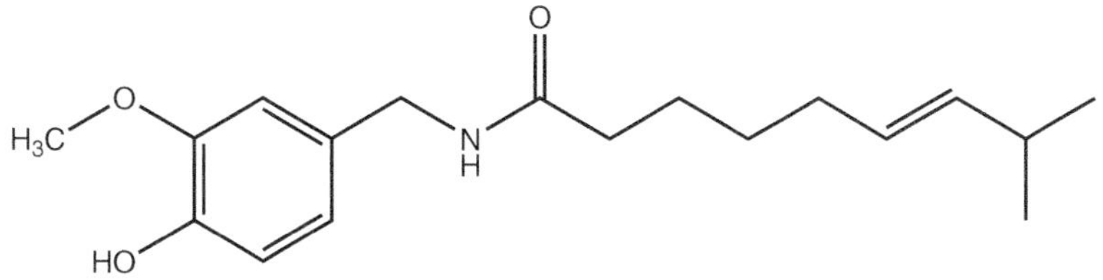

(7.1)

Capsaicin is a molecule that is often taken for granted in the UK at present. However, a little more than half a century ago, very few people in the UK ate curries on a regular basis. It is generally associated with the Indian cuisine, but many would be surprised to find that it does not originate in the Indian subcontinent.

Most spices have an Asian origin, with chillies being the principal exception. Two thousand years ago, spices were widely traded – to China in one direction and to Arabia and the Western world in another. Cardamom and pepper came from India; turmeric and ginger from Southeast Asia; cinnamon from Sri Lanka; and nutmeg (from the Banda Islands) and cloves (from the Moluccas) from small island groups.

In the Americas, Christopher Columbus and his followers encountered a pungent plant, which reminded them of the familiar black pepper (*Piper nigrum*), except that it had red pods. These became famous as red pepper, which are now classified as *Capsicum* (from the Greek *kapto*, meaning 'to bite').

The inhabitants of countries such as Mexico had domesticated chillies some eight millennia earlier, but it was kept to the Americas until European settlers took them to Europe, from where they spread to India and other parts of the world. Europeans did not take to some food imports from the New World, but chillies were an exception. In 1513, a Spanish *caballero* noted that chillies were already in regular use in Spanish and Italian cuisine. The cultivation of chillies had reached England before the end of the sixteenth century, though the cold climate was not conducive to their growth. In the twenty-first century, around three-quarters of the world's population uses chilli regularly in its diet.

Most of the heat from a chilli comes from the molecule capsaicin (7.1), which belongs to the capsaicinoid family of molecules. Chili plants use an enzyme *Capsaicin synthase* to make capsaicin from vanillylamine and 8-methyl-6-nonenoic acid (Figure 7.1), linking these two molecules in an amide group (from a carboxylic acid and a primary amine).

Capsaicin (7.1) makes up just over 0.1% of the mass of a chilli and accounts for around 70% of its capsaicinoid content, with the structurally similar molecule dihydrocapsaicin (7.2) contributing a further 20%.

DOI: 10.1201/9781003783565-7

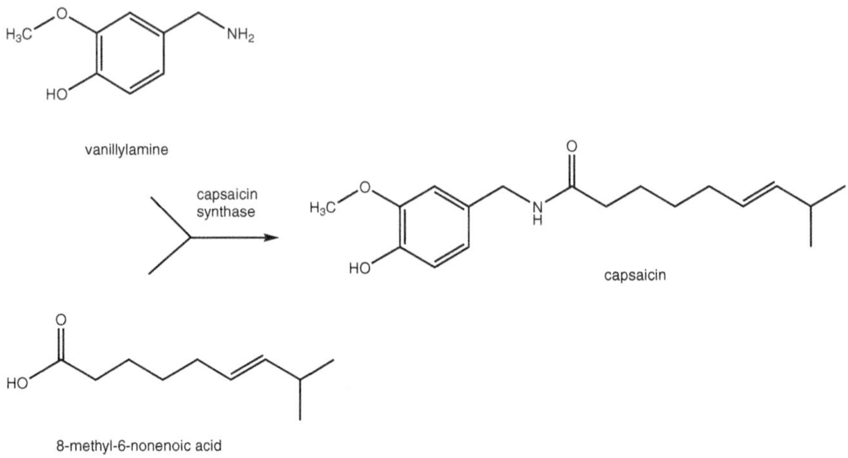

FIGURE 7.1 Biosynthesis of capsaicin from vanillylamine and 8-methyl-6-nonenoic acid.

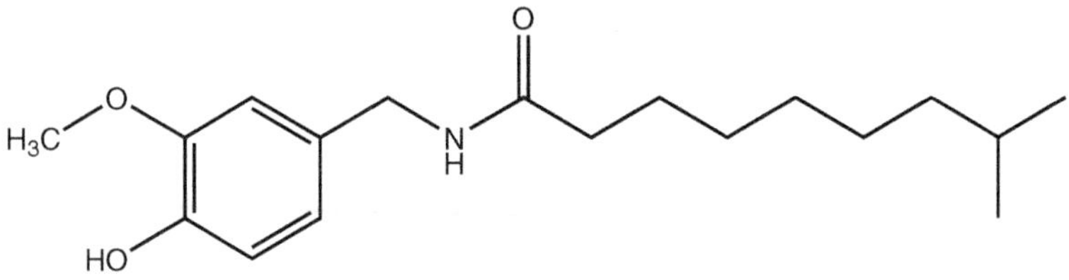

(7.2)

Nordihydrocapsaicin (7.3) and homocapsaicin (7.4) are other capsaicinoids.

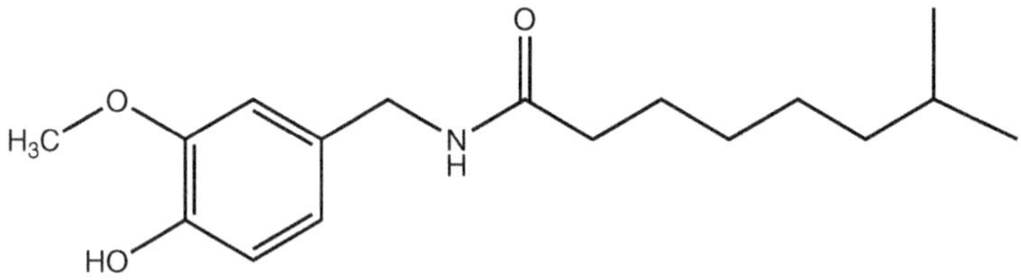

(7.3)

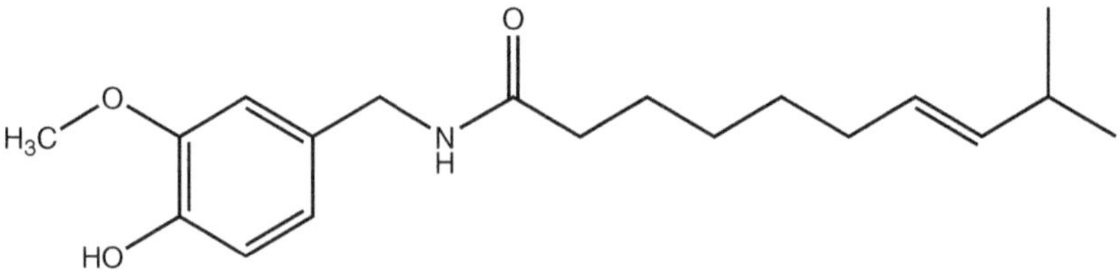

(7.4)

Chillies have long been associated with 'hot' food, notably curries and Mexican chilli con carne, but the mechanism responsible for this sensation was not discovered until in 1997. Capsaicin affects the bodily receptor, TRPV1 (*T*ransient *R*eceptor *P*otential cation channel, *V*anilloid subfamily member *1*), and its activation produces a burning sensation. These receptors are found in the mouth, lips, throat and tongue, as well as other parts of the human body. This helps to explain why people chopping chillis are advised to wash their hands thoroughly before using the toilet. The capsaicin molecule is largely hydrophobic and not water-soluble. Consequently, when a curry is too much for a customer in an Indian restaurant, drinking water is unlikely to help, as it will not dissolve the capsaicin from their tongue and other affected areas. Milk is generally recommended, as the lipophilic casein it contains is better at dissolving the lipid-like capsacinoids.

The capsaicin receptor is a protein cation channel, which contains six transmembrane helices. It is found in the outer membrane of some nerve cells. The active receptor comprises four of these subunits. The binding of capsaicin to the receptor (which appears to involve both hydrogen bonds and van der Waals interactions) causes the channel to open and allow Ca^{2+} ions to enter the cells, resulting in the transmission of a signal to the pain-processing centres in the brain. It is not just capsaicin and other capsaicinoids that activate this receptor; temperatures above about 42 °C ('noxious heat') have the same effect. The brain receives the same signal, whether the receptor is 'switched on' by the capsacinoid molecule or by heat. As a result, the brain interprets the molecule from the chilli to be 'hot'. A number of other spices contain molecules that also activate (act as 'agonists' to) the same receptor. [Another receptor, TRPM8, is activated by the 'cooling' molecule menthol, as well as by cold temperatures (see pp. 100–101)]. This 'hot' response to capsaicin is believed to trigger the release of endorphins in the brain, causing a feeling of well-being.

Capsaicin does not activate the TRPV1 channel in birds. Hence, any chilli pepper seeds eaten by birds go through their digestive tract and get dispersed. However, squirrels – like other mammals – have a response to capsaicin similar to humans. This means that if you coat the nuts intended for the birds with chilli powder (or Tabasco sauce), the squirrels will learn to give the bird table a wide berth.

Much of the research on the TRPV1 receptor that detects capsaicin (and the corresponding TRPM8 receptor for menthol) was carried out by the research group of David Julius, who in 2021 shared the Nobel Prize in Physiology or Medicine with Ardem Patapoutian for 'their discoveries of receptors for temperature and touch'.

The 'heat' of a chilli pepper is traditionally assessed using a panel of five tasters. The pepper extract is diluted until the tasters can no longer detect it. The dilution factor is taken as a measure of its 'hotness', as expressed in Scoville units. The Scoville value of undiluted capsaicin is 16 million, while that of dihydrocapsaicin is slightly lower at 15 million. Other capsaicinoids have values ranging from 8.6 million to 9.1 million. Police pepper sprays are rated at around 5 million Scoville units. The Dorset Naga is the hottest pepper bred to date, with a rating of 1.6 million Scoville units. The hottest habanero follows at around 600,000, whilst Jalapeno peppers average around 5,000. The capasaicinoid-free green bell peppers, of course, have a zero rating.

Pepper sprays can be very unpleasant, which is why some police forces use them to subdue criminals. For example, the California police have used sprays containing oleoresin capsicum (OC) since 1992. Some citizens have resorted to pepper sprays for self-protection against muggers. It appears that there may be an interaction between capsaicinoids and 'psychostimulants' such as cocaine or amphetamines, which may, in some cases, lead to fatalities.

People have been investigating capsaicin for medicinal uses for many years, particularly in pain relief. Thus, hundreds of years ago in Central America, the Mayan people used chillis to treat asthma, coughs and sore throats, whilst the Aztecs used them to treat toothache. This may be explained by the fact that, after the application of capsaicin to the skin, and initial pain sensation is followed by the desensitisation of the nociceptive neurons, so that the affected are gradually becomes less sensitive. Capsaicin is incorporated into pain-relieving creams for the same reason.

It can also alleviate the pain associated with arthritis and shingles as well as post-operative pain. Some studies suggest that capsaicin treatment can improve cerebrovascular function and cognition, thereby possibly staving off dementia. However, others have found that high chilli intake is associated with cognitive decline. The jury is out on this.

8 Cathinones and Khat

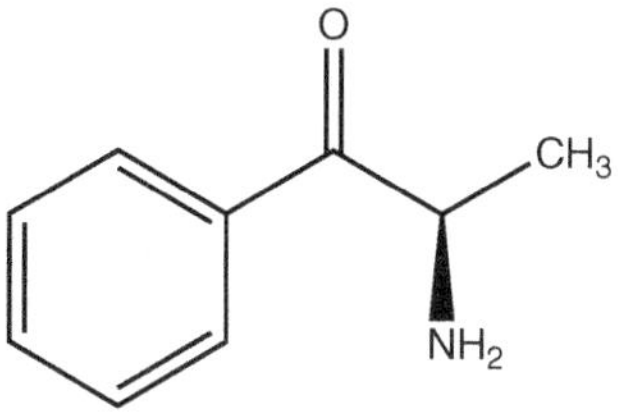

(8.1)

As ephedrine provided the structural framework for the amphethamines, cathinone (8.1) served as the forerunner of the synthetic cathinones. Cathinone is the psychoactive molecule that has, for hundreds of years, provided the stimulant effects to the users of the social drug khat (or qat). Cathinone is present in the green leaves of the shrub (*Catha edulis*), which is grown in East Africa. The leaves are widely chewed, particularly in Somalia, Ethiopia and Yemen (just as the coca leaf is chewed in the Andes), and the cathinone-laden saliva swallowed. The cathinone is oxidised rapidly after the leaves are picked, and hardly any remains three days after they are harvested. For this reason, khat is air-freighted to international destinations such as London for expatriate communities.

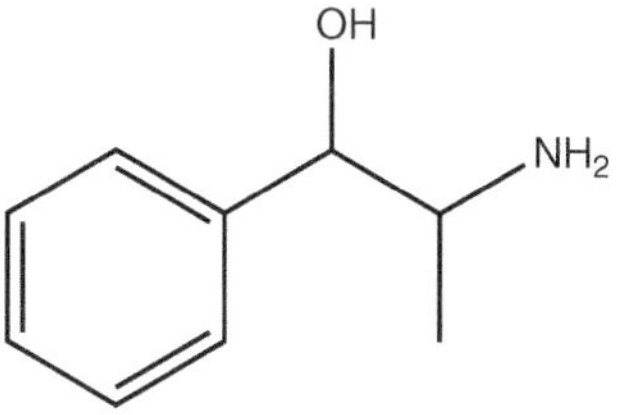

(8.2)

The structure of cathinone resembles that of amphetamine, the difference being the presence of a carbonyl group, rather than a methylene (CH_2) group, next to the benzene ring. The carbon next to the primary amine group is chiral; enzymes in the plant produce only a single isomer, (S)-cathinone, as is usually the case. In the plant, cathinone is accompanied by a similar molecule called cathine (8.2) – not to be confused with caffeine. Cathinone is a stronger stimulant than cathine, one of the latter's properties being it is an appetite suppressant.

DOI: 10.1201/9781003783565-8

Khat is a social drug used in groups. Being non-alcoholic, it is not prohibited by the Koran and therefore is acceptable to Muslims. It is particularly important in Yemen, where around 60% of men are regular consumers; chewing sessions often begin after lunch, and continue until dusk.

The stimulant properties of khat (and cathinone) have been compared to those of strong coffee. Dan Perrine described it as 'somewhere perhaps between espresso and amphetamine'. Consumers associate khat chewing with well-being and euphoria, along with improved alertness and libido. However, medical conditions arise from khat consumption. It increases the risk of myocardial infarction and may also cause liver damage. In significantly high doses, it can be associated with manic illness and delusions. And, of course, it features in fiction, notably in Evelyn Waugh's novel *Black Mischief*: 'Mahmud el Khali bin Sai'ud sat among his kinsmen, moodily browsing over his lapful of khat'. In contrast, it is an illegal drug in Europe (including the UK, where it is a Class C drug) and in many other parts of the world. In the USA, cathinone has been a Schedule I Controlled Substance, since 1993.

Though cathinones only entered public consciousness in the early twenty-first century, some synthetic methcathinones were first made in the mid-nineteenth century. Methcathinone can be prepared by oxidising ephedrine using a number of common oxidants, including potassium permanganate (Figure 8.1).

It was abused in the late twentieth century in countries such as Russia and the USA. Its effects were compared to a 'rush' associated with amphetamine, when administered intravenously.

In the first decade of the twenty-first century, numerous synthetic cathinones were marketed as street drugs, allegedly as 'legal alternatives' to Ecstasy (aka MDMA), and were reportedly synthesised largely in Far Eastern laboratories, particularly in China. These drugs went by names such as 'plant food', 'pond cleaner', 'bath salts' and 'room odoriser'. They were nothing of the kind; the names served merely as a cover. The first of these drugs to attract attention was 4-methylmethcathinone ((8.3): mephedrone), which was first synthesised in 1929.

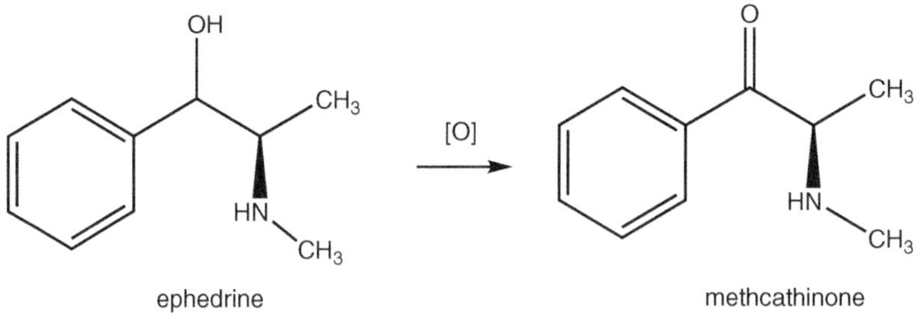

FIGURE 8.1 Synthesis of methcathinone by oxidation of ephedrine.

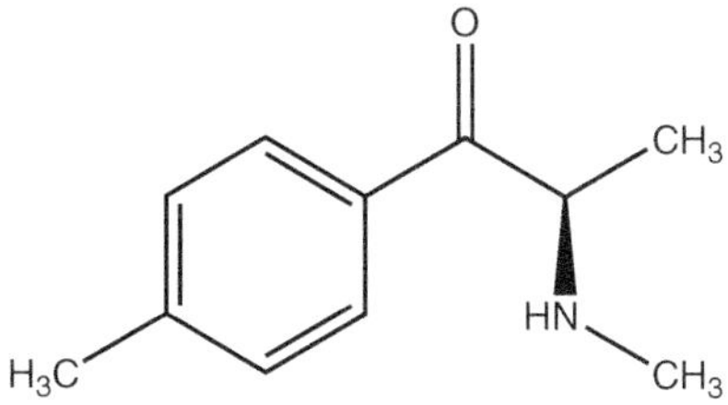

(8.3)

Mephedrone was the first substituted cathinone detected in late 2007. The media called it 'meow meow', probably because newspaper reporters had noticed the word 'cat' in its chemical name. Within three years, seizures associated with mephedrone were reported across 28 European countries. By late 2010, health issues were becoming evident, notably heart palpitations. Fatalities were reported in Sweden in 2008 and in the UK in 2010. That same year, mephedrone became a controlled drug throughout the European Union, which at that time included the UK, and was later classified as a Schedule 1 drug.

Given the structural resemblance to the amphetamines, cathinones would be expected to act in a similar manner in the body. This appears to be largely true, though mephedrone has been found to act primarily, though not exclusively, as a releasing agent at the dopamine transporter, whereas MDPV acts as an uptake inhibitor at the same transporter.

One toxicity issue is that methcathinone is often produced by the oxidation of ephedrine or pseudoephedrine, obtained from over-the-counter cold remedies, using potassium permanganate. This method can result in significant amounts of manganese being present in the methcathinone produced. In the body, this can lead to toxic effects, including parkinsonism.

Another street drug associated with cathinones is known as 'Monkey Dust'. In fact, two different molecules are sold under this street name – 3',4'-Methylenedioxy-α-pyrrolidinohexiophenone ((8.4): MDPHP) and 3',4'-methylenedioxypyrovalerone ((8.5): MDPV). They are both stimulants, with MDPV leading to stronger effects and acting mainly at the dopamine transporter. They are Schedule 1 drugs in the USA and Class B drugs in the UK. MDPV, in particular, is widely controlled around the world. As with some other cathinones, 'Monkey Dust' is associated with energy and euphoria, but is also associated with undesirable effects, including paranoia. Its 'high' lasts for up to 12 hours, followed by a crash. It has the particular attraction of being very inexpensive.

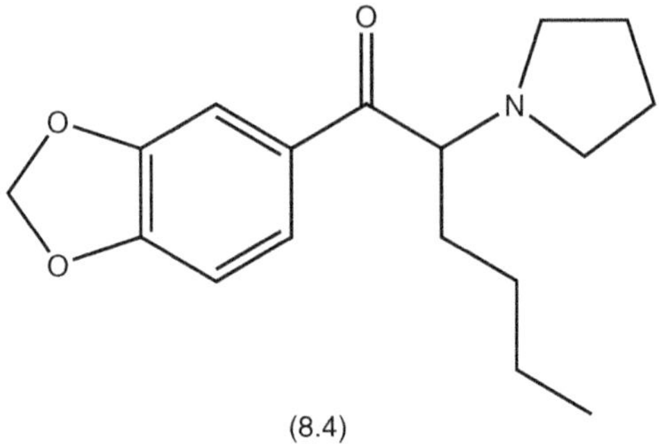

(8.4)

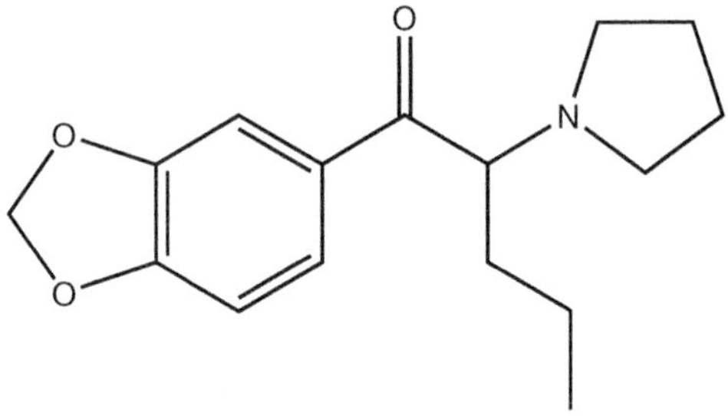

(8.5)

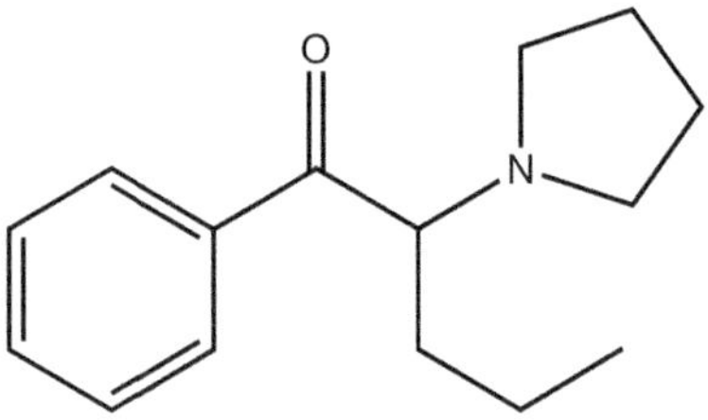

(8.6)

Flakka is the name often given to α-pyrrolidinopentiophenone (*alpha*-pyrrolidinovalerophenone, α-PVP (8.6)), another street drug from the early 2010s. Its nickname is said to be derived from a Spanish expression for an attractive woman, 'la flaca'. It is also known as 'gravel' and believed to be a norepinephrine-dopamine reuptake inhibitor.

Flakka has been regarded as a potent and inexpensive drug, producing an instant 'high' and possessing a strong potential for addiction. Its rapid onset and long-lasting euphoria is also associated with less desirable effects, including cardiac problems and sudden death. In one county of South Florida, more than 40 deaths associated with Flakka were reported in less than two years.

9 CF_2Cl_2

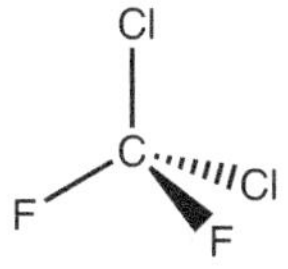

(9.1)

By itself, a molecule is morally neutral. It is only when it is put to use that it can be perceived as 'good' or 'bad'. Chlorofluorocarbons such as CF_2Cl_2 (9.1) are an example of a chemical that appeared to be a perfect solution to a problem when they were first introduced, until 50 years later they were found to contribute to an even deadlier problem. Refrigerators are a sine qua non of modern homes, but a century ago they were a luxury (even 60 years ago they were uncommon). There was no such thing as a 'weekly shop'. Fresh food was purchased and consumed promptly.

Refrigerators of the type familiar today were introduced in the 1920s America. At that time, the most generally used refrigerants were ammonia (NH_3), sulphur dioxide (SO_2) and chloromethane (CH_3Cl), all of which posed problems of toxicity. In 1928, Thomas Midgley Jr, an American mechanical and chemical engineer, was asked by his boss Charles Kettering, the head of research at General Motors, to develop a refrigerant that was safer than these substances, something that was non-flammable and much less toxic.

Midgley explored organofluorine compounds, with a boiling point between 0°C and −40°C, reasoning that such a substance would be gaseous at normal temperatures but would liquefy under pressure. However, he was concerned that they might be toxic. He discussed this with a colleague Albert L. Henne, who said that replacing some fluorines by chlorine might reduce their toxicity. They selected dichlorodifluoromethane (CCl_2F_2), a compound with a boiling point of −29.8 °C, within the desired range. Structurally, it is derived from the hydrocarbon methane CH_4 by replacing the hydrogen atoms with chlorine atoms and fluorine atoms. It therefore belongs to the class of compounds known as a chlorofluorocarbon (CFC). This compound came to be known as R-12 and later as CFC−12.

DOI: 10.1201/9781003783565-9

The numbering system for CFCs uses the format CFC-xyz, (and similarly HCFCs) and acts as follows: –

x is the number of carbon atoms in the molecule, minus 1. If x is zero, it is omitted;

y is the number of hydrogen atoms plus one;

z is the number of fluorine atoms.

You can work back from CFC-xyz, by adding 90 to xyz, the first digit representing the number of carbons, the second the number of hydrogens and the third the number of fluorines (the rest are chlorines). For example, if you do that for CFC-12, you get 102 – that is one carbon atom, no hydrogen atoms and two fluorine atoms, and the remainder being two chlorine atoms.

Nearly 40 years earlier, Frédéric Swarts, a Belgian chemist, had devised what came to be known as the Swarts reaction, a synthetic route for CFCs. It involved the reaction between tetrachloromethane and antimony(III) fluoride: -

$$3CCl_4 + 2SbF_3 \rightarrow 3CCl_2F_2 + 2SbCl_3$$

Nearly a decade later, Midgley recalled telephoning a chemical supply company and ordering five 1-ounce (28.35g) bottles of antimony trifluoride. He believed that this was the entire quantity available in the country at the time. One bottle was randomly chosen, and a few grams of dichloromonofluoromethane were produced. A guinea pig was placed under a bell jar with the compound and it survived. They took another bottle of SbF_3, prepared a few more grams of the CFC and repeated the experiment. This time, the animal died. The chemists investigated further and, on smelling the material, detected the presence of toxic phosgene. This could be removed with alkali, rendering the CF_2Cl_2 safe. They then checked the two remaining bottles of antimony trifluoride and found that they too were impure. Of the five bottles of antimony trifluoride, only one was pure.

They had selected the only pure bottle by accident for their first trial. If they had selected any one of the others, the animal would have died, as expected, and they would have abandoned the project. This is a striking example of serendipity. Famously, at a meeting of the American Chemical Society in 1930, Midgley filled his lungs with CCl_2F_2 and then exhaled it over a lighted candle, extinguishing the flame.

They proceeded with the use of CCl_2F_2 as a refrigerant material and subsequently produced other CFCs. These compounds came to be known as Freons, a trade name given by the Du Pont company. Hydrochlorofluorocarbons (HCFCs) were also made such as chlorodifluoromethane (HCFC-22) (9.2)

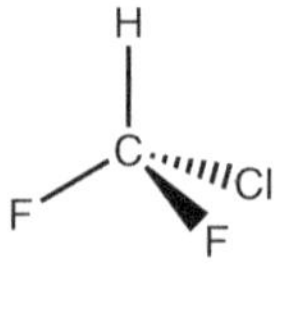

(9.2)

CFCs were associated with a range of desirable properties: –

- stable.
- non-flammable.
- non-explosive.
- non-irritating.
- non-toxic.
- noncorrosive
- does not support combustion.
- does not affect flowers, fruit, vegetables or dairy products.

The use of CFCs expanded vastly over the following 40 years, since their properties suited their use in applications such as:-

- solvents for removing grease in the electronics industry
- propellants in aerosols
- foam-blowing agents for making polyurethanes in the polymer industry
- 'inert blankets' to exclude air in fire extinguishers
- air conditioning for cars.

CF_2Cl_2 was prominent in these applications.

It was only in the 1970s that reality caught up. In 1957, the British chemist James Lovelock (who became an eco-guru) invented an electron-capture detector that proved capable of detecting tiny amounts of gas-phase halogen compounds. In 1973, he used it to show that CFCs were present in the Atlantic atmosphere, notably while travelling from England to Antarctica. The following year, independent research conducted by Sherwood Rowland and Mario Molina, along with the work of Paul Crutzen, linked the presence of CFCs to damage to the ozone layer. These three scientists shared the 1995 Nobel Prize in Chemistry for their work.

The problem with CFCs, which is linked to their very low reactivity, is that they can persist in the atmosphere for up to several hundreds of years. As a result, they survive the lower reaches of the earth's atmosphere and eventually reach the stratosphere, where their photochemical decomposition forms chlorine atoms. Naturally occurring organochlorine compounds, notably chloromethane, CH_3Cl, also decompose in the same manner.

$$CCl_2F_2 \rightarrow \bullet CClF_2 + \bullet Cl$$

$$CH_3Cl \rightarrow \bullet CH_3 + \bullet Cl$$

$$\bullet Cl + O_3 \rightarrow \bullet ClO + O_2$$

$$\bullet ClO + O \rightarrow \bullet Cl + O_2$$

The C-Cl bonds in a CFC molecule – or in CH_3Cl, for that matter – are the weakest bonds present (C-Cl bond energy ~ 346 $kJmol^{-1}$; C-H bond energy ~ 413 $kJmol^{-1}$; C-F bond energy ~ 485 $kJmol^{-1}$). Consequently, the C-Cl bond undergoes homolytic fission, forming the highly reactive • Cl radical. This radical decomposes ozone, O_3, into ordinary dioxygen (O_2) in a chain reaction. The Cl atoms are regenerated in the process and can therefore continue to decompose other ozone molecules. It has been estimated that one chlorine atom can decompose over a thousand ozone molecules.

Apart from the effect of CF_2Cl_2 and other CFCs on the ozone layer, scientists realised that CFCs contributed approximately 20% of the man-made 'greenhouse effect'. CO_2, CH_4, N_2O and tropospheric ozone contribute nearly 50%, 15%, 5% and 7%, respectively.

Within a few years, two developments took place. Several countries, including the USA, outlawed CFCs as aerosol propellant gases during the late 1970s. Then, between 1981 and 1983, Norfolk-born Joe Farman and his research team at the British Antarctic Survey located at Halley Bay, Antarctica, located what became known as a 'hole' in the ozone layer over Antarctica. Their research was first published in 1985 and subsequent studies confirmed the damage.

For once, the international community agreed on the need for action. They reached an agreement generally known as the Montreal Protocol in 1987 and subsequently strengthened in 1990. This was designed to eliminate the use of CFCs. It was agreed that CFC production in industrial countries would be completely banned by 2000, with the developing countries following ten years later.

9.1 AFTER THE CFCS

Obviously, replacements for CFCs had to be found: compounds that would perform the same functions without damaging the ozone layer. The first type comprised of hydrochlorofluorocarbons (HCFCs). Of these, chlorodifluoromethane (HCFC-22) has been widely used as a substitute for CFC–12 (9.2).
The presence of the C-H bond makes HCFCs more reactive than CFCs and consequently have shorter atmospheric lifetimes. However, these compounds contain a C-Cl bond and therefore continue to contribute to ozone depletion. They are scheduled to be phased out by 2030 by industrial countries and by 2040 in developing countries. An exception to this is $CHClF_2$ (9.2), as it is used to manufacture $F_2C{=}CF_2$ (1,1,2,2-tetrafluorethene), the monomer used (Figure 9.1) to produce Teflon (poly(tetrafluorethene)).

Hydrofluorocarbons (HFCs) do not contain chlorine and therefor do not contribute to ozone depletion. Examples include CF_2H_2 (HFC-32, (9.3)) and CHF_3 (HFC-23, (9.4)).

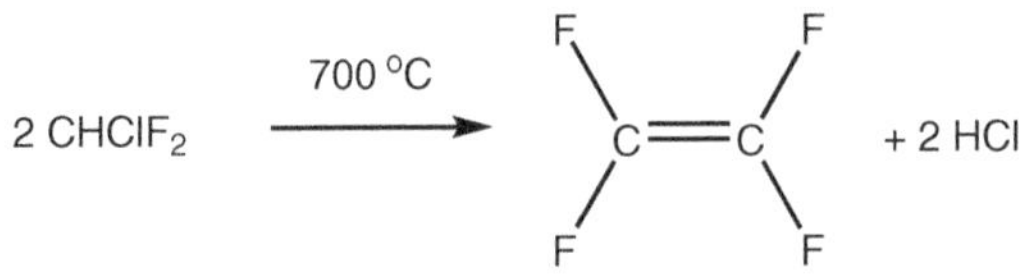

FIGURE 9.1 Synthesis of 1,1,2,2-tetrafluorethene.

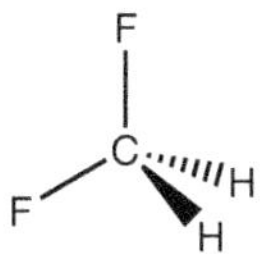

(9.3)

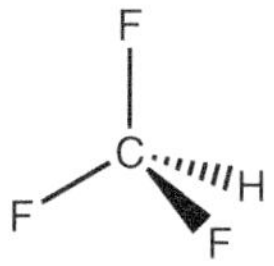

(9.4)

CF_3CH_2F, 1,1,1,2-tetrafluoroethane ((9.5), HFC-134a) is a prominent HFC that is widely used to replace CF_2Cl_2 in car air conditioners. However, it does have the disadvantage of being a greenhouse gas and therefore regarded as only a short-term solution. Consequently, it will be phased out as well.

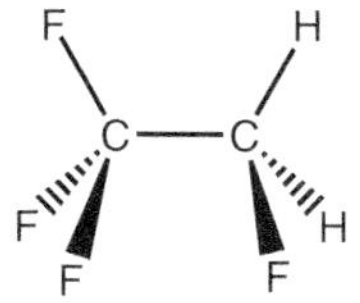

(9.5)

The Kigali Amendment to the Montreal Protocol, signed in 2016, requires its signatories to reduce production and consumption of HFCs. The long-term replacements of HFCs are expected to be hydrofluoroolefins (HFOs), which are essentially alkenes with hydrogen atoms replaced by fluorines. Their double bond makes them more reactive and short-lived in the atmosphere, resulting in a very low global warming potential (GWP100). Two typical HFOs are the isomeric compounds E-1,3,3,3-tetrafluoropropene ((9.6), E-HFO-1234ze) and 2,3,3,3-tetrafluoropropene ((9.7), HFO-1234yf).

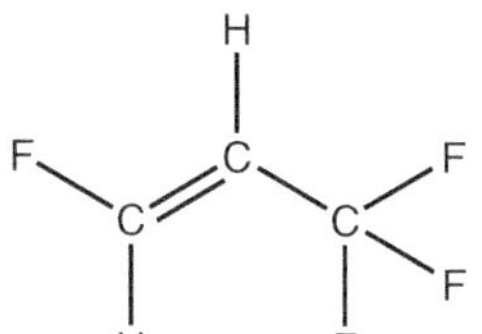

(9.6)

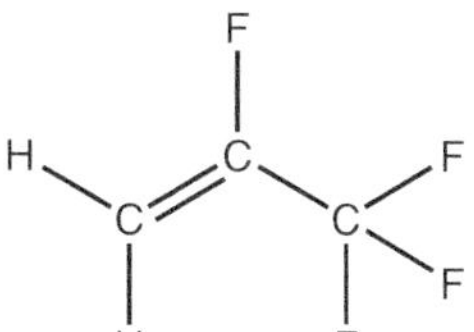

(9.7)

Although at present (2025) CFC levels have dropped since production was stopped in 2006, there have been occasional 'spikes' with $CFCl_3$ (CFC-11) levels suddenly increasing for a few years from 2012. Atmospheric levels of a few other CFCs, including CF_3Cl (CFC-13), have also risen lately. It remains to be seen whether these are just 'accidental' blips or the result of inadvertent industrial by-products.

10 Chanel No. 5 and 2-methylundecanal

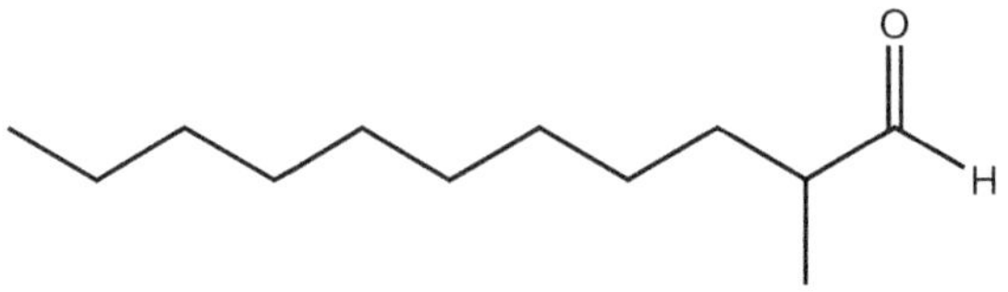

(10.1)

People have been using (and smelling) perfumes for thousands of years, dating back to the ancient Egyptians, Greeks and Romans. They very likely returned to mediaeval Europe with the Crusaders. In those days, and indeed well into the nineteenth century, perfumes were exclusively made using natural sources, such as flowers and spices. However, with the rise of synthetic chemistry during the nineteenth century, a wider range of molecules and scents became possible. The development began with Paul Parquet's *Fougère Royale* in 1882, which had synthetic coumarin. A further step was taken in 1889 by Aimé Guerlain, who not only used synthetic vanillin in *Jicky*, but went beyond floral notes in constructing his perfume. He followed a procedure that is still used today, employing a 'top note' supplied by small, volatile molecules for the initial impact; a 'middle note' of less volatile, medium-sized molecules that provides the main 'personality' of the perfume and lasts for several hours; and the bottom or base note furnished by large and rather involatile molecules that remain on the skin after all the others have evaporated.

If you asked a friend to name a famous perfume, there is a good chance that they would say '*Chanel No. 5*'. We owe this perfume to three particular people, who will emerge over the course of this chapter. *Chanel No. 5* has remained iconic for a century. Before you smell it, you notice the striking design of its bottle, which was chosen by Coco Chanel to remind her of the toiletry bottles used by Boy Capel, the love of her life, who was killed in a car crash on the St. Raphael-Cannes road on 22 December 1919, at a time when she was planning to branch out from fashion into perfumes.

Gabrielle Chanel (1883–1971), now universally known as 'Coco', had already come a long way since her illegitimate birth at Saumur, on the Loire. At 12, her father sent her to an orphanage at Aubazines, in the Corrèze, where she learned to sew. At 18, she left the orphanage and moved to Moulins in the Bourbonnais, first as a charity pupil in a Catholic school and later working at a draper's shop as a seamstress, supplementing her income by performing as a cabaret singer. Over the following years, through a combination of determination, hard work, a flair for fashion and the support of some influential boyfriends, she opened shops in Paris, Deauville and Biarritz. After

DOI: 10.1201/9781003783565-10

Capel's death, she took up with a Russian émigré, Grand Duke Dimitri Pavolivch (a cousin of the last Tsar, Nicholas II), and legend has it that he introduced Coco to a perfumer on the beach at Cannes. Ernest Beaux (1881–1961) was born in Moscow to a French father. Employed by Rallet & Co., the leading perfume and soap manufacturer in Russia, Beaux created perfumes there until World War I. The Revolution of 1917 forced him to relocate to the Riviera at Cannes, where the French perfume industry was centred on Grasse. Chanel asked Beaux to create a new perfume for her.

Beaux had already created an important perfume for Rallet & Co. in 1913–1914. It was launched as *Bouquet de Catherine* in 1913, but was rapidly renamed as *Rallet N°1*, for political reasons. It is sometimes said, incorrectly, that '*Chanel No. 5*' was the first aldehydic perfume. This is untrue. *Floramye* (1905) appears to have been the pioneer, followed by *Quelques Fleurs* (1912), and these fragrances seem to have been the inspiration for Beaux. *Quelques Fleurs* had used the aldehyde 2-methylundecanal, which had been available for less than a decade, thanks to a synthesis devised by Georges Darzens and reported in 1904. There is no single 'aldehyde smell'. Molecules with the shortest carbon chains (1 to 4 carbon atoms) tend to have pungent and unpleasant odours (methanal is also classified as a Group I carcinogen by the World Health Organization). Aldehydes containing six or seven carbon atoms are associated with 'green grass' aromas (pp. 76–80), whilst octanal, with its eight carbon atoms, has an orange-like smell (pp. 126–128). Once the chain length reaches 10–12 carbon atoms, the aldehydes are associated with odours described as fatty and citrus-like. Adding a methyl side-chain at carbon 2 of the chain, as in 2-methylundecanal, intensifies the aldehyde characteristic of the odour and also makes it more pleasant.

The contents of a surviving sealed bottle of *Rallet N°1*, probably dating back to the 1920s, were analysed recently. The results, reported in October 2007, confirmed the presence of aldehydes in the formulation, including undecanal (10.2), dodecanal (10.3) and undec-10-enal (10.4). The compounds, together with neroli and ylang-ylang, contributed to the perfume's top note and provided its initial impact. The middle note consisted of a mixture of jasmine absolute, rose de mai, orris and lily of the valley. The bottom note comprised sandalwood, vetiver, styrax, vanillin (pp. 113–120) and nitromusk (pp. 200–206).

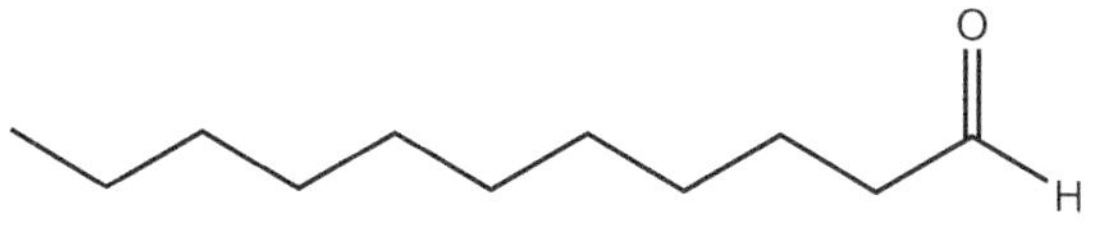

(10.2)

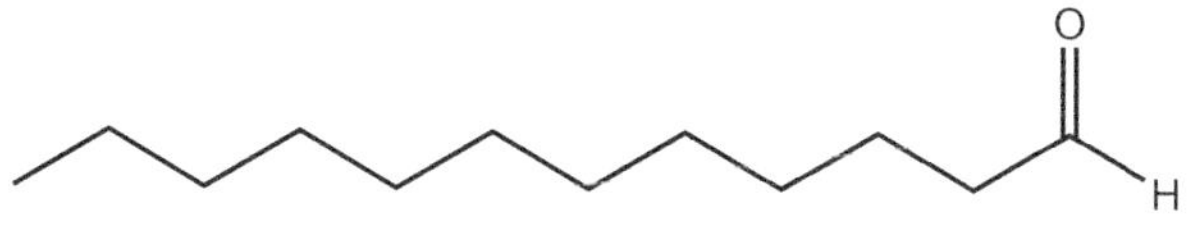

(10.3)

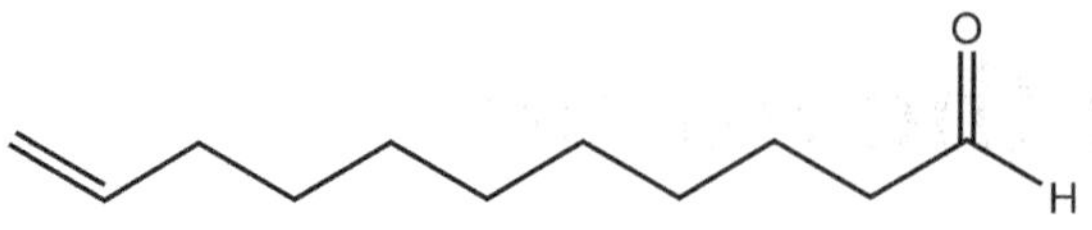

(10.4)

Ernest Beaux composed his perfume for Chanel in 1920, employing traditional natural ingredients, including jasmine, rose, lily-of-the-valley, sandalwood and vetiver. These were further accompanied by some of the newer synthetic ingredients that had recently come into use lately, such as vanillin and nitromusks. *Chanel No. 5* had a composition similar in many aspects to that of *Rallet N°1*, but Beaux appears to have used 2-methylalkanals, notably 2-methylundecanal (10.1), to reinforce the aldehydic top note. Among the many stories swirling around *Chanel No. 5* is that Beaux's assistant made an error and added far more aldehyde than his employer had instructed. Whatever the truth may be, the result gave the new perfume an immense impact. Traditionally the saying attributed to Beaux is that the effect of aldehydes on the perfume is 'like lemon juice on strawberries'.

Towards the end of 1920, Beaux prepared just one hundred bottles of the new perfume for Coco, which she distributed to her favourite customers as Christmas gifts. This was designed as part of a campaign to promote the perfume by creating demand for something that was only available in limited quantities. The strategy proved to be a success. From 1921 onwards, she sold *Chanel No. 5* from the shelves of her boutiques in Paris, Deauville and Biarritz. It was not until 1924 that she reached an agreement with the industrialists Pierre and Paul Wertheimer, whose Bourjois perfume firm could manufacture the perfume on the scale that (eventual) world-wide demand would eventually require. The company Les Parfums Chanel was established on 4 April 1924. And the rest, as they say, is history.

The aldehyde 2-methylundecanal occurs naturally in small amounts in sources such as kumquat, but the quantity available from these is insufficient to meet the needs of the perfume industry. Consequently, it is produced using synthetic routes such as the one devised by Darzens (Figure 10.1). The synthesis starts with undecan-2-one, which is reacted with methylchloroacetate in the presence of the base sodium *tert*-butoxide. The product of the reaction is a glycidyl ester. This type of reaction has come to be known as a Darzens condensation. This ester can be hydrolysed by the usual method, refluxing with alkali followed by acidification to generate the corresponding carbolic acid. Finally, when distilled under reduced pressure, the acid undergoes decarboxylation with the loss of CO_2, forming 2-methylundecanal.

But what of the man who gave us 2-methylundecanal? Georges Darzens (1867–1954) was a polymath. He studied at the elite École Polytechnique military school in Paris, in the class of 1886. His poor eyesight disqualified him from pursuing his first love, astronomy, and so he took up chemistry. He also studied for mathematics and physics degrees before embarking on medicine, obtaining an M.D. in 1899. At the École Polytechnique, he worked from 1888 to 1897 as an assistant to his mentor Édouard Grimaux (1835–1900) as a synthetic organic chemist. During that time,

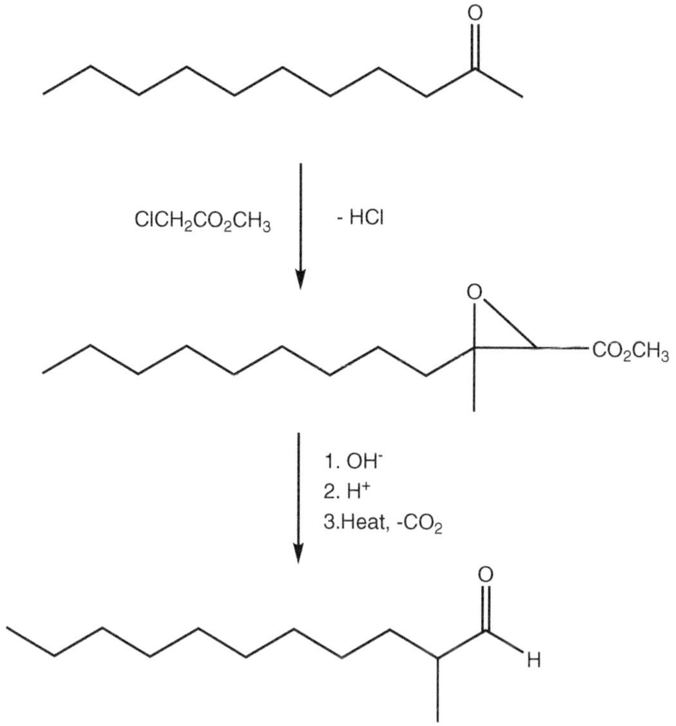

FIGURE 10.1 Synthesis of 2-methylundecanal using a Darzens condensation reaction.

Darzens and Grimaux did something heroic in supporting Alfred Dreyfus, a Jewish French Army officer, also a product of the École Polytechnique. He was exiled to the penal colony of Devil's Island in 1894, where he spent five years on trumped-up charges of treason. It was in support of Dreyfus that Émile Zola wrote his famous manifesto *J'Accuse!* Grimaux lost his chair on account of this. From 1897 to 1920, Darzens directed the research laboratory of the L. T. Piver perfumery company. He also served as a professor of chemistry at the *École Polytechnique* in Paris from 1913 until his retirement in 1939. Moreover, in the pioneering age of the automobile, he designed and built three cars.

Coco Chanel, Ernest Beaux, Georges Darzens were three very different people, yet without them we would not have one of the world's classic perfumes.

And, of course, *Chanel No. 5* has been advertised by a host of beautiful and famous women. Among them are Jean Shrimpton, Catherine Deneuve, Ali McGraw, Carole Bouquet, Nicole Kidman, Audrey Tautou, Gisele Bündchen, Marion Cotillard, Margot Robbie and most notably Marilyn Monroe, who, when asked what she wore in bed, replied: 'Nothing but a few drops of Chanel No. 5'.

11 Chloromethane

Chlorine is an element available in abundance. Everyone knows that there is a lot of NaCl, or common salt, in seawater, which contains around 3.5% of dissolved solids, of which more than 80% is NaCl. All the chlorine in nature is found in compounds. Elemental chlorine is far too reactive to be found uncombined. However, there is one particularly important application of elemental chlorine: the chlorination of water supplies to homes. This has been a practice for more than a century (Maidstone was the first town in the UK to chlorinate its water supply, in 1897). This kills harmful bacteria and viruses, preventing previously endemic water-borne diseases such as cholera, typhoid and dysentery. The amount of chlorine used is around 0.5 milligrams per litre of water. In 1991, the South American nation of Peru decided to stop chlorinating the water supply, inevitably resulting in cholera outbreaks. More than 10,000 people died during that time. Hence, Peru restarted chlorinating the water supply. Sensibly, no other country has repeated this experiment.

Chlorine is found in several organic compounds, some natural while others are synthetic. As recently as 1954, only a dozen organic compounds containing chlorine were known to occur in nature. However, continuing research and improved detection methods have increased the number of known natural organic chlorine compounds to well over 2,500, and the number continues to increase. Sadly, some peoples continue to hold an outdated view, believing that all chlorine compounds are 'dangerous' man-made materials such as CFCs and DDT.

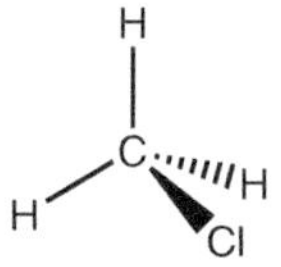

(11.1)

The organic chlorine compound that is most abundant in nature is arguably one of the simplest compounds one could find: CH_3Cl, chloromethane (11.1), formerly known as methyl chloride. Around 5 x 10^6 tonnes of CH_3Cl are released into the earth's atmosphere each year; well over 90%, and possibly as much as 99%, comes from natural sources. It was once thought that most chloromethane originated in the oceans, but more recent research now shows that no more than 10% arises from the oceans, with 90% or more originating from terrestrial sources. Tropical plants are believed to emit around half the chloromethane produced worldwide. Each year, there are some 200,000 fires around the world triggered by lightning, including forest fires, brush fires and other forms of biomass burning, and these contribute chloromethane

DOI: 10.1201/9781003783565-11

to the atmosphere, as do wood-rotting fungi. Significant quantities of chloromethane also originate from salt marshes, mangroves and peat bogs. For example, *Batis maritime*, a halophytic plant that is abundant in salt marshes, produces CH_3Cl.

Chloromethane is a gas at room temperature, with a boiling point of -24°C. Its concentration in the atmosphere is approximately 0.6 parts per billion, which is considerably higher than that of CFCs like CF_2Cl_2 (see pp. 29–33). However, chloromethane is more reactive than CFCs. C-H bonds are weaker (and hence more reactive) than C-F bonds, so chloromethane resides in the atmosphere for a much shorter period (1.4 years) than CFCs (100 years). One characteristic common to both chloromethane and CFCs is their contribution to ozone layer depletion. Of course, unlike CFCs, CH_3Cl has been causing this for a very long time, on account of the natural origin of CH_3Cl. As with CFCs such as CF_2Cl_2 (pp. 29–33), ultraviolet radiation breaks down CH_3Cl by breaking the C-Cl bond through homolytic fission, with one electron going to each atom that originally formed the bond:

$$CH_3Cl \rightarrow \cdot CH_3 + \cdot Cl$$

This process generates chlorine atoms, which are free radicals and decompose ozone molecules:-

- $Cl + O_3 \rightarrow \cdot ClO + O_2$
- $ClO + O \rightarrow \cdot Cl + O_2$

Chlorine radicals are then regenerated, allowing the process to repeat many times. Consequently, one chlorine atom can destroy many hundreds of ozone molecules. About 20 years ago, it was estimated that chloromethane accounted for approximately 15% of ozone-depleting emissions. This proportion is anticipated to increase as man-made emissions of CFCs continue to decline. Unlike CFCs, however, CH_3Cl has a very low global warming potential and is not considered a 'greenhouse gas'.

To produce chloromethane in the laboratory, the classic method dates back to 1835 and was developed by Dumas and Peligot. Their method involves the reaction of methanol with HCl, which may be generated in situ by reacting NaCl with concentrated sulphuric acid.

$$CH_3OH + NaCl + H_2SO_4 \rightarrow CH_3Cl + NaHSO_4 + H_2O$$

First, the OH group is protonated, creating an incipient water molecule, which is a much better 'leaving group' than hydroxide $O\text{-}H^-$. The resulting cationic species then undergoes nucleophilic attack by a Cl^- ion, displacing the water molecule and thereby producing chloromethane (Figure 11.1).

One important plant compound involved in the natural synthesis of chloromethane is pectin (Figure 11.2). This polysaccharide is abundant in plant cell walls and contains methyl ester groups in some of its side chains, which are believed to be the source of the methyl group in chloromethane.

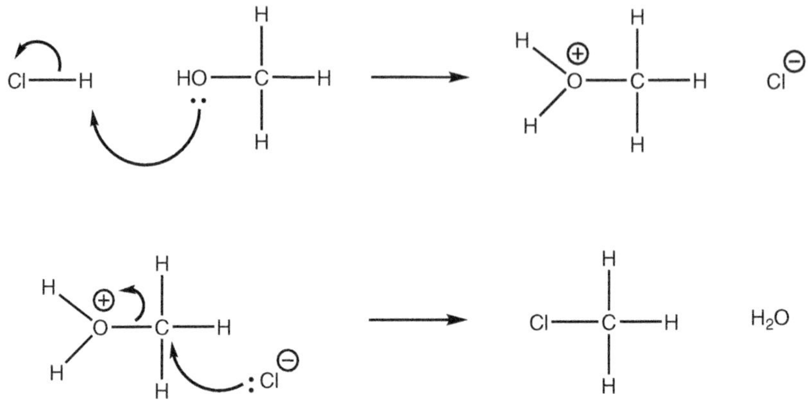

FIGURE 11.1 Mechanism of the synthesis of chloromethane from methanol.

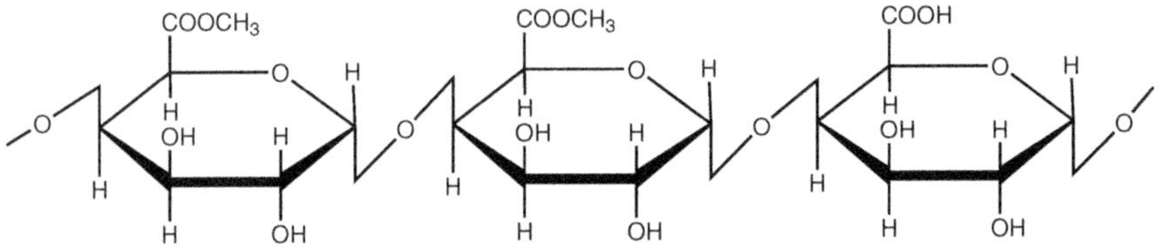

FIGURE 11.2 Structure of pectin.

Very small amounts of chloromethane are also generated in volcanic eruptions. Three volcanoes where it has been detected are Santiaguito, Guatemala (1969); Mount St. Helens, Washington, USA (1980); and Mount Etna, Italy (2012).

Reportedly, there are non-terrestrial sources of chloromethane, which do not affect the earth. Chloromethane is the first organic halogen compound to have been discovered in outer space. It was detected in 2017 in the coma of comet 67P/Churyumov–Gerasimenko (67P/C-G). More than that, both $CH_3{}^{35}Cl$ and $CH_3{}^{37}Cl$ were detected. Roughly four hundred light years away, CH_3Cl was also found in several young star clusters.

Chloromethane was once used in refrigerators, before Freons were discovered. Today, it plays an essential role in industrial synthesis. It is used in the production of methylchlorosilanes, such as $(CH_3)_2SiCl_2$, which are important in the manufacturing of silicone polymers. Chloromethane is also employed as a methylating agent for OH groups, for example in production of methylcellulose. In addition, it serves as a key agent in the organic synthesis of a range of substances like methanethiol (CH_3SH) and methylamines, CH_3NH_2, $(CH_3)_2NH$ and $(CH_3)_3N$.

12 Cinnamaldehyde

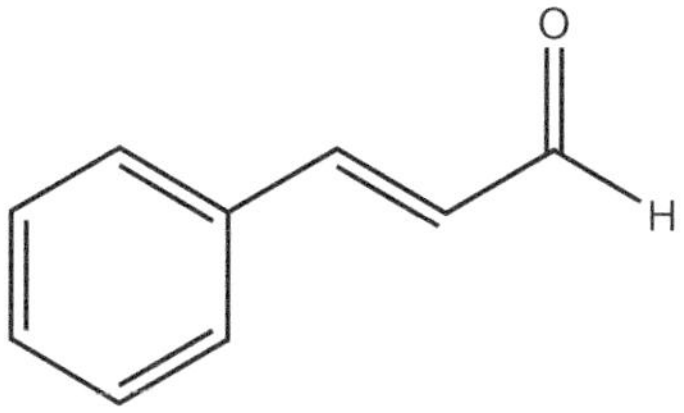

(12.1)

Some four thousand years ago, a world existed, comprising China and almost all of Asia, extending as far south as India and Ceylon (now Sri Lanka) and westwards into Arabia. The people of this region appreciated spices and traded them extensively along routes such as the Silk Road. The Chinese were familiar with cinnamon, which became the most prized spice of the ancient world, by 2700 BC. However, it was not until around 500 BC that these spices reached Greece and Rome, as their spheres and trade network expanded. At this time, the Greek historian Herodotus (c. 484–425 BC) wrote about cinnamon, citing Phoenicians who claimed that large birds carried cinnamon sticks to Arabia. Other accounts claimed that cinnamon grew on bushes guarded by poisonous snakes.

Cinnamon is obtained from the inner bark of the small evergreen tree, *Cinnamomum zeylanicum*, grown particularly in Sri Lanka and southern India. The volatile oil from the bark constitutes about 0.4 to 2.8% of the bark, of which cinnamaldehyde (12.1), whose systematic name is (2*E*)-3-phenylprop-2-enal, is the major component, comprising around 75%. Cinnamaldehyde is always found as the (E-) (trans-) isomer, rather than the (Z)-isomer. It is an oily yellow liquid (b.p. 246°C) that has a strong odour of cinnamon. The related ester *(E)*-cinnamyl acetate (12.2) is the major compound present in the volatile oils of cinnamon fruits (42–54%), which is evidently manufactured within the plant from cinnamyl alcohol.

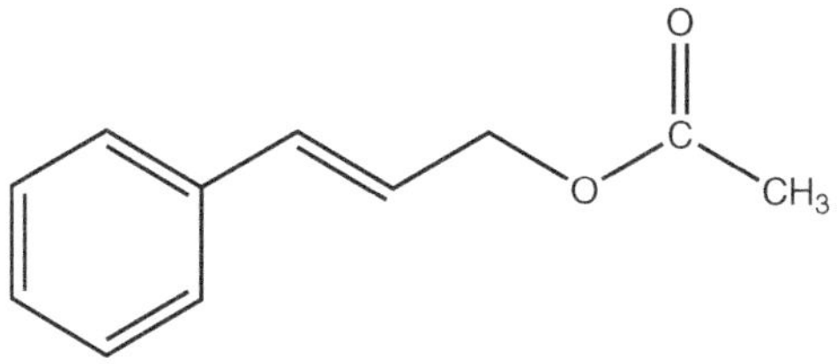

(12.2)

DOI: 10.1201/9781003783565-12

In contrast, eugenol (12.3), a compound more closely associated with cloves, is the major constituent of the oil obtained from the leaves.

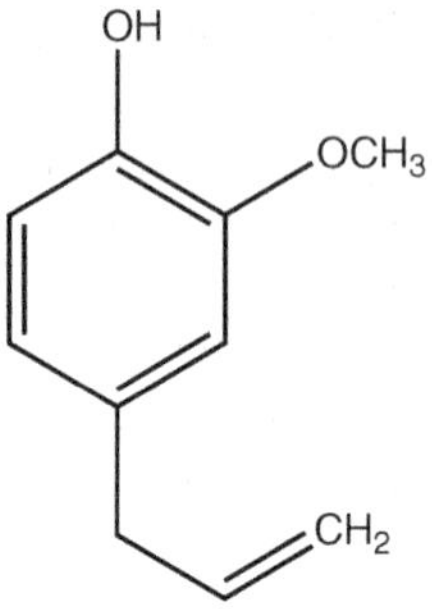

(12.3)

Cinnamaldehyde has a considerably more complex structure than the aliphatic aldehydes (e.g. octanal pp. 126–128). However, its biosynthesis is fairly straightforward (Figure 12.1). This begins with the amino acid phenylalanine, which undergoes deamination by the enzyme phenylalanine-ammonia lyase (PAL), producing cinnamic acid. This is reduced to cinnamaldehyde through a two-step process. First, it reacts with Coenzyme A (CoA), producing cinnamoyl CoA. The resulting compound is then converted into cinnamaldehyde by the enzyme cinnamoyl-CoA reductase (CCR).

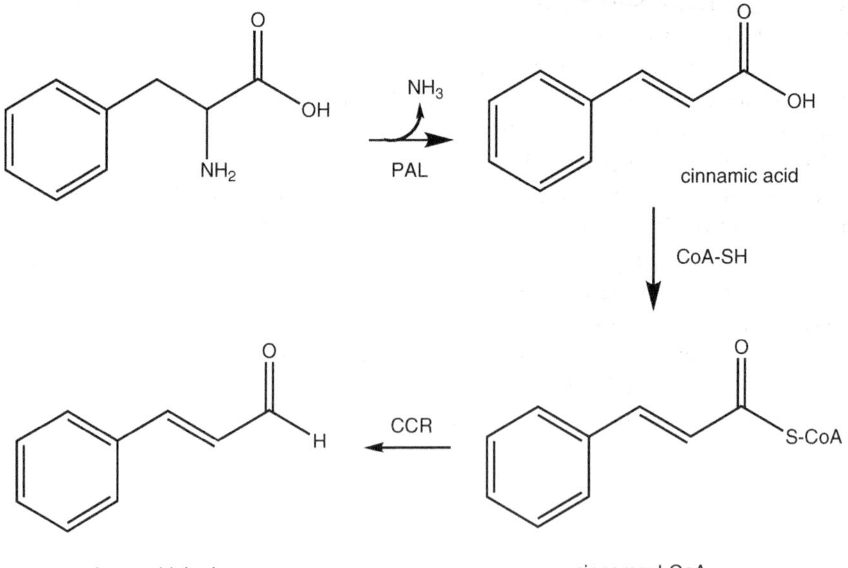

FIGURE 12.1 Simplified biosynthesis of cinnamaldehyde.

Cinnamaldehyde is widely recognised as a safe additive to foods, with nearly 180,000 kg used every year. Apart from being a flavouring agent, it has antimicrobial and antioxidant properties. It also inhibits the growth of bacteria and fungi. Cinnamaldehyde is reported to promote some anti-cancer activity and show protective effects against Alzheimer's disease and Parkinson's disease. It also exhibits anti-mosquito activity, which may represent one of its ecological functions in the cinnamon tree. In animals, it has also been proved to help manage diabetes.

One area of concern surrounds the use of cinnamaldehyde in e-liquids, alongside other flavouring agents such as menthol and vanillin. Analysis of 39 e-liquids revealed the presence of cinnamaldehyde in 20 of them. Although cinnamaldehyde is regarded as safe for use in foods, the inhalation of heated vapours resulted in other toxicological conditions, as the compound has been shown to impair the function of respiratory immune cells. For example, cinnamaldehyde has been reported to perturb bone, cartilage and vascular development in zebrafish embryos. In addition, cinnamaldehyde is toxic to certain white blood cells, particularly those present in the lungs. One likely explanation is that thermal decomposition of cinnamaldehyde (along with other volatile constituents of e-liquids) generates the vapours of toxic aldehydes, notably methanal (formaldehyde), as well as other toxic substances such as benzene.

13 Cisplatin and Other Platinum Anticancer Drugs

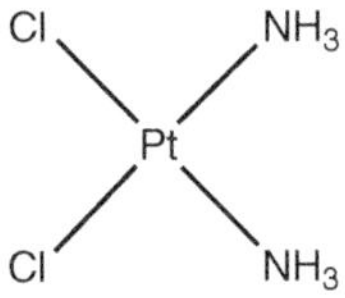

(13.1)

Scientific discoveries cannot be made to order. Some scientific research involves following existing knowledge and established theory. However, at other times, scientists make a leap in the dark and a serendipitous discovery emerges, as the story of cisplatin illustrates.

Barnett Rosenberg (1926–2009) was a physicist who became a biophysicist at Michigan State University. In 1964, he decided to investigate whether electric currents affected cell division, prompted by the thought that mitotic spindles in a dividing cell reminded him of the way in which iron filings align themselves in a magnetic field. His very talented research technician, Loretta Van Camp, carried out the experiments that Rosenberg devised. Platinum electrodes were placed in an ammonium chloride buffer containing *Escherichia coli* (*E. coli*) cells, and an electric current was passed through the solution. The cells began to grow without dividing, becoming increasingly long filaments, reaching lengths of up to 300 times their normal size. Cell division had ceased.

After a while, it was established that this change to the cells was not caused by the electric current. Rather, it was platinum that was dissolving from the electrodes, despite being a highly unreactive metal. Consequently, researchers decided to investigate if any platinum complexes led to the observed effect in the cells. Van Camp subsequently found that certain simple platinum salts, particularly $(NH_4)_2[PtCl_6]$, caused this elongation. In 1966, Andrew Thomson, an English postdoctoral researcher (who later became a professor at the University of East Anglia), produced *cis*- and *trans*-$[Pt(NH_3)_2Cl_2]$ and *cis*- and *trans*-$[Pt(NH_3)_2Cl_4]$ for testing, reasoning that these could plausibly have been formed in the ammonium chloride buffer. He demonstrated that the *cis*- complexes, instead of the *trans*-isomers, were biologically active. Soon it became clear that *cis*-$[Pt(NH_3)_2Cl_2]$, now commonly known

DOI: 10.1201/9781003783565-13

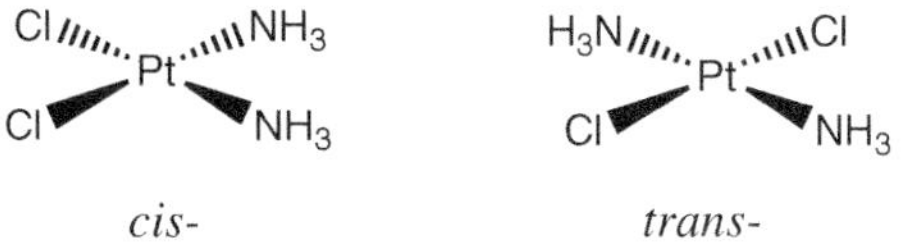

FIGURE 13.1 Isomers of $[Pt(NH_3)_2Cl_2]$; *cis*- (left) and *trans*- (right).

as cisplatin, was by far the most effective substance in promoting the filamentous growth of the cells.

In a moment of inspiration, Rosenberg suggested that, since cisplatin inhibited the division of bacterial cells, there was a possibility that it could stop the growth of cancerous cells. Further experimental studies, published in 1969, demonstrated that *cis*-$[Pt(NH_3)_2Cl_2]$ was indeed effective in treating Sarcoma 180 and Leukaemia L1210 in mice, whereas *trans*-$[Pt(NH_3)_2Cl_2]$ had no effect. Treatment with *cis*-$[Pt(NH_3)_2Cl_2]$ led large solid tumours to shrink, and the mice became healthy and survived for months.

Testing on humans began in the UK in 1971, rather before trials in the United States, as the USA was not keen on testing metal-based drugs. Cisplatin was found to be most effective against patients with testicular cancer. However, there were problems particularly involving toxicity in the kidney, which were addressed with hydration therapy as well as the use of other drugs. In 1978, cisplatin was approved by the FDA, just nine years after the first reports from animal trials. Many people have benefitted from cisplatin's success, notably those suffering from testicular cancer. An early patient treated with cisplatin was the British jockey Bob Champion, a familiar personality in the UK. He was diagnosed with Stage-3 testicular cancer in July 1979. Within a few months of starting a course of cisplatin-based therapy, he was back in the saddle, winning the 1981 Grand National Steeplechase at Aintree, riding a horse named Aldaniti. The event is also celebrated in the 1983 film '*Champions*', starring John Hurt.

Perhaps the most famous recipient of cisplatin was the American cyclist Lance Armstrong. Having ignored symptoms for several months in the year 1996, he finally consulted a friend who was a doctor in October after coughing up streams of blood and noticing a significantly huge swelling in his right testicle. He immediately visited a urologist. X-rays revealed testicular cancer with extensive metastasis to the lungs. Armstrong underwent surgery to remove the testicle at 7 a.m. the following day. However, within a week, a brain MRI scan revealed two spots, indicating that the cancer had spread beyond the lungs. Armstrong was diagnosed with Stage 3 testicular cancer.

His treatment at the Indiana University Medical Centre in Indianapolis commenced with a surgery to remove the two lesions on the brain, followed by chemotherapy, which was a combination of vinblastine, etoposide, ifosfamide and cisplatin. As he wrote in his book *It's Not about the Bike*, 'I thought, "Bring it on,

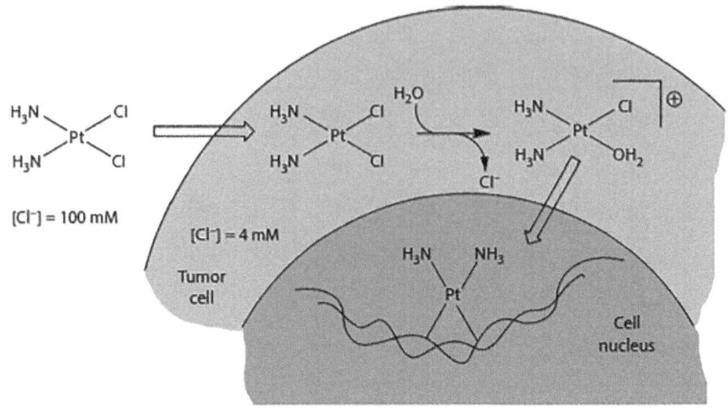

FIGURE 13.2 How cisplatin crosslinks DNA.

Source: Based on R. A. Alderden, M. D. Hall and T. W. Hambley, *J. Chem. Educ.*, **2006, 83, 728–734**

give me the platinum"'. He completed chemotherapy on 13 December that year and was cured. Within three years, Armstrong won the 1999 *Tour De France*. He continued to win the race for seven consecutive years. However, he later admitted to using performance-enhancing drugs and was stripped of these (and other) titles in 2012.

As mentioned earlier, the cis-geometry is important in understanding how cisplatin acts as an anticancer drug when it enters cancer cells (Figure 13.2). Generally, *cis*-$[Pt(NH_3)_2Cl_2]$ is administered intravenously in a saline solution, where the high concentration of chloride ions, approximately 100 mM (milliMolar), prevents the substitution of chloride ions in the cisplatin molecule by water. Once inside the cell, where the chloride ion concentration is much lower (around 4mM), water molecules can substitute chloride ions, producing *cis*-$[Pt(NH_3)_2(H_2O)Cl]^+$ ions.

$$[Pt(NH_3)_2Cl_2] + H_2O \rightleftharpoons [Pt(NH_3)_2(H_2O)Cl]^+ + Cl^-$$

$$[Pt(NH_3)_2(H_2O)Cl]^+ + H_2O \rightleftharpoons [Pt(NH_3)_2(H_2O)_2]^{2+} + Cl^-$$

These water-substituted cisplatins lose one water molecule upon binding to a purine base in one strand of the cellular DNA. Once the platinum-containing species is bound to the DNA, a base in the second strand replaces the remaining chloride, thereby cross-linking the two DNA strands. This causes the DNA to distort, so that it is not recognised by the cell's repair mechanism, resulting in cell death. The (E)-isomer of $[Pt(NH_3)_2Cl_2]$ is not active against cancers, as its *trans*-geometry means that it cannot cross-link DNA.

Cisplatin has a bit of a pedigree. It was first synthesised in 1844 by the Italian chemist Michele Peyrone (1813–1883). However, its structure was not clearly explained

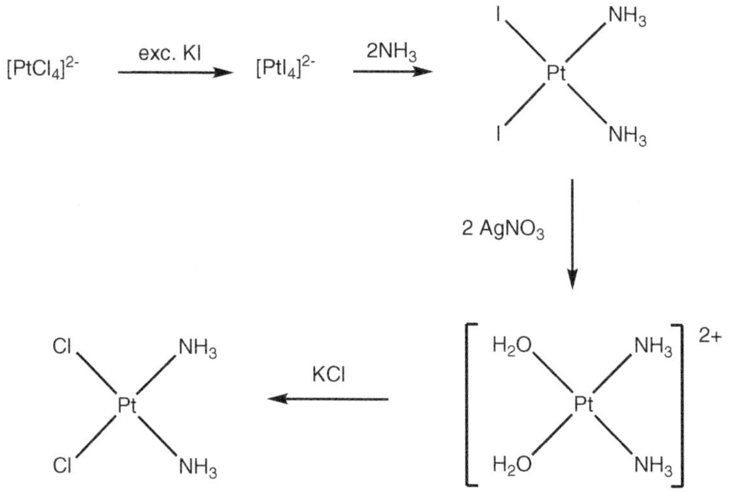

FIGURE 13.3 The synthesis of cisplatin.

until almost half a century later, mostly through the work of Alfred Werner. The synthesis most commonly used today to prepare cisplatin is a method reported in 1970, which begins with an aqueous solution of K_2PtCl_4.

The steps are: –

(i) addition of potassium iodide to K_2PtCl_4, forming K_2PtI_4.
(ii) $K_2[PtI_4]$ is reacted with NH_3 (aq), which affords a precipitate of slightly soluble *cis*-$[Pt(NH_3)_2I_2]$
(iii) On addition of $AgNO_3$ (aq) to a solution of *cis*-$[Pt(NH_3)_2I_2]$, AgI precipitates; this is filtered off, leaving a solution of *cis*-$[Pt(NH_3)_2(H_2O)_2]^{2+}$
(iv) When KCl(aq) is added to the solution of *cis*-$[Pt(NH_3)_2(H_2O)_2]^{2+}$, yellow *cis*-$[Pt(NH_3)_2Cl_2]$ precipitates.

This route avoids complications due to the precipitation of green Magnus' salt $[Pt(NH_3)_4]^{2+}$ $[PtCl_4]^{2}$ if ammonia is added to K_2PtCl_4.

Cisplatin was not a perfect anticancer drug, particularly because of resistance problems and side-effects associated with the drug. Hence, scientists began synthesising and evaluating other platinum-based compounds.

The other platinum-based compounds included carboplatin, which certainly has fewer side effects, and oxaliplatin, which is used in the treatment of colorectal cancer.

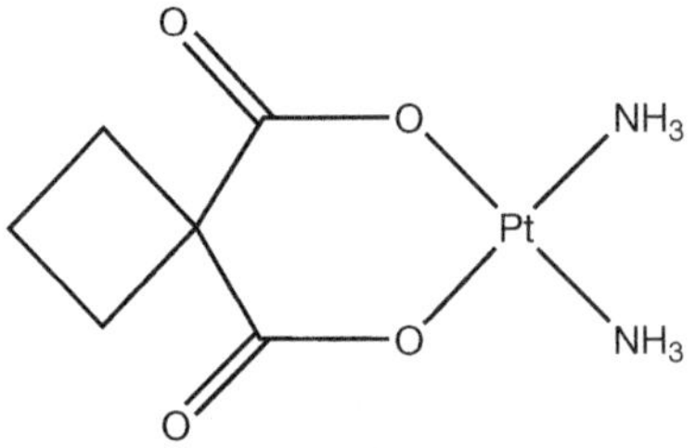

(13.2)

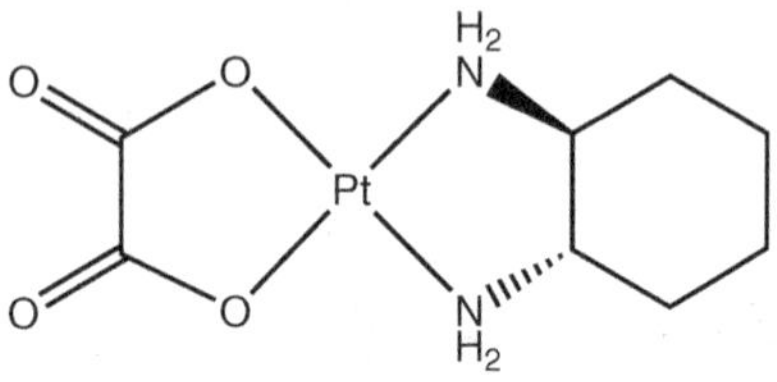

(13.3)

Carboplatin (13.2) received FDA approval in 1989 for the treatment of lung, head and neck cancers, as well as ovarian carcinoma. Its mechanism of actions is similar to cisplatin, but with significantly fewer side effects, as it is more stable in the blood and less toxic to the kidneys. Oxaliplatin (13.3) has a different activity spectrum to cisplatin, which circumvents resistance. It has been approved both in Europe (1999) and the USA (2002) for secondary treatment of metastatic colorectal cancers.

Satraplatin (JM216: (13.4)) is an example of a Pt(IV) compound as an anti-cancer drug. It has been under investigation for some years, undergoing Phase III trials as a treatment for advanced prostate cancer. It is the first platinum-based drug to be taken orally, while all the others require intravenous administration.

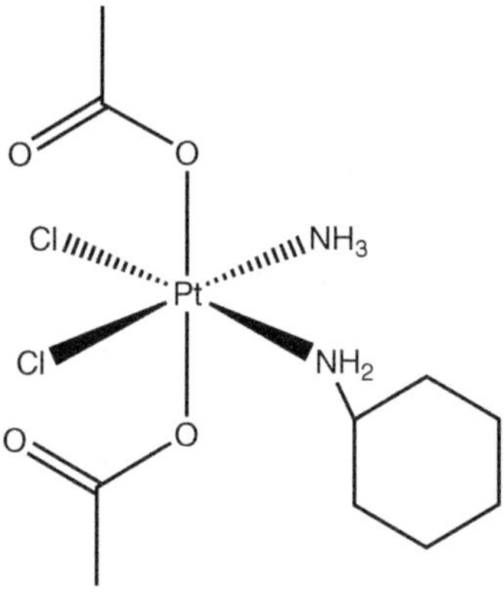

(13.4)

And of course, one should always have in mind the words attributed to Pasteur, 'In the field of observation, chance only favours the prepared mind' (*le hasard ne favorise que les esprits préparés*).

14 Cocaine

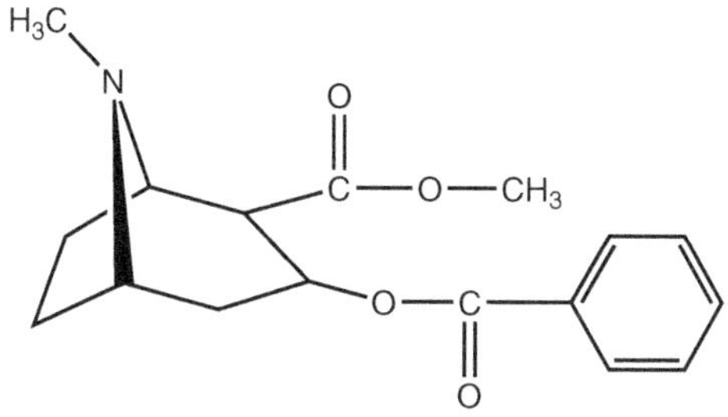

(14.1)

Cocaine (14.1) is synthesised in the leaves of the coca plant, *Erythroxylum coca,* which grows wild in the Andes and northern Amazon region. It is biosynthesised from the amino acids glutamine and orthinine. The plant produces it as a natural insecticide, as evidenced by the rarity of insect damage to the plant compared with other plants in the area. The cocaine molecule contains four chiral carbon atoms and can therefore exist in 16 isomeric forms. However, the enzyme-catalysed reactions in its biosynthesis ensure that only one isomer is formed, namely (R)-(-)-cocaine.

The leaves, which contain nearly 0.3–0.7% cocaine, have been chewed by the people of Bolivia, Peru, Ecuador and Chile for thousands of years. In addition to drinking coca tea, the Incas prepared pellets of the leaves by mixing small amounts of alkali (lime is used today) and holding them in their mouths. This practice led to the gradual release of cocaine, which was absorbed through the saliva. Cocaine enhances strength and endurance, while suppressing appetite. It enabled the relays of messengers (*chasqui*) covering distances up to 300 km a day, carrying messages throughout the Inca Empire. In the mountains, cocaine helped counter altitude sickness. Following the Spanish conquest of Peru, the conquistadors found that the natives they had enslaved could work harder for longer periods at jobs such as silver mining when they were provided with coca leaves to chew. During the second half of the nineteenth century, cocaine entered the public consciousness in Europe. First isolated in 1855 by Friedrich Gaedcke, cocaine was obtained in its pure state in 1859 by Albert Niemann, a student of the great Friedrich Wöhler.

It was deemed useful in medicinal purposes, including dentistry, anaesthesiology and ophthalmology. The American surgeon William Stewart Halsted used cocaine injections to invent nerve-block anaesthesia in 1884 but unfortunately he became a cocaine addict. Sigmund Freud, inventor of psychoanalysis, pioneered the use of cocaine as a local anaesthetic for eye surgery. In fact, for a significant period, cocaine was an over-the-counter medication for the treatment of toothache and catarrh. In a parallel study, Angelo Mariani, a chemist from Corsica, discovered that the ethanol

DOI: 10.1201/9781003783565-14

in wine extracted the cocaine from coca leaves. He subsequently marketed the preparation as *Vin Mariani*. He sent bottles of his wine to famous people in hope of securing endorsements. These were promptly provided by Thomas Alva Edison, Jules Verne, H.G. Wells and even Pope Leo XIII. On the other side of the Atlantic, 'French Wine Cola' was being prepared by John Styth Pemberton, a Georgian pharmacist. Undaunted by the declaration of prohibition in Atlanta in 1886, Pemberton launched an alcohol–free version of the wine, but continued to used coca leaves and kola nut. A further change occurred in 1903. Following the outlawing of cocaine across the USA, cocaine was replaced by sugar and caffeine as stimulants in the drink, which evolved into the drink known to us as *Coca-Cola*.

The stimulant effects of cocaine were generally considered undesirable by this time, leading to its ban in North America and Europe. During the 1880s, Arthur Conan Doyle, an ophthalmologist-turned novelist, described his fictional detective Sherlock Holmes as the first celebrity with a cocaine problem in *A Scandal in Bohemia* and *The Sign of Four*. Its stimulant properties were also exploited by the polar explorer Sir Ernest Shackleton during his 1909 expedition. He carried 'Forced March' coca tablets to do what was said on the tin.

Since then, cocaine has been popular as an illegal stimulant, with a demand later fuelled by gangs from South America, famously the Cali and Medellin cartels of Colombia. Pablo Emilio Escobar Gaviria (1949–1992) was the founder of the Medellin cartel, who was once reckoned to be the seventh richest man in the world with a fortune estimated at approximately 30 billion American dollars and died in a shoot-out with Colombian National Police. Today, cocaine is cultivated in a number of South American countries, notably Bolivia, Colombia and Peru. Its cultivation has expanded to other parts of the continent, including Belize, Guatemala and Honduras. The cocaine coming from different soils and climate has slightly different isotopic compositions. The isotopic ratios ^{13}C:^{12}C and ^{15}N:^{14}N, as well as the levels of trace alkaloids truxilline and trimethoxycocaine, have been used to pinpoint where a particular type of cocaine originates. More recently, ^{2}H and ^{18}O levels have been added to refine the method. Sadly, cocaine has proved to be dangerously addictive. Its use and abuse has been linked with numerous fatalities, including several popular entertainers such as Whitney Houston, John Belushi, River Phoenix and Carrie Fisher.

For many years, the only form of cocaine used by its 'consumers' came in the form of the hydrochloride salt, but in 1984–1985 'crack' cocaine arrived – a mixture of cocaine hydrochloride and baking powder ($NaHCO_3$) which, when heated for smoking, emits a crackling noise (hence its name) and generates the free cocaine base, which is quite volatile above c. 90°C. This was a cheaper and fast-acting form of cocaine. So traditional cocaine is a salt, whilst 'crack' cocaine is molecular in structure.

Cocaine production is illicit, of course, but certain aspects of it illustrate a number of important chemical principles. It is extracted from coca leaves in three steps, often in three different laboratories which can be hundreds of miles apart. Solvent extraction is involved, and this depends upon a fundamental difference between neutral cocaine molecules (the free base), which have low solubility in water but are very soluble in non-polar organic solvents; in contrast, protonated cocaine ions are soluble in water, whilst having very low solubility in organic solvents. Both ionic and molecular forms are involved in cocaine processing.

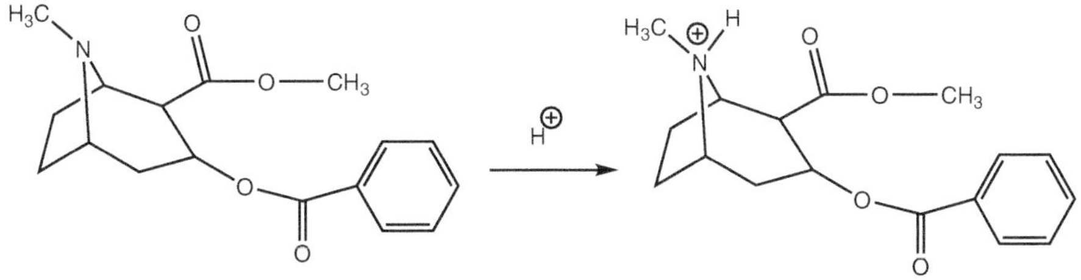

FIGURE 14.1 Extraction of protonated cocaine from coca leaves.

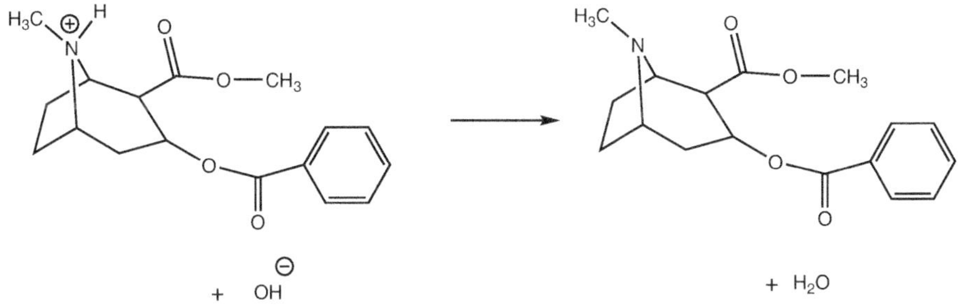

FIGURE 14.2 Formation of neutral cocaine molecules by deprotonation of protonated cocaine.

The first step of cocaine processing involves agitating coca leaves with a base (usually lime), the base ensuring that all the cocaine is present as the neutral molecular form, which can then be extracted using a non-polar solvent like kerosene. The kerosene solution of cocaine is then mixed with dilute sulfuric acid; the acid protonates the cocaine (Figure 14.1), and these ions are extracted into the polar aqueous layer as the salt cocaine sulphate in a 'back-extraction' step.

The aqueous layer containing cocaine sulphate is then separated and an alkali is added to it, usually calcium hydroxide, which deprotonates the protonated cocaine ions (Figure 14.12), resulting in the precipitation of neutral cocaine molecules – the crude 'coca paste', which can be anything between 30% and 80% pure.

This impure cocaine is moved to another laboratory for purification (step 2). The base is dissolved again in dilute sulphuric acid, and potassium manganate (VII) solution ($KMnO_4$) is added slowly until the mixture is just colourless (rather than purple), in an unusual 'permanganate titration'. The permanganate oxidises the alkaloid impurities present and itself is reduced to a brown precipitate of insoluble MnO_2, which is filtered off from the aqueous layer. This aqueous layer containing protonated cocaine is then treated with alkali (normally ammonia solution), which neutralises the cocaine sulphate; the precipitated cocaine is filtered off and dried. This is 'coke base'.

'Step 3' follows at a 'crystal lab'. The cocaine molecules are dissolved in an organic solvent like ethoxyethane (ether), with any insoluble impurities being filtered off. The filtrate, still containing neutral cocaine molecules, is mixed with a mixture of concentrated HCl and propanone, so that cocaine hydrochloride crystallises out. Other solvents like butanone ('methyl ethyl ketone') are also used.

As already noted, two forms of cocaine are currently consumed, with normal cocaine powder being the (ionic) salt cocaine hydrochloride, traditionally injected or snorted. Smoking it destroys most of the cocaine. Treating the hydrochloride salt with baking soda makes 'crack' cocaine, which is much more volatile, so many more cocaine molecules survive to be absorbed. If cocaine is injected or smoked, then, like nicotine, it reaches the brain in seconds; by contrast, if swallowed, most of it is destroyed in the liver before it can get to the brain. Inhaling crack generates a vapour which is easily absorbed and gives a rapid sensation. Cocaine is a central nervous system stimulant which prevents the reuptake of the body's neurotransmitters like dopamine, serotonin and noradrenaline (notably the first-named), thus affecting the pleasure centre of the brain, which is why cocaine is a stimulant. Cocaine is very soluble in water, yet can cross the blood-brain barrier, a property ascribed to an intramolecular hydrogen bond pulling the molecule into a conformation which shields its hydrophilic regions.

Cocaine's anaesthetic effects are caused by its ability to act as a sodium-channel blocker, interfering with the transmission of nerve impulses. It appears to affect neurons in the *nucleus accumbens* region of the brain, causing loss of grey matter. Cocaine is a stimulant, increasing pulse rate and blood pressure. Of concern is the risk of cardiac arrest, even for very healthy people, for whom numerous deaths have been documented, famously including the University of Maryland basketball star Len Bias, in 1986. Some deaths in the USA have been associated with cocaine users being pepper-sprayed by police, the suggestion being that exposure to capsaicin increases the toxic effects of cocaine.

The body uses two enzymes in its liver to metabolise cocaine, carboxylesterase enzymes hCE1 and hCE2, with just 1% of the cocaine excreted unchanged. Hydrolysis of an ester linkage affords the readily detectable benzoylecgonine, which shows that the subject has consumed cocaine and not accidentally picked it up on their hands (Figure 14.3). Benzoylecgonine can similarly be detected in a user's sweat, notably in their fingerprints, again indicating that the person concerned has taken cocaine into their body. Such metabolites have implicated sportsmen such as Diego Maradona, Kieron Fallon and Martina Hingis in cocaine abuse. That's not all that hCE1 does.

If a person has consumed alcohol at the same time as cocaine, then hCE1 carries out a transesterification reaction (Figure 14.3) forming cocaethylene, which is more toxic than either cocaine or ethanol on its own. Mixing cocaine and alcohol has been credited with the deaths of the American baseball pitchers José Fernández and Tommy Hanson, the novelist Jacques d'Adelswärd-Fersen, and the musician Jay Reatard, among many others.

Another metabolite of cocaine, methylecgonidine ((14.2) -anhydroecgonine methyl ester), is a pyrolysis product of smoking crack cocaine, though it is not formed in 'normal' ingestion of cocaine. It can thus be used to distinguish between someone who has been smoking crack and another who was merely snorting cocaine.

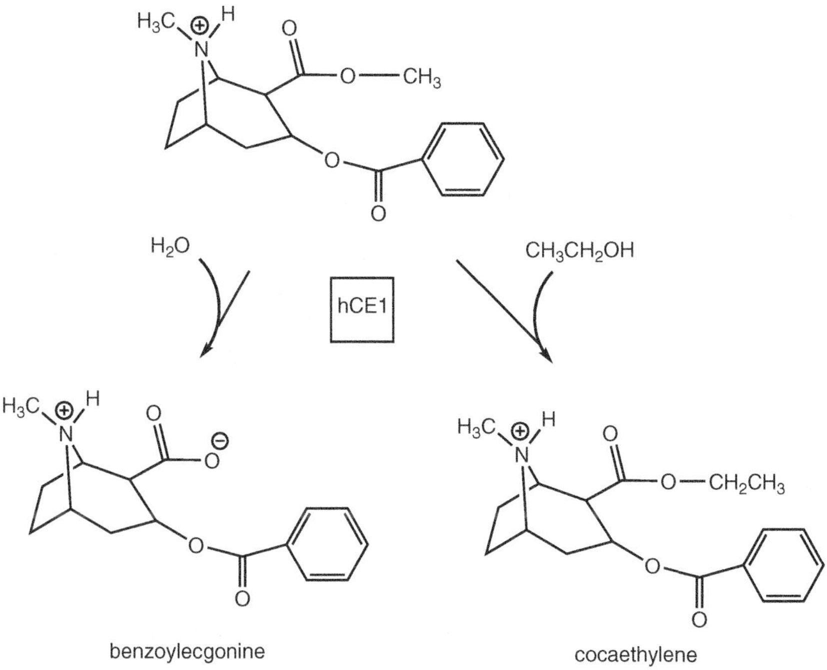

FIGURE 14.3 Metabolism of cocaine by the enzyme hCE1.

Cocaine and its metabolites including benzoylecgonine were found in hair samples from ancient Peruvian coca-leaf chewers dating back to AD 1000.

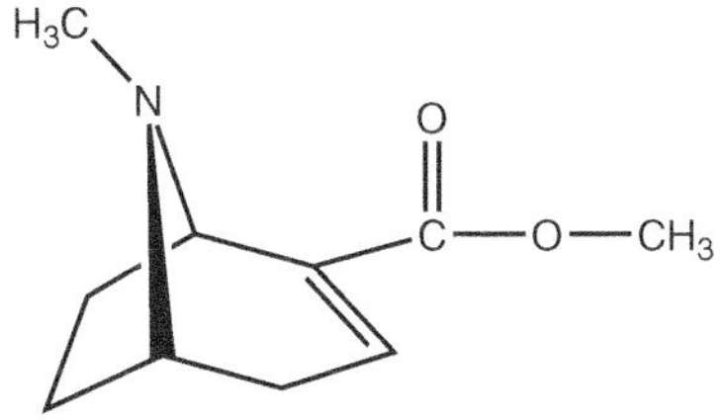

(14.2)

Cocaine doesn't just contaminate people. Researchers at Dublin City University found that nearly all the euro banknotes they examined in Dublin in 2007 had detectable levels of cocaine. Similar findings have been made for £10 and £20 notes in the London area, and for dollar bills in the USA. The cocaine can be removed from the notes by solvent extraction, but a quicker way of detecting it involves tandem mass

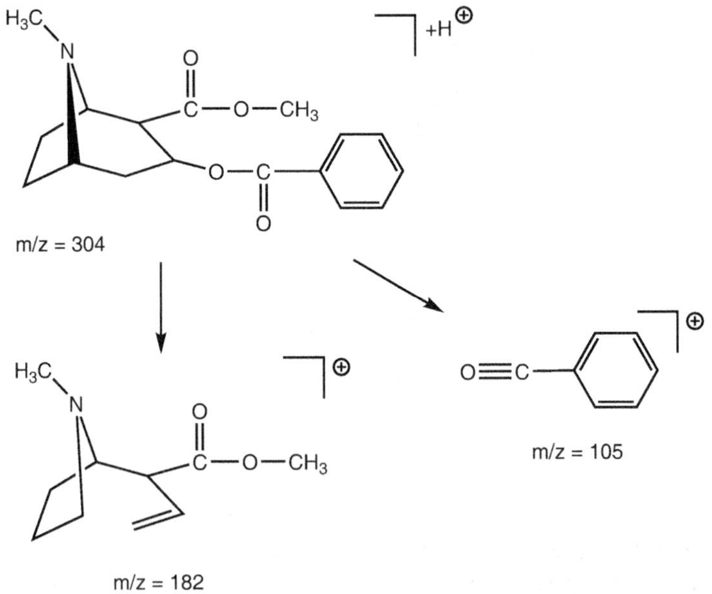

FIGURE 14.4 Fragmentation of cocaine in a mass spectrometer.

spectrometry. The banknote is heated up to 285°C for a second, the cocaine molecules get released, then carried on a stream of air into the spectrometer, where they are ionised and protonated. This ion is fragmented when it bumps into other ions, and the spectrometer can detect these fragment ions with m/z values of 105 and 182 (Figure 14.4).

Although the vast majority of the banknotes examined by the Irish researchers had just a trace of cocaine – around 2 nanograms per note – some 5% of them had 200 ng or more. Most notes owed their cocaine contamination to contact with seriously contaminated notes in counting machines at a bank, but the notes with the high level of cocaine had probably been rolled up and used to 'snort' the drug. Just because drugs can be detected on notes in your wallet does not mean that you are a drug user.

Beyond that, scientists can detect cocaine – or its metabolites – in the environment. It turns up in the wastewater produced by cities in countries all over Europe, such as England, Ireland, Belgium, Italy, Spain and Switzerland, as well as in the air of cities including Milan, Rome, Porto, São Paolo in Brazil and Santiago in Chile. Its use is so widespread, with an estimated 20 million or more people consuming it.

15 Codeine and Oxycodone and Desomorphine

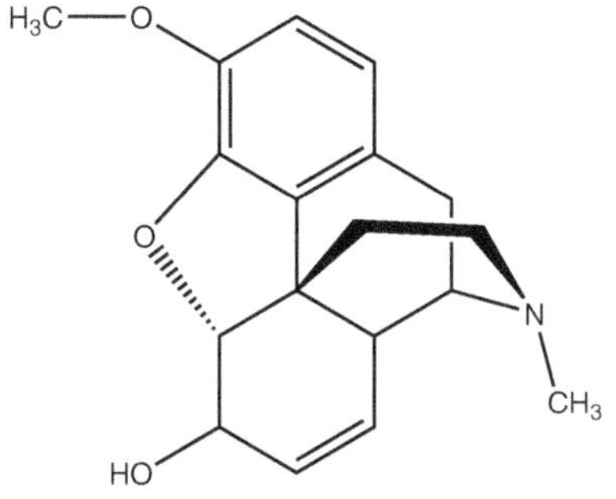

(15.1)

This chapter discusses codeine (15.1), a naturally produced medicinal molecule, and two synthetic opiates made from it. Codeine is one of only two painkilling molecules produced naturally by the opium poppy (*Papaver somniferum*), along with morphine, though there is much more morphine in the poppy. The poppy makes codeine using a series of enzyme-catalysed reactions, starting from L-tyrosine, an amino acid. One molecule of tyrosine is turned into dopamine, and another into 2–(4-hydroxyphenyl) acetaldehyde; these two molecules are then joined together (in another enzyme-catalysed reaction), forming (S)-norcolaurine (Figure 15.1), and generating the very outline of the skeleton of the codeine molecule.

Further steps generate more rings, eventually forming the alkaloid thebaine (15.2). Although thebaine is another alkaloid present in the opium poppy, it is not a painkiller. Thebaine can be turned into codeine, and also morphine, in a few more steps.

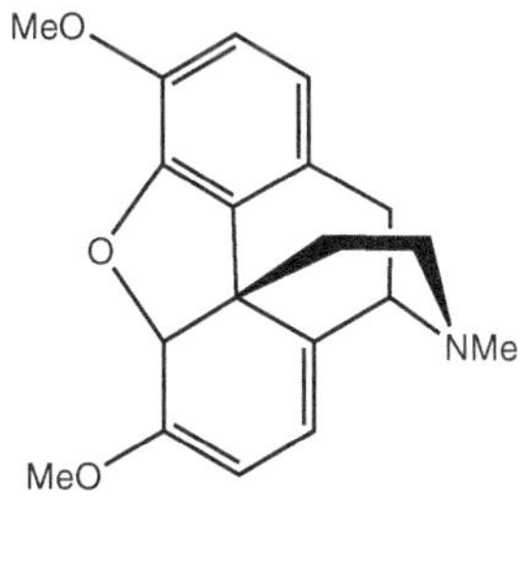

(15.2)

DOI: 10.1201/9781003783565-15

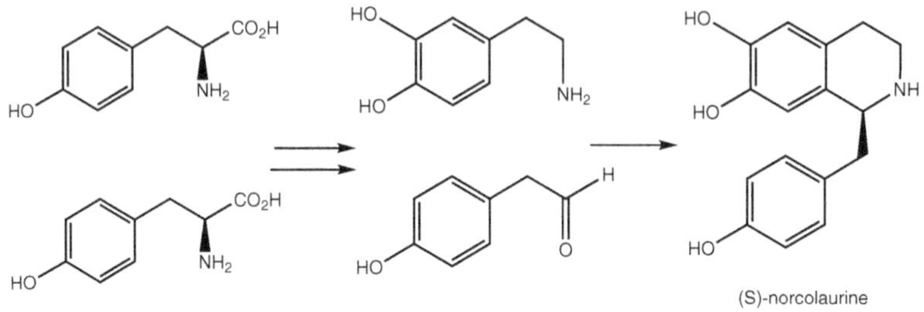

FIGURE 15.1 Biosynthesis of (S)-norcolaurine from L-tyrosine.

Codeine is a milder painkiller than morphine. It is in considerable demand by the pharmaceutical industry, far exceeding the amount that can be obtained directly from poppies, so much codeine is made from morphine by methylating its phenolic -OH group, with the other –OH group unaffected. This route to codeine was first reported by the Russian chemist V. M. Rodionov (1878–1954).

Although codeine was not isolated until 1832, the painkilling ability of codeine (and morphine) has been known for several millennia in the Near East through the medium of opium. In the early sixteenth century, Paracelsus (aka Phillipus Aureolus Theophrastus Bombastus von Hohenheim) created the opium-derived drug known as laudanum; in various formulations, this was widely used as a medicine, along with opium. From Victorian times onwards, codeine has been used as a mild painkiller, sometimes mixed with aspirin, paracetamol (acetaminophen) or ibuprofen to produce both analgesic and fever-reducing effects, or with caffeine, which adds stimulant power. Possibly the most well-known codeine-containing medication, a hundred or so years ago, was Dr J. Collis Browne's Compound, sold widely as an over-the-counter medicine for the treatment of complaints such as diarrhoea; it was also used in cough syrups. Over the years, the codeine content was reduced, and for many years it has been codeine-free. Nowadays, codeine is only available in prescription drugs.

Codeine acts as a prodrug, because metabolism in the liver turns about 10% of the codeine taken into morphine, which is the substance responsible for the pain relief experienced when codeine is taken. Morphine binds to the μ-(mu) opioid receptors in the central nervous system.

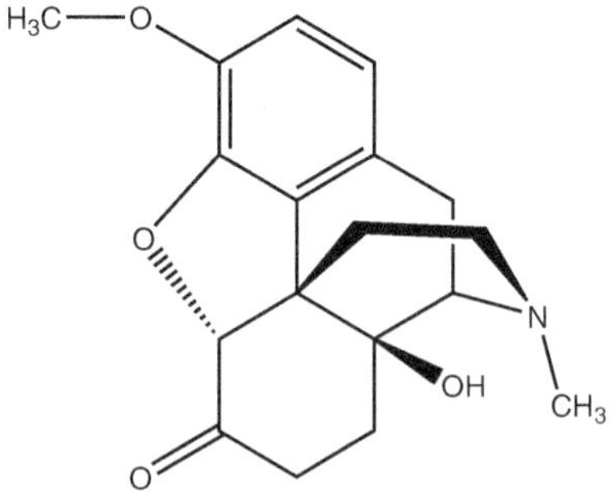

(15.3)

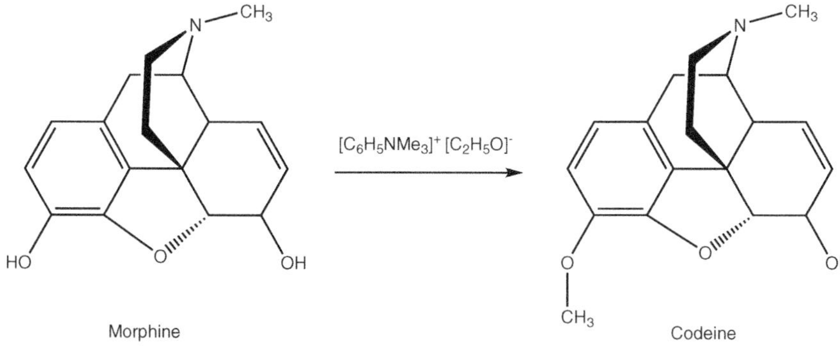

FIGURE 15.2 Synthesis of codeine from morphine.

Oxycodone (15.3) is a synthetic painkiller; it does not exist in nature. It was first made in 1916, not very long after heroin was first synthesised. At that time, pharmaceutical companies were trying to 'invent' synthetic painkillers that were better than morphine but less addictive than heroin.

Oxycodone is made from thebaine or codeine. This is much easier – with fewer steps – than 'starting from scratch'. For one thing, it gives a better yield, so it is more efficient. The process is known as 'semisynthesis' – you get nature, in a plant, to do the hard work of building the complicated core, then you just add the 'finishing touches'.

If you start from thebaine, obtained from opium poppies, you first oxidise thebaine to 14-hydroxycodeinone. This can be done simply with hydrogen peroxide, but the results are better with m-chlorobenzoic acid (m-CPBA) in acetic acid/trifluoroacetic acid. Then the conjugated C=C bond in 14-hydroxycodeinone is reduced to a C-C bond by catalytic hydrogenation, forming oxycodone (Figure 15.3).

Following its discovery, oxycodone was used in painkillers in Central Europe and then after World War II in the USA. So, oxycodone had been used for some time before Oxycontin® was introduced in 1995; it was a 'controlled-release' formulation of oxycodone, in which, once the medication was taken, the oxycodone content was released gradually. So, the patient would need to take a new dose every 12 hours or so to get continuous pain relief. However, American doctors started to prescribe it widely as a painkiller, not for the kind of severe pain in terminal cancer patients for which morphine is used, but for milder pain relief. Oxycontin became the leading painkiller in the USA.

Unfortunately, along with the painkilling effects, oxycodone brings euphoria with it. Patients began liking the feelings they got when taking OxyContin and became addicted. Oxycontin abuse soon began, as some people discovered that if they crushed the medication into a powder, all the oxycodone was released immediately; they then began snorting the Oxycontin or dissolving it in water and injecting it for the 'high' it gave them. OxyContin addiction became a serious problem, linked with thousands of deaths in the United States, as well as leading to heroin addiction.

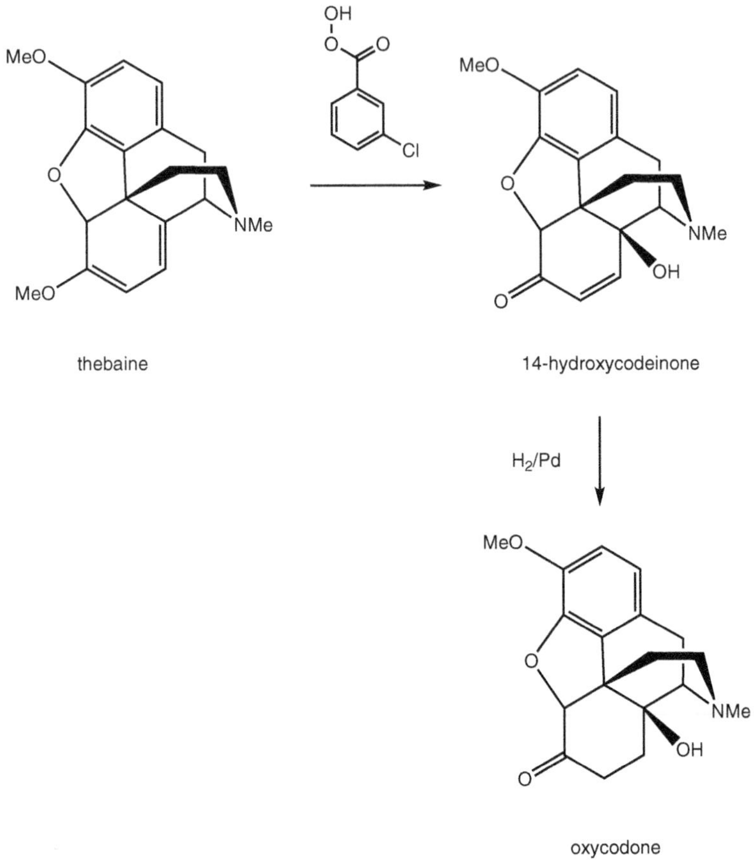

FIGURE 15.3 Synthesis of oxycodone from thebaine.

How oxycodone exerts its painkiller effects is uncertain. Some have suggested that it acts through a mixture of μ-(mu) and κ-(kappa) opioid receptors. It may be that its metabolite, oxymorphone (15.4), which is also a painkiller, is involved.

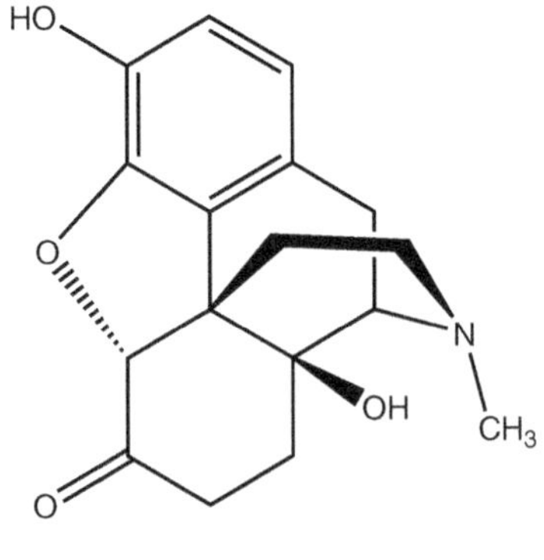

(15.4)

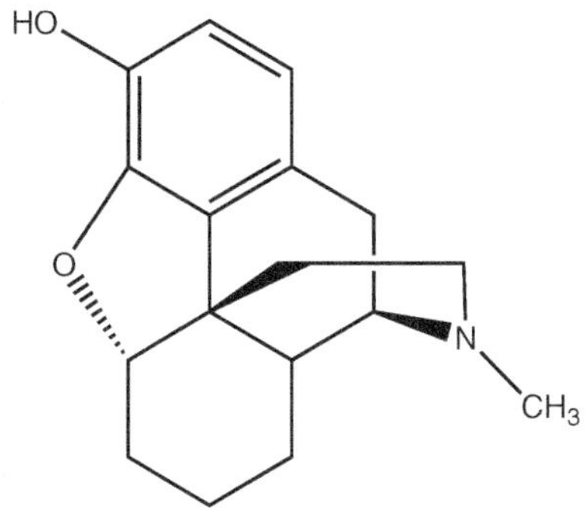

(15.5)

Desomorphine (15.5), with a rather similar structure to morphine, was first reported nearly 20 years after oxycodone, in 1932. Compared to morphine, it has one C=C double bond fewer, as well as lacking one –OH group and a methyl substituent. Its synthesis (Figure 15.4) generally begins with codeine. The –OH group

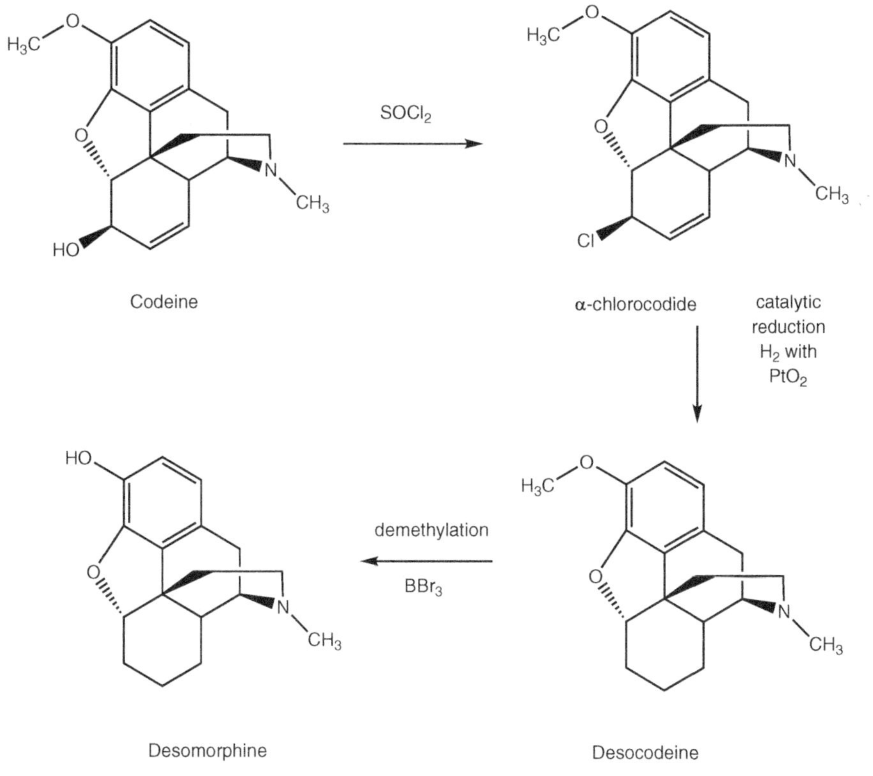

FIGURE 15.4 Synthesis of desomorphine from codeine.

is chlorinated, then hydrogenation removes both the chlorine and the C=C bond. Finally, the methoxy group is demethylated.

Once again, desomorphine's synthesis was part of the ongoing quest to devise a painkiller with fewer side effects than morphine. It turned out to be a stronger painkiller than morphine because having one fewer –OH group than morphine meant that desomorphine was more lipid-soluble and more likely to cross the blood-brain barrier. Thus, desomorphine is predicted to access the brain better than morphine. So, desomorphine is roughly ten times as strong a painkiller as morphine and its effects are felt sooner. It has fewer side effects than morphine (such as depressing respiration), but sadly it has a greater potential for dependence. Thus, it was never adopted as a medicine, but it has become an addictive drug, notably in Russia.

When Russian heroin addicts found it more difficult to get heroin in the early 2000s, at a time when supplies from Afghanistan were cut off, someone came up with the idea of using desomorphine. Because, at that time, codeine was an over-the-counter drug which was easy to obtain (and remained so in Russia until 2012), and because other chemicals needed, such as phosphorus (from matchboxes) and iodine (a common antiseptic), were easily available, desomorphine was more inexpensive to prepare than heroin. It could also be made in an hour by people with no chemical training; this was important, as addicts needed to inject it every few hours, on account of the short-lived 'high'.

Desomorphine addiction comes with very real problems. The addicts don't take the time to purify the desomorphine they have made – they inject (intravenously) a very impure substance, full of toxic and corrosive impurities. These impurities cause problems including thrombosis and gangrenous damage, subcutaneous abscesses and ulceration. The addicts' flesh rots, and bones and muscles become infected, so it has come to be called a 'flesh-eating' drug. They develop scaly skin with a green-black colour, resembling a crocodile, which is where the name 'Krokodil' comes from. And intravenous injectors often share syringes, with the usual concomitant risks of AIDS/HIV and hepatitis C. Krokodil addicts are often dead within a year or so.

16 Dimethylmercury

$$H_3C—Hg—CH_3$$

(16.1)

Dimethylmercury (16.1) can best be described as a supertoxic compound. Given what is now known about its toxicity, it is remarkable that any of the early chemists who worked with it survived, given the absence of the protective materials that we take for granted today. It was first prepared in 1858 by George Bowdler Buckton (1818–1905), who worked for some time as an assistant to A. W. Hofmann at the Royal College of Chemistry, now part of Imperial College, London.

$$2\ CH_3HgI + 2\ KCN \rightarrow Hg + Hg(CH_3)_2 + 2\ KI + (CN)_2$$

Buckton reported it as a colourless volatile liquid. The boiling point is variously reported as 92 or 96 °C. A few years later, the research group of the pioneering organometallic chemist Edward Frankland, investigated the compound. In February 1865, Dr Carl Ulrich inhaled some $Hg(CH_3)_2$, was admitted to hospital, and was dead within a fortnight. A technician in the group was admitted to hospital the following month with milder symptoms but soon became demented and died a year later. That same year, 1865, Buckton married Mary Ann Odling, sister of William Odling, who had also worked with Frankland on dimethylmercury. Buckton promptly gave up chemistry and moved back to his earlier interest in entomology. Buckton did live to the age of 87, though. In the succeeding years, occasional chemists worked with dimethyl mercury, including Henry Gilman, the leading U.S. organometallic chemist in the early twentieth century. The next fatality occurred in a 28-year-old Czech chemist, who had synthesised 6 kg [*sic*] of dimethylmercury over a three-month period in 1971.

By this time, the toxicity of mercury compounds was familiar. Two hundred years ago, mercury nitrate solution was used as a preservative for some types of felt hat, causing some of the workers handling these hats to develop 'hatter's disease', a mercury poisoning that is believed to have inspired the character of the Mad Hatter in Lewis Carroll's *Alice in Wonderland*. Organomercury compounds were used as fungicides on some seeds, and poisonings occurred when people ate these seeds, in Iraq in 1956 and again in 1960, whilst around 100 people were poisoned through eating treated seed wheat in West Pakistan in 1961. Even worse was to happen in Iraq in 1971–1972. Methylmercury compounds had been used as a fungicide on grain that was meant for planting. Instead, someone sold it to be used for milling and making bread. The grain was dyed red and the sacks containing it had warning labels in Spanish and English, which no one could read. A total of 459 people died. In general,

DOI: 10.1201/9781003783565-16

the compounds used were monomethylmercury derivatives, of the general formula CH_3MgX.

In parallel to these events, strange things happened around the seaside town of Minamata, on Kyushu Island in Japan in the early 1950s. First, it was cats that exhibited nervous behaviour; next, birds dropped from the sky. Then around 1956 humans started showing symptoms and the following year, fishing was banned – and fish are important in the Japanese diet. It transpired that the factory of the Chisso Chemical Company, which was using a mercury-containing catalyst to make acetaldehyde (ethanal), was discharging wastes into the Minamata River and later directly into Minamata Bay. By 2005, the number of certified cases of mercury poisoning (Minamata disease) in the area had risen to 2,955 cases, with around 700 still alive at that time.

People were eating fish contaminated with methylmercury compounds. It is believed that what became known as 'Minamata disease' was caused by bacteria in sediments on the seafloor or in riverbeds methylating 'inorganic' mercury ions, and possibly by the presence of methylmercury ions in effluent discharges from the industrial plant. The methylmercury species are absorbed by plankton, which are eaten by small fish, which in turn are eaten by larger fish, which eventually are eaten by humans, with the concentration of mercury increasing along the food chain. This 'biomagnification' means that the final concentration in animals higher up the food chain, including humans, can be – at least – a thousand times larger than that present in the original water.

But it was the events of 1996–1997 that led to a serious reassessment of mercury poisoning. Karen Wetterhahn was a senior professor at Dartmouth College, a leading American university based in New Hampshire. An expert on the role of heavy metals in living systems, she was working at the lab bench on 14 August 1996. She was well aware of the toxicity of dimethyl mercury and was thus doing a manipulation that she did not want her students to risk. Her group was studying how mercury ions interacted with DNA repair proteins; carrying out ^{199}Hg NMR measurements, she had decided to use $[Hg(CH_3)_2]$ as a reference standard, the equivalent of using $Si(CH_3)_4$, tetramethylsilane, in ^{1}H NMR. Wetterhahn took the approved safety precautions, donning safety glasses and latex gloves and working in a fume cupboard. She took a sealed glass vial of dimethylmercury, which had been cooled in ice-water to reduce the volatility of the mercury compound. Her colleague David Lemal cut the top off the glass vial, so that Wetterhahn was able to pipette a small sample of $[Hg(CH_3)_2]$ into an NMR tube. She sealed and labelled the tube, put the rest of the dimethylmercury into a storage container, then disposed of the gloves and cleaned up. Then she went home to her husband and children.

Five months later, in January 1997, Wetterhahn started to notice some strange symptoms. She was getting tingly feelings in her fingers and toes, and her speech started to slur. Her field of vision began to shrink. She was losing her hearing abilities and started to have balance problems. She had lost a stone in weight. She went for tests and, on 28 January, she was diagnosed with mercury poisoning. The normal level of mercury in the blood of a healthy person would be 1 to 8 micrograms per litre; the toxic threshold was 50 micrograms per litre. Wetterhahn's test results came back with a level of 4000 micrograms per litre, at least 400 times 'normal' and 80 times the toxic threshold. Chelation therapy – using succimer to remove mercury from the body – was begun immediately and succeeded in gradually reducing the amount of mercury in her body, but to no avail. Wetterhahn lapsed into a coma on 6 February 1997. Karen Wetterhahn never recovered from the coma, dying on 8 June 1997.

In January 1997, Wetterhahn recalled spilling possibly one or two drops of dimethylmercury on her protective gloves on that August day. It is now known that latex offers no protection – the compound would have penetrated straight through the glove and entered her skin within 15 seconds. A safety reassessment concluded that the only safe course in handling dimethylmercury is to wear highly resistant laminated gloves underneath a pair of long-cuffed neoprene (or other heavy-duty) gloves.

As already stated, dimethylmercury is 'supertoxic', one of the most lethal neurotoxins in existence. It easily crosses the blood-brain barrier, most likely as a methylmercury-cysteine complex. The large Hg^{2+} ion is a 'soft acid' and therefore prefers to bind to easily polarisable donor atoms in 'soft bases', so that it has a liking for sulphur and sulphur-containing ligands. One consequence of this is that it will bind to the thiol groups of enzymes and thus inhibit them. On a historical note, it was the Danish chemist W. C. Zeise who made the first mercaptan (R-SH) ligands and coined their name, based on the Latin expression *mercurium captans*, meaning 'capturing mercury'. Mercury causes ataxia (lack of coordination), sensory disturbance and changes in mental state. It inhibits several stages of neurotransmission in the brain. It is a cumulative poison that is very slowly removed (excreted) from the body, and by the time its effects are noted, it will be too late to do anything about it. It spreads throughout the body, reaching the brain relatively quickly. One (disputed) theory is that a transporter that takes amino acids into the brain mistakes the methylmercury-cysteine complex (16.2) for the amino acid methionine (16.3), because they have similar shapes. This phenomenon is called 'molecular mimicry'.

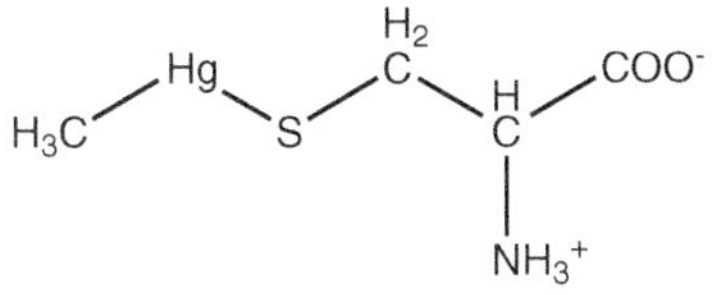

(16.2)

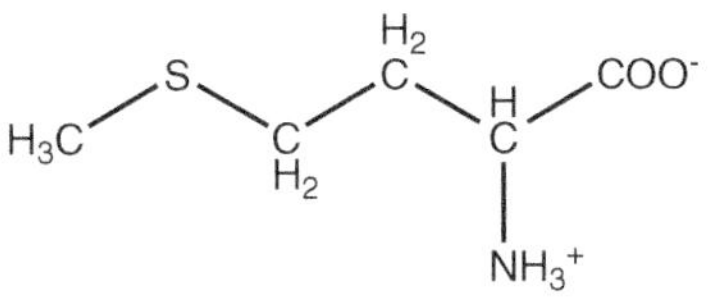

(16.3)

Study of the X-ray scattering spectra (XAS) of human hair, taken from people in the Seychelles known to have mercury poisoning, has shown that CH_3Hg (cysteinate) was present. EXAFS (Extended X-ray Absorption Fine Structure) spectra indicated that it has bond lengths of Hg-C of 2.06Å and Hg-S 2.36 Å, similar to the values known for CH_3Hg (cysteinate), which are 2.10 and 2.352 Å, respectively. The Hg-C distance in $Hg(CH_3)_2$ is 2.083 Å. Chemical analysis showed that 80% of the mercury in the hair was present as CH_3Hg (cysteinate). XAS and EXAFS examination of the brain of Karen Wetterhahn identified the dominant mercury species as CH_3Hg (cysteinate) too.

Karen Wetterhahn's death led to a drastic reassessment of the toxicity of dimethylmercury and of mercury poisoning. But there is still concern about mercury in the environment. Both dimethylmercury (DMHg) and monomercury (MMHg) species are ubiquitous in the world's oceans, a matter of concern, not least because of the exposure of humans to MMHg through eating seafood, whether shellfish or fish. MMHg is formed in the oceans both by methylation of mercury and by demethylation of DMHg. DMHg is volatile, so it can escape into the atmosphere, affecting environmental levels of methylated mercury. In the California Current System (CCS), a fishery generating some $0.5 billion a year, it appears that the upwelling of DMHg from nutrient-enriched deeper waters generates a good deal of MMHg.

17 Dimethylsulfide

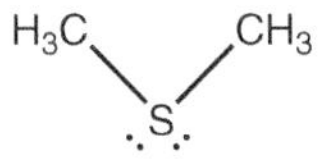

(17.1)

Dimethyl sulphide (DMS (17.1)) is a versatile molecule, with many roles. You could call it a bigger brother of hydrogen sulphide, which is associated with the smell of bad eggs, of course, and similarly DMS has a characteristic smell. Like H_2S – and indeed H_2O – dimethyl sulphide has a V-shaped structure (Figure 17.1). All three of these molecules have eight outer-shell electrons surrounding the central atom, comprising lone pairs as well as two bond pairs of electrons. Because repulsions involving lone pairs are stronger than those involving bonding pairs of electrons, the H-O-H bond angle is reduced from the 'tetrahedral' angle of 109.5° to 104.5° in water.

Compared to water, the bond angle is much smaller in H_2S (92.1°), because the sulphur atom is significantly larger than oxygen, so that the bond pairs are further from the central atom than in H_2O, thus reducing the repulsion between them. The 99.1° angle in $(CH_3)_2S$ is due to the greater size of the methyl groups, which prevents them from approaching each other as closely as the two hydrogens do in H_2S.

Whilst H_2S is a gas at room temperature (b.p. – 60.3°C), $(CH_3)_2S$ is a volatile liquid (b.p. 37°C), because of the stronger Van der Waals forces associated with the larger molecule.

$(CH_3)_2S$ also has an unpleasant smell, detectable at levels perhaps as low as 1 part per billion. 'Rotting cabbage' or 'cooked cauliflower' are common descriptions of that smell. One place where you may meet it is in the Brussels sprouts on your Christmas dinner plate, as dimethyl sulphide can be generated by the decomposition of methionine or S-methylmethionine, both present in large quantities in brassicas.

A number of carrion-smelling *Amorphophallus* plants, notably the dead horse arum (*Helicodiceros muscivorus*) use dimethyl sulphide among the molecules that attract insects to pollinate them. The plants produce an aroma like rotting flesh to attract these insects. By the way, $(CH_3)_2S$ is a major component – together with H_2S and CH_3SH – of human flatus.

Dimethyl sulphide does not always smell so unpleasant. It occurs in a number of foods, where it is mixed with other compounds that have odours, so it contributes to the 'overall' smell of the foodstuff. It's found in some wines, where below a certain level it can enhance fruity flavours, and at low concentrations it can smell pleasant. Dimethyl sulphide is also found in some beers; it is generated by more than one process, such as the reduction of dimethyl sulfoxide $((CH_3)_2S=O)$ to dimethyl sulphide by yeast in the brewing process or it is formed from molecules

DOI: 10.1201/9781003783565-17

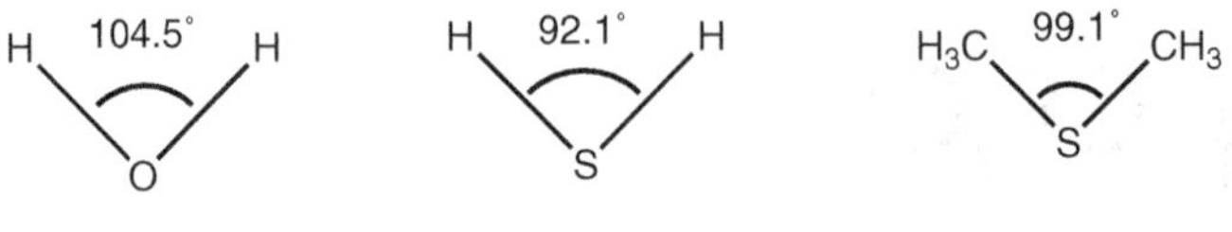

FIGURE 17.1 Bond angles in H_2O, H_2S and $(CH_3)_2S$.

like S-methylmethionine in germinating barley and drying malts. Some English breweries, notably those in Burton, are celebrated for brewing beer with an 'eggy' note, possibly due to a high sulphate content in the water supply that those breweries use. Likewise $(CH_3)_2S$ contributes to the smell of certain cheeses including Cheddar and Camembert. Enzymes contribute to the ripening of cheeses in various ways, including proteolysis, the breakdown of proteins to their constituent amino acids. In this way, methionine is converted into a number of sulphur compounds, particularly $(CH_3)_2S$.

But the more celebrated foodstuff containing dimethyl sulphide is truffles, and there is quite a story here. Truffles are formed by a subterranean fungus, notably in France (black truffles) and Italy (white truffles). In the Dordogne, black truffles (*Tuber melanosporum*) are traditionally hunted out by their smell; in the past pigs were used, whereas nowadays it is dogs that pick up the truffle's scent. For many years, it was believed that the pigs were attracted by a steroid that could be a sex pheromone, such as 5α-androst-16-en-3α-ol (17.2). Nearly 40 years ago, the French chemist Thierry Talou decided to test this theory. He composed a synthetic aroma containing $(CH_3)_2S$ – but without the steroid. He then buried samples of this 'synthetic aroma' at some locations; while at others, he buried real black truffles; and at yet other sites, he buried samples of 5α-androst-16-en-3α-ol. He then unleashed the pigs, finding that the pigs ignored the steroid, unearthing either the 'synthetic aroma' or the black truffles themselves.

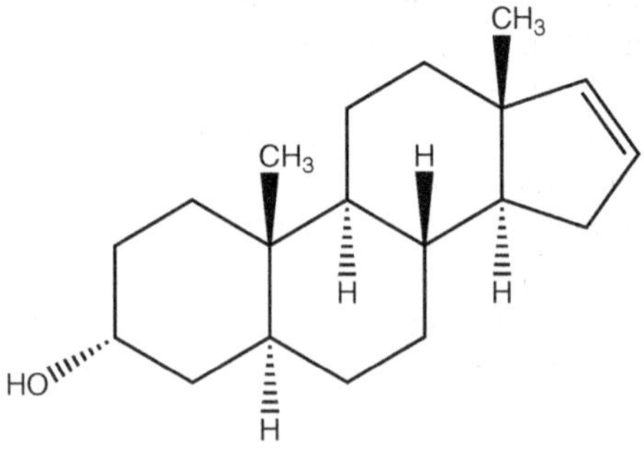

(17.2)

Some 20 years after Talou published his research, a European research team published the results of its five-year investigation of the genome of *Tuber melanosporum*

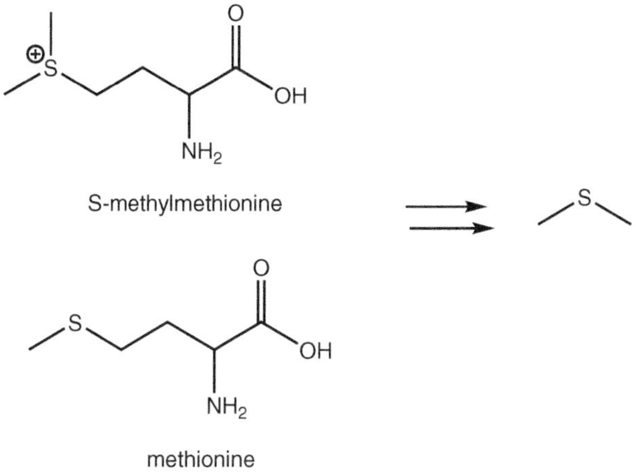

FIGURE 17.2 Formation of dimethyl sulphide from sulphur-containing amino acids.

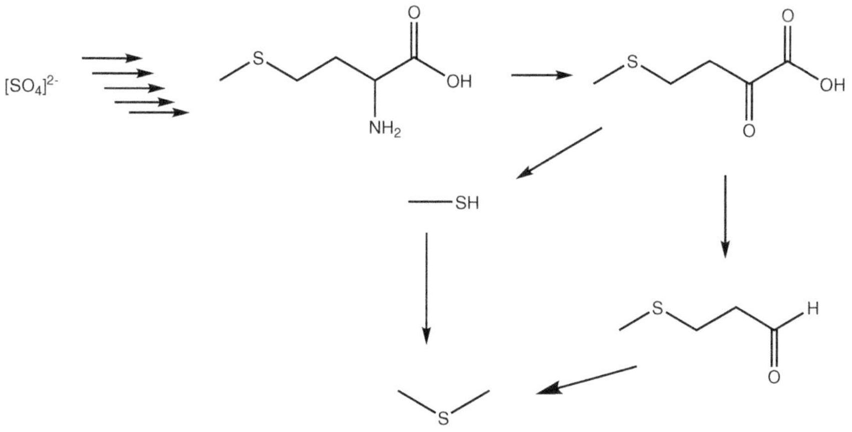

FIGURE 17.3 Steps in the synthesis of dimethyl sulphide from sulphate in the soil.

in 2010. They confirmed that the truffle did indeed make $(CH_3)_2S$, identifying the gene responsible. This was important, as some scientists had believed that the dimethyl sulphide was made by soil microbes. The truffle uses sulphate ions from the soil as the starting material for a sequence of enzyme-catalysed reactions, creating a series of organic sulphur compounds which includes the amino acid methionine, then methanethiol and 3-(methylthio)propanal, leading to dimethyl sulphide itself (Figure 17.3).

Incidentally, the dimethyl sulphide aroma is not retained by the truffle because of its volatility; it is replaced by the smell of other molecules, notably that of 1-octen-3-ol (the main compound responsible for the smell of mushrooms), so that truffles, after a while, smell 'mushroomy'. $(CH_3)_2S$ is also an important volatile of the Italian white truffle (*Tuber magnatum*), along with $CH_3SCH_2SCH_3$ (2,4-dithiapentane).

By far the most important application of dimethyl sulphide is something quite different. It was originally James Lovelock who suggested that dimethyl sulphide is key to regulating the whole global climate cycle (Figure 17.4). In fact, $(CH_3)_2S$ is responsible for 75% of the global sulphur cycle. It starts with phytoplankton, at the bottom of the oceanic food web. Phytoplankton are ultimately the food supply for the whole population of the oceans, all the way from tiny zooplankton to whales. The dimethylsulfoniopropionate (DMSP) molecule is formed by phytoplankton; when zooplankton or small fish eat the phytoplankton, dimethyl sulphide is formed. In its turn, this attracts more zooplankton to feed. In fact, $(CH_3)_2S$ is an attractant for many other members of the oceanic population. Although most of the dimethyl sulphide – 90% – is broken down in the ocean, before reaching the atmosphere, it is calculated that 50 million tonnes of $(CH_3)_2S$ are released into the air each year, where it is oxidised successively to SO_2, $(CH_3)_2SO_2$, CH_3SO_3H (sulfonic acid) and sulphuric acid. It is the sulphuric acid that is the precursor to the cloud-forming aerosols over oceans – the clouds which reflect solar energy back into space, cooling the earth.

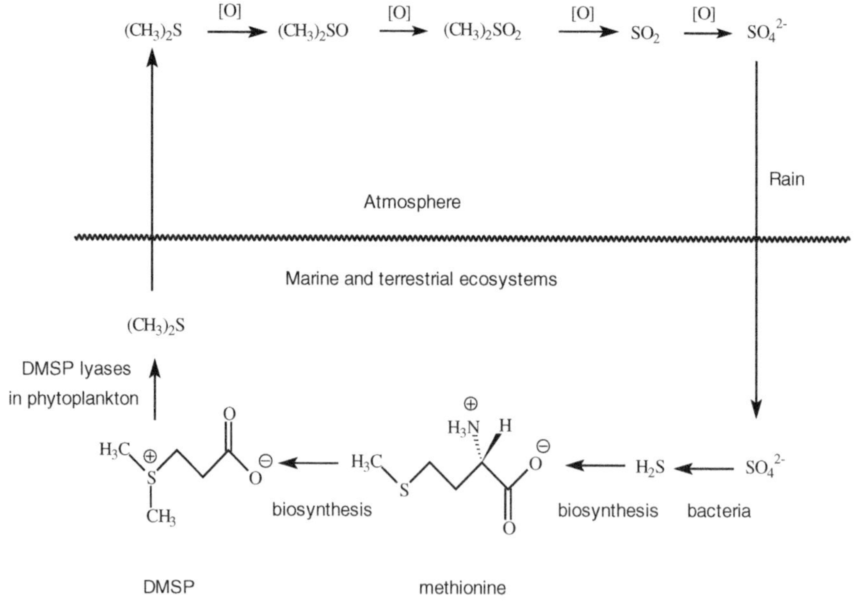

FIGURE 17.4 Dimethyl sulphide and the global sulphur cycle.

Source: Based on Figure 1 of T. G. Chasteen and R. Bentley, *J. Chem. Educ.*, 2004, 81, 1524

Dimethyl sulphide's attraction brings marine predators to areas rich in zooplankton; apart from fish, there are numerous seabirds, and also penguins and seals. When you see a seal poking its nose out of the water, it is sniffing for dimethyl sulphide.

Some 15 years ago, Professor Andy Johnston's research group at the University of East Anglia in Norwich (UEA) isolated a gene which turned DMSP into dimethyl sulphide. After inserting this gene into *E. coli*, they found that, when grown on DMSP, these bacteria could produce the characteristic 'rotten cabbage' smell of $(CH_3)_2S$. During their studies, these intrepid researchers took a bottle of the $(CH_3)_2S$-producing bacteria to a salt marsh on the Norfolk coast, whereupon they were dive-bombed by hungry seabirds who smelt 'food'.

One last thing. The bracing smell of sea air is not due to ozone, the title that it was once given. Dimethyl sulphide is the 'smell of the seaside'.

18 Fentanyl and Its Homologues

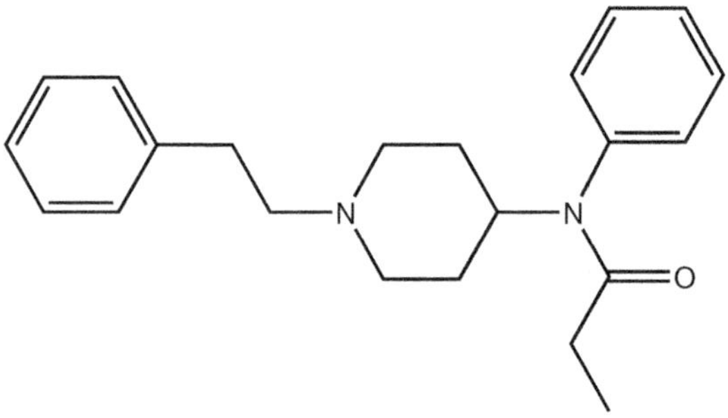

(18.1)

In some parts of the world, notably the United States, fentanyl (18.1) is overwhelmingly linked with pathways to addiction. It has been stated that in the 20 years from 1999 to 2018, nearly 450,000 people died from opioid overdose in the United States. It is hoped that this account will supply some perspective on the history of fentanyl.

It is a human instinct to want to reduce the pain that others suffer. For many years, pain relief was provided through molecules derived from the opium poppy (*Papaver somniferum*), first as opium, then using morphine (pp. 108–110), codeine (pp. 55–57) and their derivatives. As synthetic chemistry developed over the course of the nineteenth century, chemists produced diacetylmorphine (better known as heroin) but, though this was an advance on morphine, it proved to be a mixed success, and so the search for better analgesics went on.

Fentanyl was first synthesised in 1960 by the Belgian pharmacist Paul Janssen (1926–2003). Not just a brilliant synthetic chemist, he was an entrepreneur – his company *Janssen Pharmaceutica* pioneered numerous medicines. Fentanyl was his biggest breakthrough. Janssen was looking for a molecule that was more lipid-soluble, giving faster access to the central nervous system and resulting in more rapid analgesic action. Pethidine ((18.2), aka meperidine) was already known but was not as effective a painkiller as morphine; Janssen used this structure as a basis for his new painkillers.

Fentanyl was found to be 75–100 times more potent than morphine; its advantages over morphine included its greater speed of onset, along with a short duration of action, as well as its greater potency (as already noted). It also had fewer cardiovascular risks compared to morphine.

DOI: 10.1201/9781003783565-18

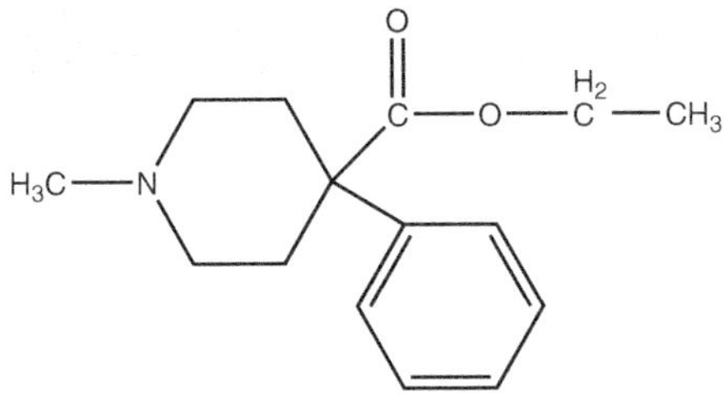

(18.2)

Within three years, fentanyl was in use as a fast-acting painkiller in much of Europe, and within a decade it was in regular use in the USA. as an intravenous anaesthetic (*Sublimaze*). Around a decade later, fentanyl transdermal patches (*Duragesic*) were approved for use in 1990 as a treatment for patients with chronic pain, giving slow release of the drug into the bloodstream over 48 hours for palliative use. Following the success of fentanyl, Janssen developed other fentanyl derivatives in the succeeding decades, with sufentanil and carfentanil being the most successful and even stronger than their parent. Like morphine, they all target the mu-opioid receptor. A recent paper showed that fentanyl forms a hydrogen bond to a histidine residue, moving deeper into the μ-opioid receptor.

The success of fentanyl compounds as medicines, particularly fentanyl itself, was followed within a decade or so by their abuse, initially by some of the surgeons and anaesthesiologists using them. When fentanyl became more widely available after patents lapsed in 1980, not only did its sales increase steeply, but its abuse increased too. Medical professionals abusing it would not realise how powerful it was, leading to fatalities. Soon afterwards, people found a way of abusing fentanyl patches by applying multiple patches simultaneously, again leading to fatalities.

The first major fentanyl drug outbreak was in California; some 112 deaths due to alpha-methylfentanyl (18.3) abuse were reported between 1979 and 1988. On the streets, it became known as 'China White'; often supplied mixed with heroin, users again underestimated its potency, with α-methylfentanyl said to be about 5000 times stronger than morphine.

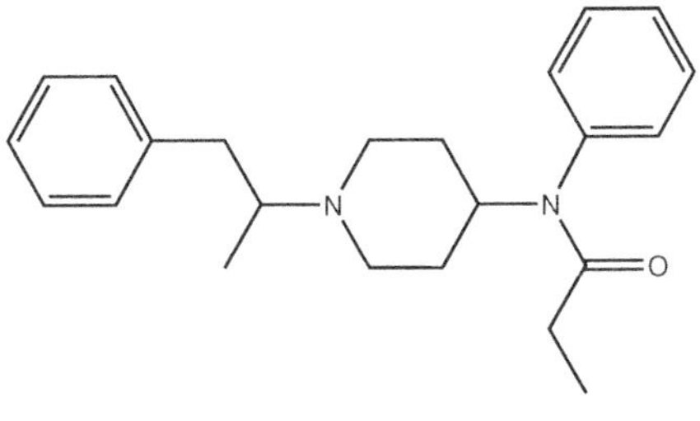

(18.3)

An isomer of α-methylfentanyl, 3-methylfentanyl (18.4), is another painkiller.

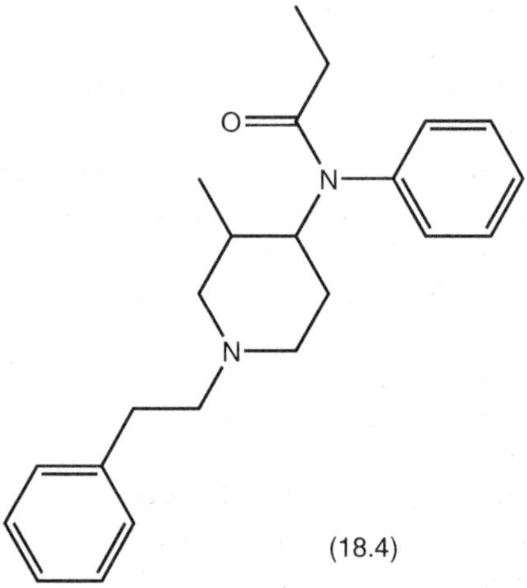

(18.4)

As time has gone on, fentanyl-related deaths have spread out of America to countries including the UK, Germany, Sweden and Eastern Europe, with fentanyl and its derivatives often marketed as heroin substitutes.

Paul Janssen went on looking for more potent fentanyls, molecules that would be better painkillers with higher safety margins and fewer side effects, notably molecules with higher lipid solubility, conferring faster access to the central nervous system and thus more rapid analgesic action. The two most successful were sufentanil (18.5), which is 5–10 times stronger than fentanyl, and notably carfentanil (18.6), a good 100 times stronger than fentanyl. First made in 1974, carfentanil is some 10,000 times stronger than morphine. It has no approved uses in humans; that is how toxic it is.

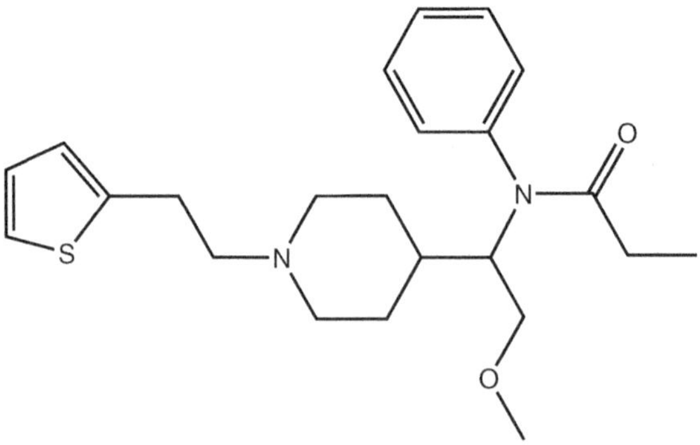

(18.5)

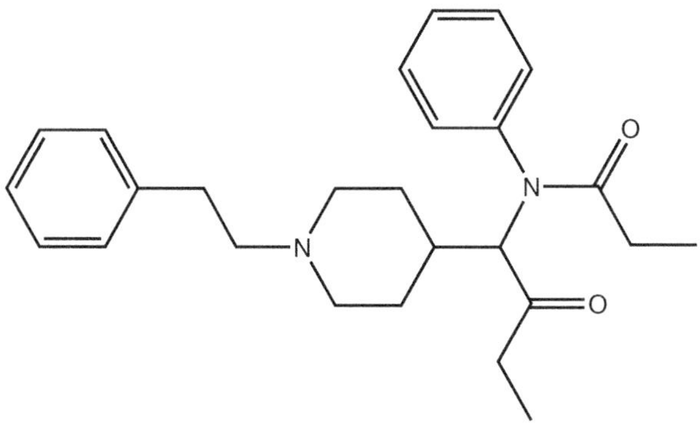

(18.6)

The literature does not quote a lethal dose of carfentanil in humans, but it must be less than 20 micrograms (20 μg), too small to be seen. Carfentanil is used to anaesthetise large animals, such as 'big cats', rhinoceroses or elephants. The anaesthetic is administered by firing darts at the animals, the darts essentially being syringes fitted with a hypodermic needle, which is equipped with a 'flight' to make it fly straight after firing (using compressed gas, of course, not an explosive propellant). Because of the toxicity of carfentanil, the vets have to take precautions to avoid exposing themselves to the chemical. First, they should not work alone, but must have at least one colleague in case of an accident. This must be followed by full facial covering, avoiding inhalation or even contact between mucous membranes and the carfentanil; latex gloves must be worn to avoid skin contact (e.g. with cuts); and clothing must cover the whole body. Last of all, they must carry several doses of the opioid antidote naloxone (Narcan).

A case history, published in 2010, explains the reasons for all these precautions. A 42-year-old (male) vet. was trying to sedate an elk. One dart missed and lodged in a tree trunk, so he tried to remove the dart from the tree. In the process, he got splashed by the carfentanil solution on his mouth, eyes and face. He washed his face with water as soon as possible, but within two minutes he became drowsy, so his colleagues administered naltrexone, the 'reversal agent' (aka antidote) used in animals. The vet was feeling better by the time he reached the emergency department an hour later. The hospital kept him under observation for 24 hours before discharging him.

A much more serious case of the effects of carfentanil on humans occurred on 23 October 2002 in Moscow. The Dubrovka Theatre was packed for a sold-out performance of the musical '*Nord-Ost*', when a band of more than 30 armed Chechen terrorists suddenly occupied the theatre, taking the audience – over 900 people from all over the world – as hostages. The armed terrorists – additionally equipped with explosive devices strapped to their bodies – demanded the immediate withdrawal of Russian troops from Chechnya. There was a stand-off for three days, until a Russian Special Forces unit pumped a 'dense, white and cloudy aerosol' (to quote survivors) into the building, then stormed the building, deactivating the bombs. They killed the terrorists, but over 120 hostages died too. The identity of the chemical in the aerosol was not divulged. The Russian emergency services had been told to bring opioid antagonists to the theatre, but they did not bring enough naloxone or naltrexone. Two days after the siege ended, clothing, urine and blood samples from three British survivors held in the theatre arrived at the Defence Science and Technology Laboratory (Dstl) at Porton Down, near Salisbury, in the United Kingdom. They used chemical analysis, particularly mass spectrometry, to show that a mixture of carfentanil (18.6) and remifentanil (18.7) was used.

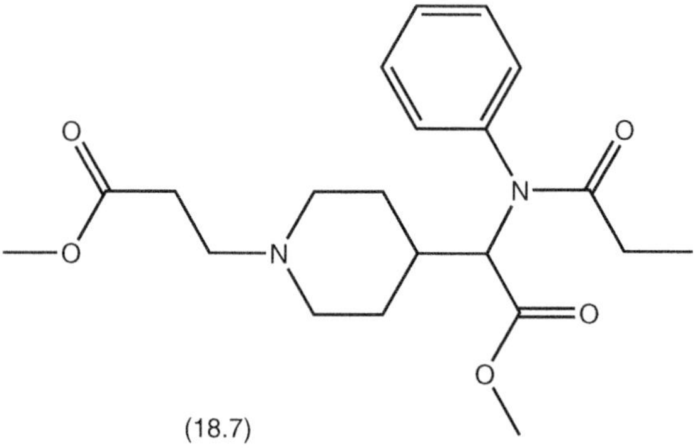

(18.7)

Fentanyls have been a big part of the 'painkiller crisis' in the USA, which has involved a number of chemicals. Overdose deaths there started to rise around the year 2000, at which time this was due to significant increases in the prescription of natural and semisynthetic opioids – neither heroin nor fentanyl use contributed to this. Suddenly, around 2010 there was a very steep rise in deaths due to heroin – these peaked around 2016. Then deaths due to fentanyl (and other synthetic opioids) rose sharply – within three years these exceeded other causes of death. A possible contributing factor may have been the use of fentanyls to 'spike' heroin (or even them being marketed as heroin) so that people would take an overdose without realising it. Media attention was attracted by the death of the musician Prince, who died on 21 April 2016 from an accidental fentanyl overdose, and by the death on 2 October 2017 of the singer and guitarist Tom Petty, from an accidental overdose of prescription painkillers, including fentanyl, acetylfentanyl and despropionylfentanyl, which he was taking for a medical condition. From around 2020, further deaths have been caused by abuse of mixtures of fentanyls and stimulants, notably methamphetamine and cocaine. The overall picture in the United States is that overdose-caused deaths in the USA have risen from less than 20,000 in 1999 to over 106,000 people in 2021.

Carfentanil has become a significant part of the problem since around 2016. Its use was first noticed in the region of Cincinnati, Ohio, but is now found across the country and, beyond America, has spread to countries including Canada, Australia, the United Kingdom and Latvia.

Much of the carfentanil is 'imported' from abroad, notably the Far East, though not all of it gets through. On 27 June 2016, a team of Royal Canadian Mounted Police officers, dressed in hazmat suits with respirators and facemasks, confiscated a consignment that had arrived from China at Vancouver Airport. Labelled as 'printer accessories', the consignment of laser jet printer toner cartridges contained a kilogram of carfentanil, reportedly sufficient for 50 million lethal doses. The addressee, a 24-year-old man in Calgary, was arrested. Although China introduced regulations which restricted fentanyl and its derivatives, the problem of 'underground' laboratories remains, aided by the 'Dark Web'.

The fentanyls, notably carfentanil, contribute to overdoses because of their high toxicity, since drugs like heroin are often supplied with the addition of one of the fentanyls. A heroin user, accustomed to taking a certain dose of the drug, can unwittingly take a lethal overdose.

19 Hexenal and Green Grass Smell

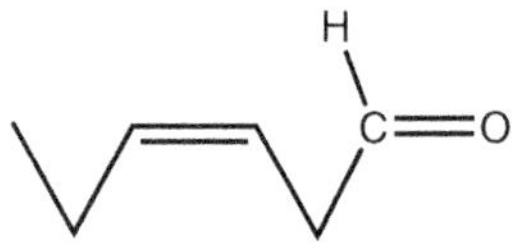

(19.1)

The green colour of grass looks wonderful, but when you cut it, the grass smells even better. This evocative smell is due to a number of connected enzyme-catalysed reactions, and particularly to the compound (Z)-3-hexenal (19.1).

Grass contains esters of long-chain unsaturated organic acids, notably linoleic acid ((9Z,12Z-octadeca-9,12-dienoic acid) and linolenic acid (9Z,12Z,15Z-octadecatrienoic acid). When grass is cut – either mechanically or when damaged by an animal – cell membranes are broken. One particular enzyme, a hydrolase, first catalyses the hydrolysis of the esters, freeing the parent acids. They are then free to attach to another enzyme, a lipoxygenase, which holds the enzyme in such a way that a reactive double bond in the acid molecule is positioned near the enzyme's active site, and an O_2 molecule (from the atmosphere) can oxidise the chain at this point. Figure 19.1 shows how this happens, with linoleic acid as the substrate, first oxidised to a hydroperoxide, before the chain is broken by another enzyme, a hydroperoxide lyase, forming the saturated C_6 aldehyde hexanal.

The hexanal undergoes further enzyme-catalysed reactions (Fig. 19.2), being first reduced by an alcohol dehydrogenase enzyme to hexan-1-ol, which can then be esterified by an acetyl transferase enzyme to hexyl acetate. These volatile compounds form part of the so-called green leaf volatiles (GLVs), a mixture of C_6 compounds, aldehydes, alcohols and esters which make up the smell associated with cut grass.

Linolenic acid possesses an extra C=C bond in its carbon chain but is otherwise similar to linoleic acid. When it is held in the active site of the lipoxygenase enzyme, it is held in the same place, resulting in the backbone being split at the same point (between carbon atoms 12 and 13). The 'extra' double bond in linolenic acid is retained in the C_6 product, which in this case is an unsaturated aldehyde, (Z)-hex-3-enal (Figure 19.3). This can undergo the two subsequent reactions, similar to those of hexanal, forming an unsaturated alcohol, (Z)-hex-3-en-1-ol, which can then be acetylated to generate the ester (Z)-hex-3-enyl acetate.

Additionally, a further sequence of reactions is possible, the work of isomerisation factor enzymes (IFs). (Z)-hex-3-enal is unstable, quickly changing into its geometric

DOI: 10.1201/9781003783565-19

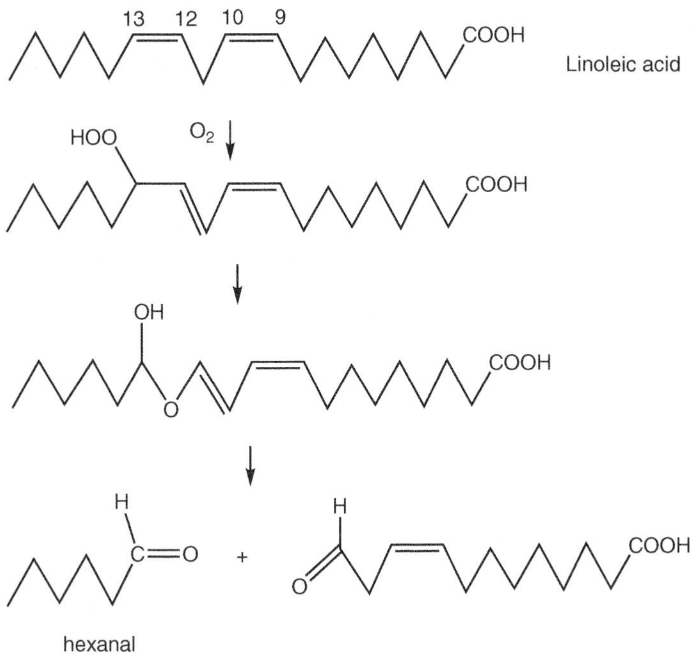

FIGURE 19.1 Steps in the formation of hexanal from linoleic acid. The first step involves a lipoxygenase enzyme, using atmospheric oxygen to form a hydroperoxide adduct. This is subsequently broken down by a hydroperoxide lyase enzyme.

isomer (*E*)-hex-3-enal, which then can isomerise to (*E*)-hex-2-enal, with the double bond shifting one place nearer to the carbonyl group. As in the earlier cases, these can undergo reduction to the corresponding unsaturated alcohols; these can then be esterified. The compounds shown in Figure 19.3 are also found among the GLVs.

This may seem complicated – so, what does it mean? First, (Z)-hex-3-enal is the main cause of the evocative smell of freshly cut grass. (Z)-hex-3-enal has an incredibly low odour threshold – the level at which it can be detected – of 0.25 parts per billion (ppb). That means that it can be detected in even tiny amounts. Secondly, as already noted (Z)-hex-3-enal undergoes two successive isomerisations – one to its geometric isomer (*E*)-hex-3-enal and then to (*E*)-hex-2-enal. These compounds have less 'fresh' smells than that of (Z)-hex-3-enal but can still be detected in the smell of cut grass after it has been standing for a while, an odour change which you may have noticed. Both (*E*)-hex-2-enal and (*Z*)-hex-3-en-1-ol are stable. This alcohol is known as 'leaf alcohol', and its green smell is used to good effect in some perfumes.

These aldehydes do seem to serve some important functions. In grass, the plants probably use them to heal cut ends – unsaturated aldehydes have bactericidal and antifungal properties. Both (*E*)-hex-2-enal and (*E*)-hex-3-enal are active against various bacteria, including *Salmonella choleraesuis*, the most common cause of septicaemia (blood poisoning).

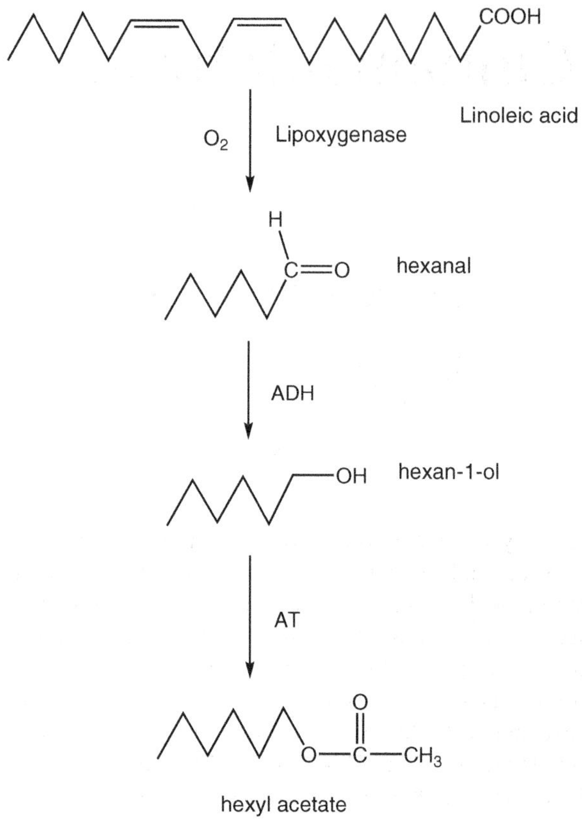

FIGURE 19.2 Formation of hexanal and its successors from linoleic acid.

Note: ADH = alcohol dehydrogenase enzyme; AT = acetyl transferase enzyme

Being volatile, these C_6 compounds spread through the air and prompt neighbouring plants to switch on their 'defence mechanisms'. Sometimes these defences are unusual, issuing a summons to 'friendly' insects to attack insect herbivores. Scientists examined the GLVs produced by tobacco plants (*Nicotiana attenuata*). When they are 'mechanically' wounded, they produce large amounts of (Z)-GLVs such as (Z)-hex-3-enal and rather smaller amounts of (E)-GLVs, such as (*E*)-hex-2-enal. However, when the plants were attacked by their specialist herbivore *Manduca sexta*, the amount of (*E*)-hex-2-enal emitted increased sharply and resulted in the appearance of 'insect cavalry', *Geocoris spp*., which are predators of *Manduca sexta*. Obviously, these insects can recognise isomerism at work.

Birds are not usually associated with a sense of smell, but a study reported in 2021 is intriguing. It was found that European white storks were attracted to freshly mown meadows, where the insects and rodents that they feed on were most exposed. The investigators covered unmown fields with freshly cut grass from elsewhere. This

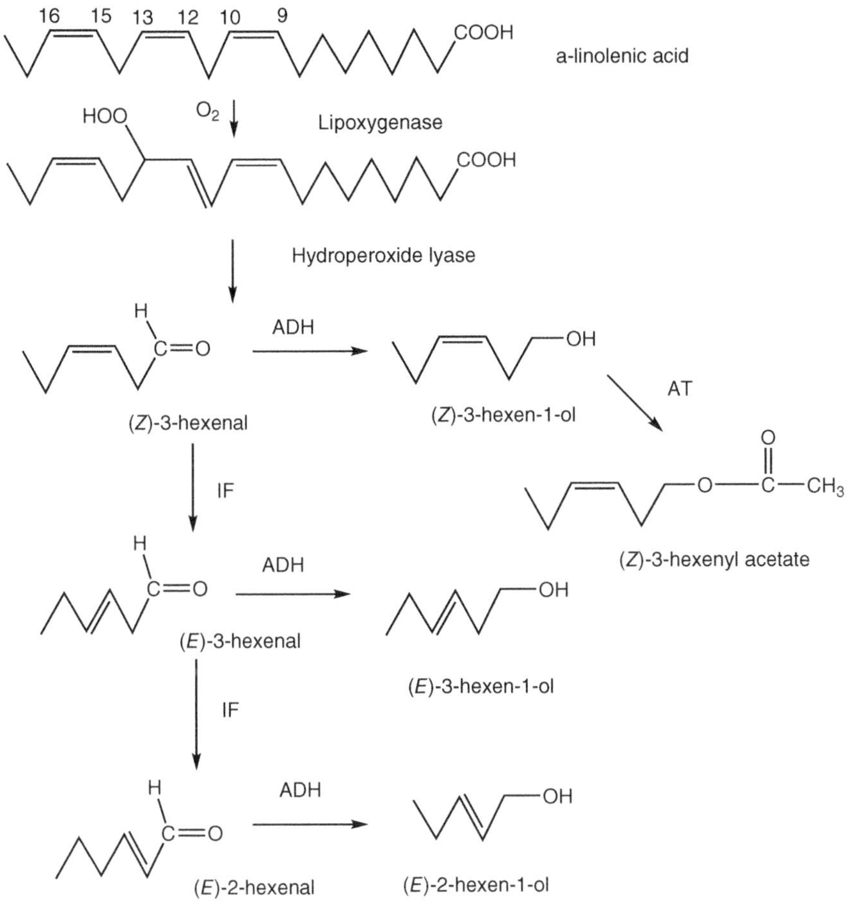

FIGURE 19.3 Formation of (*Z*)-3-hexenal and its successors from linolenic acid.

resulted in the immediate arrival of storks. They then tried spraying unmown fields with a GLV mixture containing ((Z)-3-hexenal, (Z)-3-hexenol and hexenyl acetate; this, too, immediately attracted the storks.

Hexenals are found in other plants, such as apples, as well as in apple juice. (*E*)-hex-2-enal is one of the molecules responsible for the freshness of apple juice; thus, hexenals can be used as additives to give a 'greener' note to food flavours. Food scientists have suggested that our bodies recognise some of these GLVs as a sign of food being good to eat.

(*E*)-hex-2-enal is one of several aldehydes found in the herb coriander. All parts of the coriander plant are used in cooking, and its dried fruits – coriander seeds – are used as a spice. Spices have long been associated with preserving food, due to their antimicrobial properties, which appear to be linked with alkenals, but not alkanals.

More than that, there is evidence that GLVs have a role in human health. Studies involving PET (positron emission tomography) have shown that a 'green odour' mixture of (*E*)-2-hexenal and (*Z*)-3-hexen-1-ol has a healing effect on the psychological damage caused by stress in monkeys. The 'green odour' activates regional blood flow in the primary olfactory cortex (like other smelly molecules) and also in the anterior cingulated gyrus, the latter being linked with the 'healing effect'. A green odour mixture reduces stress in rats and has a similar effect on humans. The green grass smell really could be good for you.

20 Ivermectin

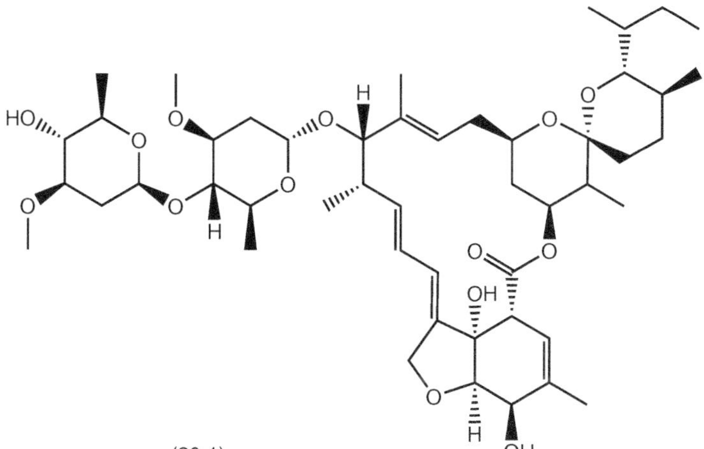

(20.1)

Organisms living in the soil must protect themselves against their enemies, so they use chemicals for their defence. There is a long tradition of scientists searching the soil for possible medicines, examples being the tuberculosis drug streptomycin (pp. 162–164) and the tetracycline family of antibiotics (pp. 183–187).

In the 1970s, a group of Japanese researchers at the Kitasato Institute in Tokyo, working in partnership with the American pharmaceutical company Merck Sharp & Dohme, discovered a bioactive compound, avermectin, in a soil sample from a golf course. The research team was led by Professor Satoshi Ōmura. (Remarkably, this microorganism remains the only source of avermectin ever discovered). Further research revealed that there was a family of very similar avermectins that were potent antiparasitic medications; the most active of the avermectins is B_1, shown below.

Further research found that selective hydrogenation of one of the five C=C double bonds in avermectin (Figure 20.1) led to 22,23-dihydroavermectin B1a, known as ivermectin (20.1), which is a better antiparasitic agent. This reduction was achieved using gaseous hydrogen, with a catalytic amount of the compound $RhCl(PPh_3)_3$, Wilkinson's catalyst (page 223). The reason for this selectivity is that, of the five C=C double bonds in avermectin, three are in the *E*- (*trans*-) configuration and two are in the more open *Z*- (*cis*-) configuration. One of these two is much less 'hindered' than the other, making it easier for the catalyst to bind to it than to any of the others.

Ivermectin was introduced onto the market in 1981. It is an effective medication for treating many parasites, including GI roundworms, lungworms, mites, lice and hornflies, as well as bed bugs and ticks. Among the last-named is the ixodid tick, Rhipicephalus (Boophilus) microplus, which plagues cattle in the tropical and

DOI: 10.1201/9781003783565-20

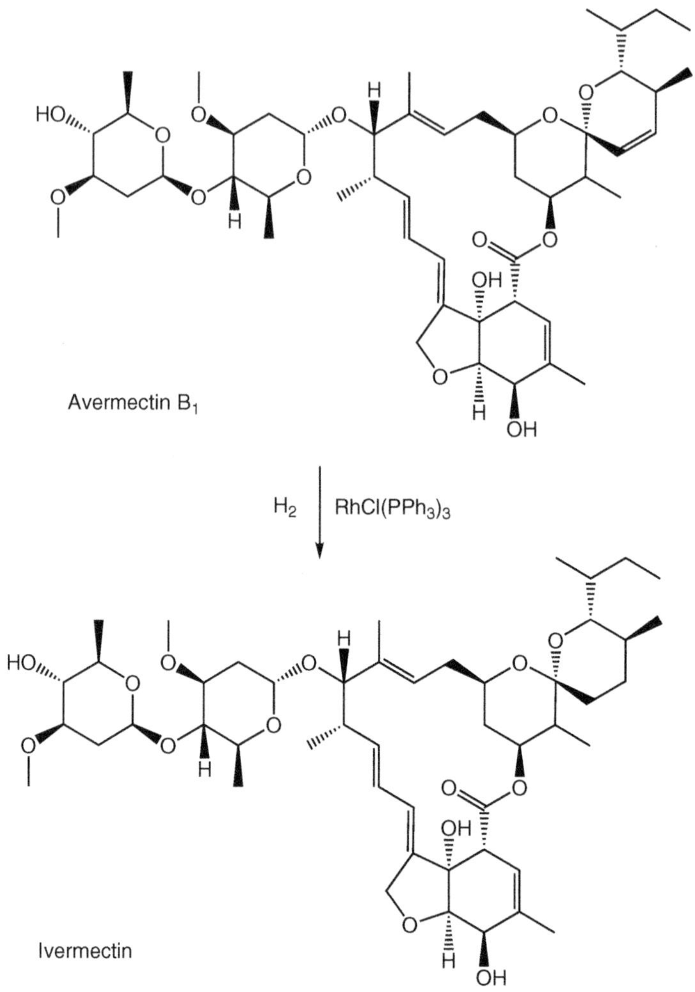

FIGURE 20.1 Synthesis of ivermectin by catalytic hydrogenation of avermectin.

subtropical regions of the world, notably Brazil. Apart from livestock, ivermectin is important in treating pets, such as cats and dogs, including heartworm infections in the latter.

The success of ivermectin in treating parasitic infections in animals prompted thoughts that it could also be an important medication against parasitic infections in humans. One such infection that was immediately considered was onchocerciasis; this is a parasitic infection responsible for blindness in millions of people. This is caused by the filarial worm, Onchocerca volvulus, which infects human beings living in fertile river valleys in regions of Africa, Yemen, southern Mexico, Guatemala, Ecuador, Colombia, Venezuela and the Brazilian Amazon. The vector for this disease is the blackfly, which breeds in rapid rivers; its bite transmits the larvae of the worm, Onchocerca volvulus. The larvae develop into worms living under the skin, and as adults can grow to over a foot long. In the final stages of the disease, the worms invade the eyes, causing visual impairment and even blindness, leading to the common name, 'river blindness'. It is thought that around 100 million people are at risk of contracting the disease. In Africa, there are 40,000 new cases of blindness every year. Half a century ago, this disease had no known cure.

It was the parasitologist William Campbell who suggested to Merck that ivermectin should be tested on human subjects. He pointed out that these people could not afford the medication. Initial trials took place in Senegal in 1981 and were successful, and were followed up in other countries. In 1987, Merck, the developers of ivermectin (its brand name is *Mectizan*) promised to provide it free of charge to anyone for as long as it was required. Patients receive a single dose of ivermectin every six months to keep the parasite at bay. Each year some 25 million people receive treatment.

In October 2015, the 2015 Nobel Prize in medicine was awarded to William C. Campbell and Satoshi Ōmura 'for their discoveries concerning a novel therapy against infections caused by roundworm parasites' and to the Chinese scientist Youyou Tu 'for her discoveries concerning a novel therapy against malaria'. This latter discovery was artemisinin (pp. 9–12). Thus, ivermectin has been a highly successful medicine, transforming the lives of millions of people in the developing Third World countries.

When the COVID-19 viral infection hit the world in 2019–2020, there were claims that ivermectin could be an effective treatment. Some of these claims appeared in print, though not as peer-reviewed papers in academic journals. Instead, they were posted on 'preprint servers' and reported findings that had not been subjected to peer-group review; that is, they had not been refereed prior to publication in a recognised academic journal, which is the standard procedure. Scientists who reviewed these claims found that they often contained muddled or misleading claims, errors and, in some cases, fabrications.

A number of properly conducted academic studies examined the effect of treating COVID patients with ivermectin. One such appeared in a leading medical periodical, the *Journal of the American Medical Association*: it was a randomised clinical trial carried out in Colombia on patients with mild COVID–19. These patients received a five-day treatment with ivermectin; when compared with placebo, no significant improvement was found in the time taken to resolve the symptoms of COVID. A careful review of trial data from several sources, published in the respected *Cochrane Database of Systematic Reviews*, found that many of these trials left a lot to be desired. Among the conclusions was the following: 'For the outpatient setting, there is currently moderate-to high certainty evidence that ivermectin has no beneficial effect on risk of death and quality of life for people with COVID–19'. As far as 'the authorities' are concerned, ivermectin was not approved as a treatment for COVID in the United States, the United Kingdom or the European Union. The World Health Organisation has strongly recommended against the use of ivermectin as a COVID-19 treatment. Furthermore, Merck, the developers of ivermectin, has stated that there is no good evidence that ivermectin is effective against COVID-19.

More recently, there have been some reports suggesting that ivermectin may have a role to play in the treatment of malaria. A careful investigation is called for in this case.

21 Limonene

Limonene is an important part of the 100-million-tonne citrus harvest each year. The essential oil of oranges is generally over 80% limonene, sometimes over 95%, with almost as much present in grapefruit and lemon essential oil, and a significant amount also occurring in the oil from limes. A lot of the citrus fruit, however, ends up as waste and there are good reasons – not just environmental ones – for utilising it better.

Limonene's systematic name is 1-methyl-4–(1-methylethenyl) cyclohexene, so it seems sensible to call it limonene instead. The limonene molecule contains one chiral carbon atom, so the molecule exists as two enantiomers. Of these, (R)-(+)-limonene is associated with citrus fruits.

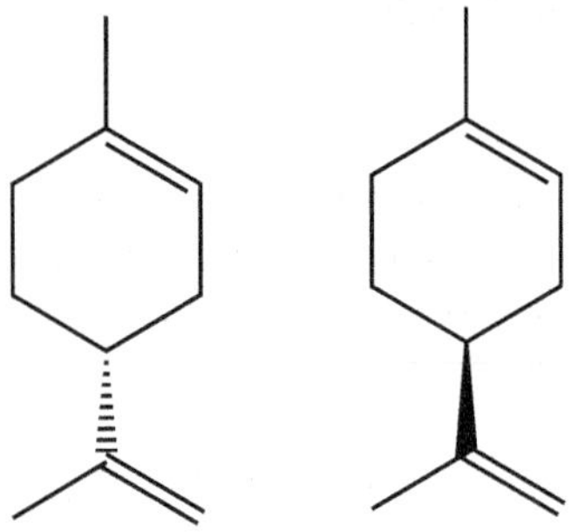

(+)-limonene (21.1) (-)-limonene (21.2)

The smell of the two isomers of limonene has been a controversial topic. A 1971 paper in the journal *Science* associated the smell of oranges with (+)-limonene (21.1) and the smell of lemons with (-)-limonene (21.2). Because it had appeared in a prestigious journal, this statement was accepted for years. After all, there are many other pairs of enantiomers with different smells, such as the isomers of carvone, which have spearmint and caraway odours, respectively.

In 1993, a paper in *Perfumer and Flavorist* suggested that the orange smell associated with (+)-limonene was due to traces of highly odorous aldehydes. This point was taken up in 2004, in a paper entitled 'Scent through the Looking Glass', by the distinguished perfume chemist Charles Sell. Sell said that the more carefully a sample of (+)-limonene was purified, the less smell it had. He also noted that the authors of the 1971 paper said that they obtained their (-)-limonene from lemons,

DOI: 10.1201/9781003783565-21

'a very odd claim since lemons, like all citrus fruits, produce the (+) enantiomer in high enantiomeric purity' (this author's italics). Others have noted that commercial samples of (+)-(*R*)-limonene, often obtained from oranges, have an orange smell, whereas commercial samples of (-)-(*S*)-limonene, obtained from turpentine (derived from pine trees), have a 'pine' smell. A paper by three Norwegian chemists which appeared in the *Journal of Chemical Education* in 2021 revisited the question, again pointing out that (S)-(−)-limonene is almost absent from in citrus fruits, though it is the major isomer (albeit in modest amounts), in the oils of Citronella (*Cymbopogan nardus*) and lemongrass (*Cymbopogan citratus*).

Orange oil consists mainly of (R)-(+)-limonene; additionally, however, it contains more than 100 other components, each present in quantities less than 2%. In an examination of hand-squeezed, non-pasteurised juice of Valencia late oranges, ethyl butanoate (21.3) and (*Z*)-3-hexenal (p. 76) emerged as the two most important aroma compounds. Subsequently, a study of freshly reconstituted orange juice from concentrate identified ethyl butanoate, linalool and octanal (p. 126) as the odourants with the highest flavour dilution (FD) values and therefore the most potent odorants.

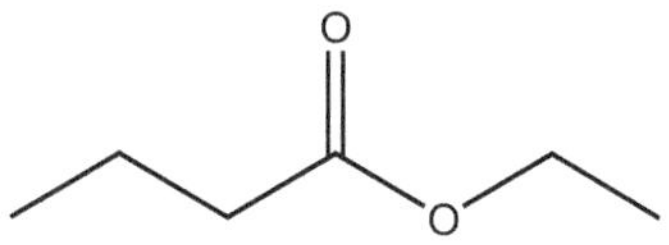

(21.3)

So, the picture that emerges is that the two enantiomers of limonene have very weak smells when pure. However, since they are commonly obtained from orange peels and pine trees respectively, commercial samples typically smell like oranges and pine, respectively, due to impurities.

Limonene is made in spearmint, *Mentha spicata*, using the enzyme (4*S*)-limonene synthase. This is a monoterpene cyclase which, starting from geranyl diphosphate, catalyses its isomerisation and cyclisation to generate the α-terpinyl cation, which leads to (-)-limonene (Figure 21.1). The enzyme's structure is known, as is that of the enzyme which makes the enantiomer, (+)-limonene. Having made limonene, a particular plant uses it to make other terpenoids, including (−)-menthol in peppermint and the carvone enantiomers in spearmint and caraway.

A component of fruit such as oranges, limonene is found in all sorts of consumer products, not just in drinks, soaps and detergents. But limonene has many other uses. It is a feedstock for making other organic molecules, such as *p*-cymene (a potential source of terephthalic acid) and *p*-menthane. Beyond that, it has 'street cred' as a biodegradable substance from a renewable source that is also non-toxic, unlike halogenated solvents or most other hydrocarbons. It is therefore a good solvent and is widely used as an industrial cleaner, especially as a good solvent for greases. It was employed as a degreaser in the clean-up after the 2011 Deepwater Horizon oil spill

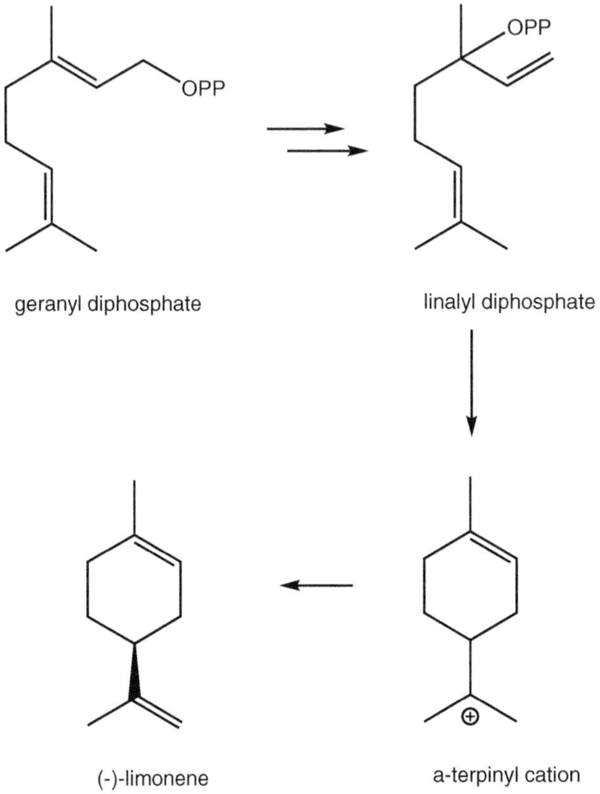

FIGURE 21.1 Sequence in the biosynthesis of (-)-limonene.

in the Gulf of Mexico. It has antimicrobial properties and is also an insecticide, most familiarly a component of the pest-control product *Orange Guard*; it is also considered friendly towards honeybees.

On the health front, limonene has been reported as a breath biomarker for cirrhosis of liver, thus offering potential for identifying early-stage liver disease. There is considerable scope for recovering – and using – large quantities of limonene from waste orange peel, generated in the production of orange juice.

22 LSD

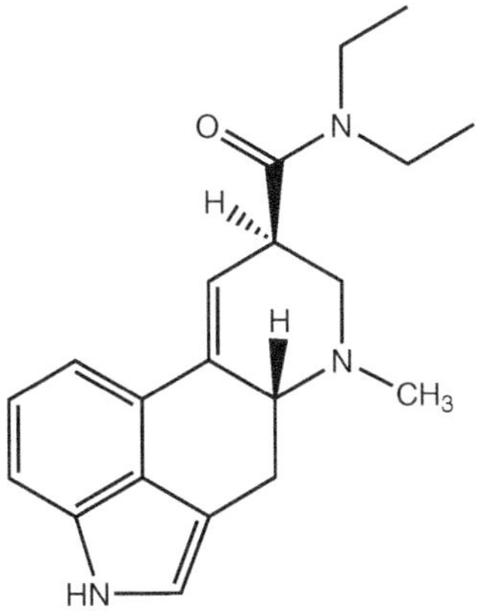

(22.1)

If a molecule embodies a decade, it is LSD (22.1) and the 1960s. But the genesis of LSD lies much further back in time. Centuries ago, life expectancy was much shorter than today, and not just for the poor. One factor explored by the American historian Mary K. Matossian is poisoning caused by infected crops. Infectious diseases like typhoid and dysentery peak in the summer, whereas others, such as influenza, pneumonia, typhus and smallpox peak in winter. Diet-related disease is another factor.

In the distant past, the majority of people, certainly in Europe, depended heavily upon cereals for their bread (and much of their diet); richer people tended to eat wheat-based white bread, whilst the browner bread consumed by most people contained high amount of rye. Whether before harvesting or in storage, moulds tended to develop in the cold, damp, weather prevalent in much of Northern Europe at that time. Among these moulds were the fungus, *Claviceps purpurea,* which develops on crops like rye. If bread made from infected grain is eaten, it can lead to epidemics, with concomitant fatalities. During the nineteenth-century Russian ergotism epidemics, the mortality rate was over 40%.

Ergot alkaloids, mainly derivatives of lysergic acid, were at the root of this. In the Middle Ages, the gangrene often associated with the ergotism led to blackened limbs that appeared to have been burned. The complaint was known as *ignis sacer*, 'St Anthony's fire', after the order of clergy who sought to minister to sufferers. Matossian summarised the arguments linking at least some of the mortality at the time of the Black Death (1348–1350), and the subsequent demographic depression, with ergotism. Similarly, she linked to ergotism the bizarre waves of panic that affected much of France in the early summer of 1789 – and which may have contributed to the French Revolution.

DOI: 10.1201/9781003783565-22

More recently, what appears to have been an outbreak of ergotism occurred in the small French town of Pont-Saint-Esprit at Assumptiontide around 15 August 1951, with seven fatalities among around 250 casualties. One enterprising writer has ascribed this to CIA experiments with LSD (see below), though this is not generally accepted.

A number of alkaloids are present in ergot, some with medicinal uses. For one, ergonovine (22.2) is an abortifacient and also possesses hallucinogenic properties. Another is ergotamine (22.3), which is a vasoconstrictor with medicinal uses including the treatment of migraines and post-partum haemorrhage.

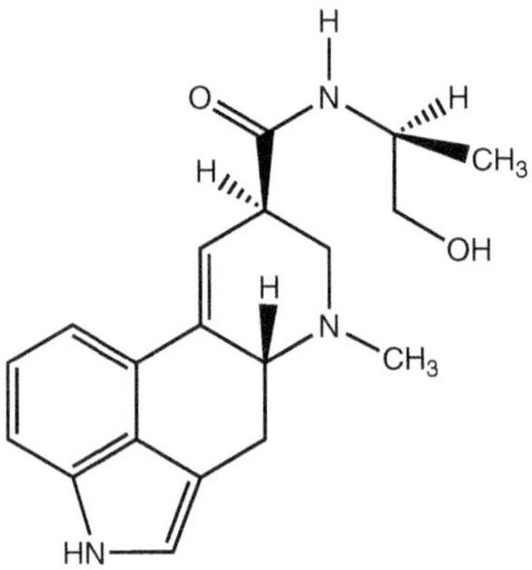

(22.2)

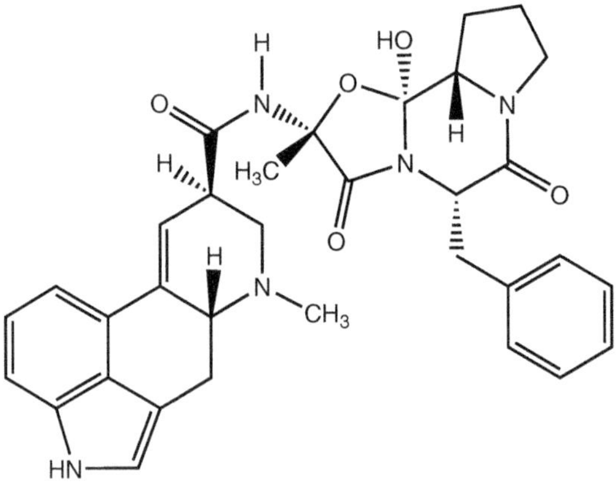

(22.3)

'Time and again I hear or read that LSD was discovered by accident. This is only half true'. These are the opening words of Albert Hofmann's book 'LSD My Problem Child'. Working at the Sandoz Laboratories in Switzerland, Hofmann first synthesised LSD (22.1) on 16 November 1938. His aim was to prepare ergot alkaloid derivatives that could serve as useful medicines. In fact, the molecule was originally called LSD-25 because it was the 25th lysergic acid derivative that he had made.

It was four years later that he accidentally ingested some LSD and discovered its psychedelic properties, a stream of fantastic pictures, extraordinary shapes and an intense kaleidoscopic play of colours. On 19 April 1943 – three days later – he deliberately ingested 250 µg (0.25 mg) of LSD, finding that it was much more potent than other ergot alkaloids; he experienced dizziness and visual distortions for several hours. He rode home by bicycle (as no car was available because of wartime fuel shortages), feeling that he was stationary and that the fields were whizzing past. His assistant, who accompanied him, later assured him that they had ridden quite rapidly.

Nevertheless, he arrived home safe and sound, though he summoned the family doctor and requested that his next-door neighbour give him some milk, as an antidote to the poison he imagined he had consumed. He did not recognise her. By the time the doctor arrived, the worst of the symptoms had disappeared, and he found nothing wrong with his patient – blood pressure, pulse and respiration were normal. So, he sent Hofmann to bed and watched over him, along with Hofmann's assistant. Hofmann awoke the next day feeling remarkably clear-headed, though his senses were still stimulated throughout the day. He survived to the age of 102, dying on 29 April 2008.

We now know that LSD is a very powerful psychoactive drug – a 'normal' dose is in the range of 50–100 µg (much less than Hofmann had taken in his experiment), though as little as 10 µg can produce symptoms. On account of its small active dose, LSD is normally dissolved in water, and the resulting solution is absorbed into blotting paper, which is then administered as small pieces of paper. Alternatively, it may be administered as microdots. LSD has two chiral centres but only (+)-LSD with the absolute configuration (5*R*,8*R*) is psychoactive.

Research on the properties of LSD began after World War II, with hundreds of psychologists and psychiatrists looking for possible uses in psychoanalysis throughout the 1950s. The American Central Intelligence Agency (the CIA) carried out tests on both young servicemen and members of the public to see whether it was a potential mind-control agent. One of these psychologists became a great advocate for LSD; his name was Timothy Leary, a Harvard academic. The 1960s saw LSD enter the counterculture of the day as a recreational drug and entheogen, and its use reached a peak during that decade. For the last decade of his life, Aldous Huxley, the English writer and philosopher ('*Brave New World*'; '*The Doors of Perception*') experimented with psychedelic drugs. Suffering from terminal oesophageal cancer, he took an intravenous dose of 100 µg of LSD on the day of his death, 22 November 1963 (President John F. Kennedy and C. S. Lewis also died on that same day).

People taking LSD shared Hofmann's experiences of the effects of the drug, as well as others. Thus the psychological effects of LSD depended upon various factors – the person involved; the environment during the 'trip'; and even their state of mind. A sunny blue sky could bring on euphoria, but all that could change when a cloud covered the sun, causing a quite tremendous 'down'. Inanimate objects could be distorted – buildings could 'come alive' and colours intensify. People could undergo

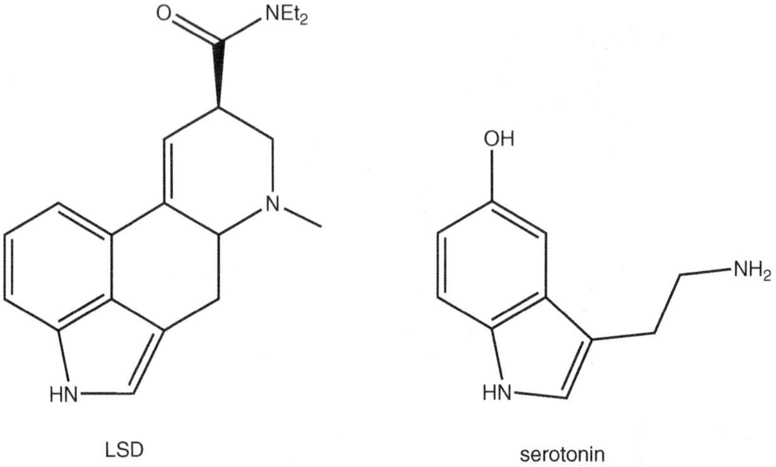

FIGURE 22.1 Comparison of the structures of LSD and serotonin.

a crossover in their senses, by 'hearing colours' or 'seeing sounds'. Furthermore, 'flashbacks', years after the actual events, have been linked with taking LSD.

How LSD works is unclear, though many think that LSD interferes with the body's levels of the neurotransmitter – one of the body's natural 'messenger molecules' – serotonin (5-hydroxytryptamine, 5-HT), through competing for some kinds of serotonin receptors on the surfaces of neural cells in the brain, particularly due to agonist effects at the 5-HT_{2A} receptors. Serotonin and LSD share an indoleamine core (Figure 22.1).

During the San Francisco 'Summer of Love' in 1967, the subculture in the Haight-Ashbury area came under the influence of LSD-induced psychedelic experiences and psychedelic performers, as well as groups whose music reflected this philosophy. Just look at the albums of that time from groups like the Byrds and Jefferson Airplane. But 'Love, Joy, Peace' did not last, as LSD was replaced by less peaceful drugs such as methamphetamine that took its place. The Summer of Love passed into history. At the end of that decade, the increased use of LSD as a recreational drug caused LSD to be placed into Schedule 1 of the Controlled Substances Act of 1970. The world had changed.

It has sometimes been stated that there have been no human fatalities from LSD overdoses, but Nichols (2018) records two, arising from massive LSD overdoses (there is also a reported case of an elephant dying from a dose of 0.3 g. of LSD). Additionally, LSD may well have caused deaths through impairing the judgement and capacity for decision-making of the people involved, as they may have been prompted to try rock climbing or jaywalking, not to mention jumping from a sky-scraper. It was probably a good thing that Albert Hofmann was riding a bike whilst under the influence of LSD that day in 1943, instead of driving a car.

Interest in the medical properties of LSD continues. A report published in September 2025 describing a clinical trial using 100-microgram doses of LSD to treat generalised anxiety disorder (GAD) indicated encouraging results that will inform Phase 3 trials.

One last thing. The chemical name of LSD is (6a*R*,9*R*)-*N*,*N*-diethyl-7-methyl-4,6,6a,7,8,9-hexahydroindolo[4,3-fg]quinoline-9-carboxamide. Now you know why it is easier to refer to it as LSD.

23 Mefloquine and Some Other Synthetic Antimalarial Drugs

As the nineteenth century unfolded and synthetic chemistry advanced, scientists started looking for ways to make molecules that were alternatives to natural medicines such as quinine (p. 150).

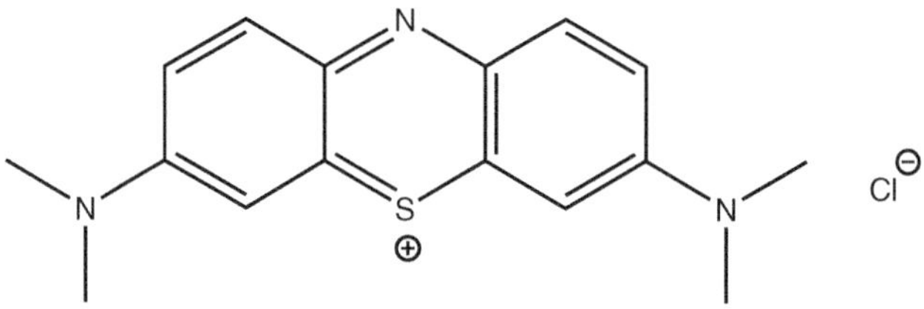

(23.1)

The first synthetic antimalarial to be discovered was the dye methylene blue (23.1). The eminent German immunologist Paul Ehrlich (1854–1915) had found that this dye could be used to stain malaria parasites, so in 1889 he administered 100 mg of it to two patients suffering from mild malaria and found that it cured them (though it was subsequently found to be ineffective against the more virulent *Plasmodium falciparum* strain). Molecules emulating methylene blue in having two or three fused six-membered rings formed part of several of the synthetic antimalarial drugs subsequently developed, which often featured a quinoline system. Methylene blue had the side-effect of turning the patient's urine and faeces bright blue.

Plasmaquine ((23.2); 1925) was found to be a cure for malaria in some patients, but there were problems with side-effects, so it was not widely used. However a related compound primaquine (1952) was subsequently found to be an effective drug.

DOI: 10.1201/9781003783565-23

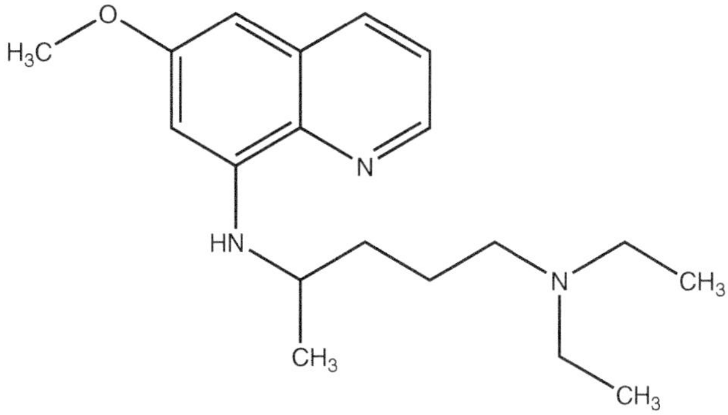

(23.2)

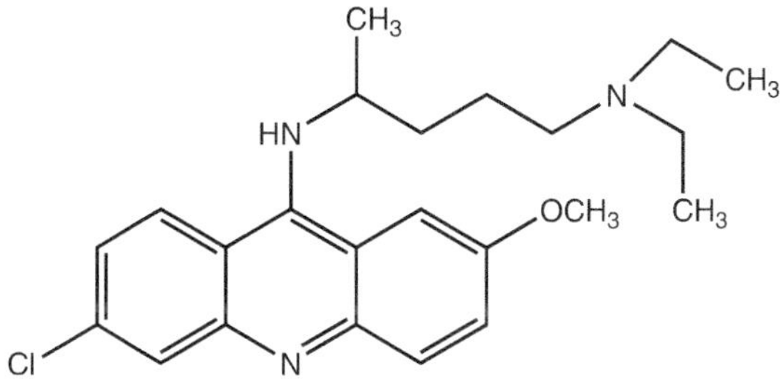

(23.3)

The next widely used antimalarial drug was quinacrine, or mepacrine ((23.3); Atabrine®), first synthesised in 1932. Soon after the attack on Pearl Harbour on 7 December 1941, Japanese forces occupied the island of Java, hitherto the source of around two thirds of the world's supply of quinine, meaning that the Allies (USA, UK and the British Empire) had to rely upon synthetic substitutes for quinine. In fact, there was also a logistical issue, as hitherto the USA had obtained its supplies of Atabrine from I.G. Farben in Germany, which had declared war on the USA on 7 December. Therefore, the Winthrop Chemical Corporation, hitherto an I.G. Farben-owned company, severed ties with Germany and was ordered by the US Government to manufacture Atabrine for use by its military. In Britain, it was manufactured by Imperial Chemical Industries (ICI).

This drug had disadvantages. One obvious one was that it turned the skin of its users yellow. There were rumours – unfounded but widely believed – that taking it rendered the user impotent. Consequently, many troops did not take their mepacrine tablets and thus caught malaria. At one point in the war, General William Slim, the Allied commander on the Burma front, whose forces consisted largely of troops from Britain, India and Australia, found that for every wounded soldier evacuated, over a hundred troops were evacuated with malaria. After he enforced the use of mepacrine, malaria rates dropped below 5%. It took a lot of persuasion, including a campaign featuring the cartoon character 'Annie the mosquito', to get troops across the Pacific theatre to take the drug.

The Japanese picked up on the impotence rumour and promoted it in propaganda broadcasts. One peripheral story concerns Iva Ikuko Toguri (1916–2006), an American-born and educated graduate (UCLA, BS in Zoology) of Japanese parentage, who was visiting relatives in Japan when war broke out, and was unable to return home. She was forced into working as a DJ and presenter on English-language radio programmes broadcast to Allied troops (she came to be known to Americans as 'Tokyo Rose'), though she secretly smuggled food to American prisoners of war. After the war she was tried on charges of treason (1949) and convicted on one count based on perjured testimony. She served over six years in gaol of a ten-year sentence. She was subsequently exonerated and received a Presidential pardon from President Gerald Ford in 1977.

One side effect of mepacrine was that, in extreme cases, it could cause toxic psychoses, the most famous victim being Major General Orde Wingate (1903–1944), commander of the British jungle long-range penetration unit, known as the Chindits, in the Burma campaign. Wingate had earlier attempted suicide while taking higher-than-recommended doses of mepacrine for malaria. Such side-effects led to mepacrine being supplanted by later drugs, such as chloroquine, but it was both an effective therapeutic and a prophylactic antimalarial medicine during World War II.

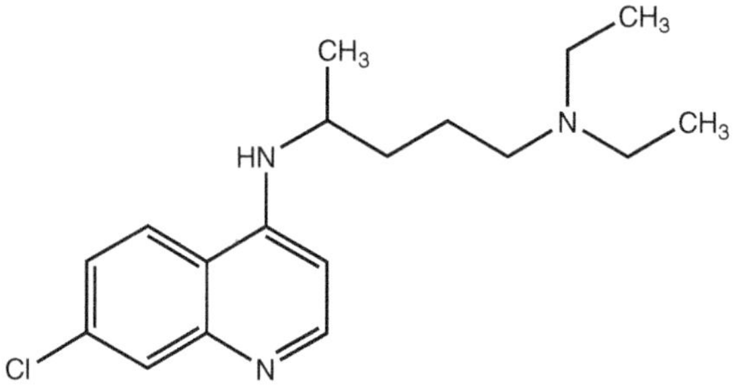

(23.4)

Mepacrine was succeeded by chloroquine (23.4). Chloroquine has a rather similar structure to mepacrine, differing primarily in the absence of one benzene ring. This compound had been discovered in 1934, and during World War II it was extensively tested in the United States during the screening of some 14,000 compounds, before coming into general use in 1947. Its advantages over mepacrine included requiring lower doses and having lower toxicity, with none of the undesirable side-effects of mepacrine. Unfortunately, cases of *P. falciparum* resistance to chloroquine started to appear within a decade, and these resistant strains became widespread across much of South America, notably Brazil; East and West Africa; and Southeast Asia. It may be that resistance is lost when chloroquine is withdrawn for a few years, so that it can subsequently be reintroduced in combination therapy with another antimalarial drug.

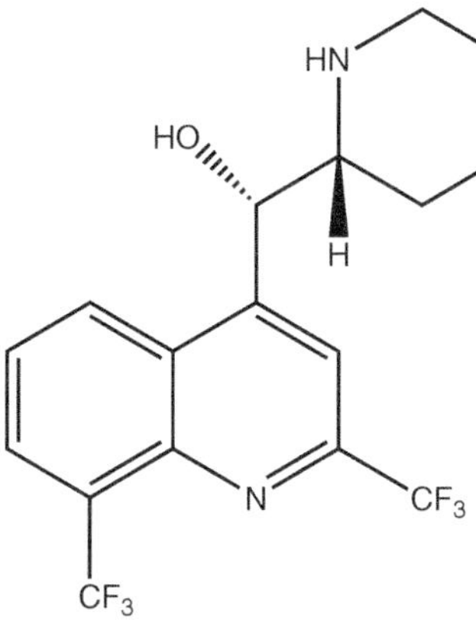

(23.5)

From 1965 onwards, the United States troops were actively fighting a war in a part of the world where malaria was endemic: Vietnam. By that time, resistance to chloroquine was only too prevalent, and in Vietnam the deadly *P. falciparum* parasite had to be combated. Chloroquine had a failure rate of between 80% and 89%; using quinine as an alternative was only slightly better, with a failure rate between 50% and 74%. In 1969, there were 12,000 cases of malaria among the US troops in Vietnam; there were 26,000 acute cases of falciparum malaria over the years 1965–1969, with 53 deaths. Something had to be done.

A screening programme for new antimalarial agents had been launched in 1963 by the Walter Reed Army Institute of Research, and some 400 4-quinolinemethanol derivatives – compounds with an aminoalcohol sidechain attached to a quinolone ring – were screened. These compounds were tested against mice and monkeys infected with the malaria parasite, and several were found to be effective against chloroquine-resistant malaria parasites. One in particular stood out. Originally designated WR-142490 (first synthesised in 1969), it later became known as mefloquine (23.5), or Lariam. Hofmann-La Roche subsequently developed an improved (and therefor inexpensive) synthetic route to the drug and marketed it under the name Lariam.

It was first tested (on volunteers) in two prisons in Illinois and Maryland, USA, and then in various parts of the world. It was in 1971 that mefloquine was first used to treat US troops in Vietnam. At the time, it was suggested that it be used in combination with another antimalarial, but this course was not adopted. Resistance started to be noticed in the early 1980s, so more recently it has been used in combination with other medications, notably artesunate.

Side-effects such as dizziness, nightmares, headaches and insomnia can be a problem with mefloquine, which makes it a controversial drug. Some 20 years ago, a fit and healthy young woman, who was at that time a colleague of mine, took mefloquine when on holiday overseas; she told me that from then on, she suffered from epileptic fits. Serious neuropsychiatric reactions can occur in a very small number of cases, perhaps one in 10,000. Dramatically, in 2002 four American soldiers at Fort Bragg, North Carolina, killed their wives, with two of the men immediately committing suicide. Three of the men had recently returned from duty in Afghanistan, where they were reportedly taking mefloquine; a military panel refused to cite the drug as a cause, instead blaming the stress of the foreign deployment, as well as marital problems.

On 11 March 2012, Staff Sergeant Robert Bales of US Special Forces murdered 16 people in two villages in a poor region of Kandahar Province, Afghanistan. During his trial, on 5 June 2013, Bales was asked by the trial judge 'What was your reason for killing them?' and replied that he had asked himself that question 'a million times', adding 'There's not a good reason in this world for why I did the horrible things I did'. Bales was sentenced to life in prison without parole. Bales' lawyers claimed that taking mefloquine had contributed to his behaviour, but this was not accepted.

At the time of the Vietnam War when mefloquine was developed, the other forces in the field, the North Vietnamese Army, was encountering problems of their own with malaria and appealed to its Chinese allies for help. The Chinese launched a research programme, and artemisinin was the result (see pp. 9–12).

Atovaquone (23.6) was discovered in the 1980s and came into use in 1992. For over 25 years, together with proguanil, it has been used as the antimalarial drug Malarone. It appears to be tolerated better than alternatives, with fewer side-effects. It has been very expensive compared with other medications, but it has now been off-patent for some years and is now significantly cheaper.

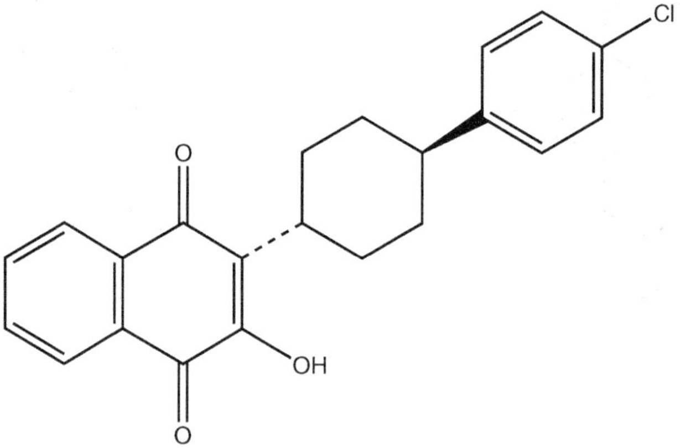

(23.6)

And in 2025 a new drug, ganaplacide (23.7), showed promise in clinical trials as a component of ganaplacide–lumefantrine (23.8), for the treatment of drug-resistant malaria in Africa.

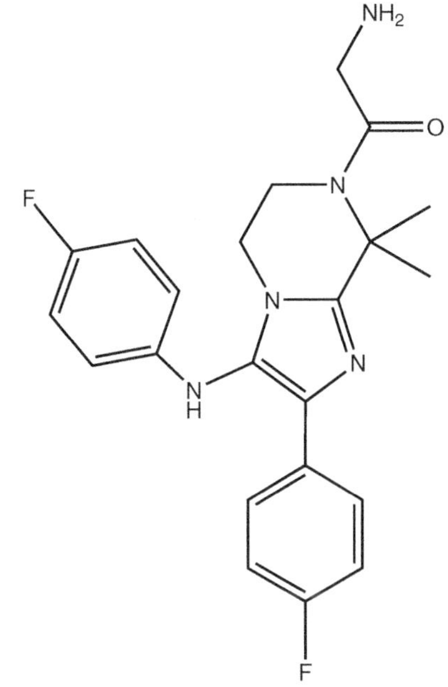

(23.7)

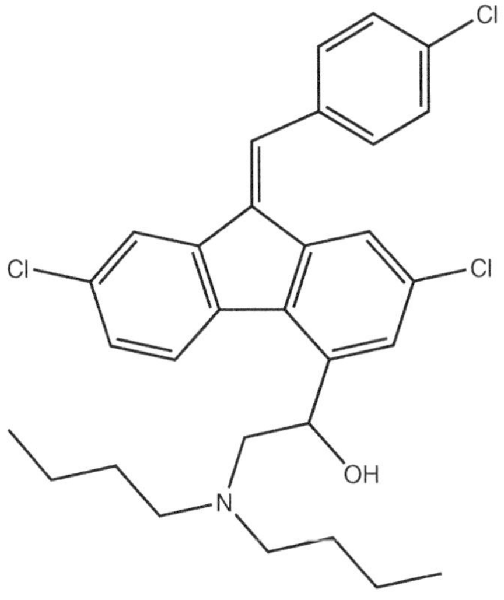

(23.8)

24 Menthol (and Its Relatives)

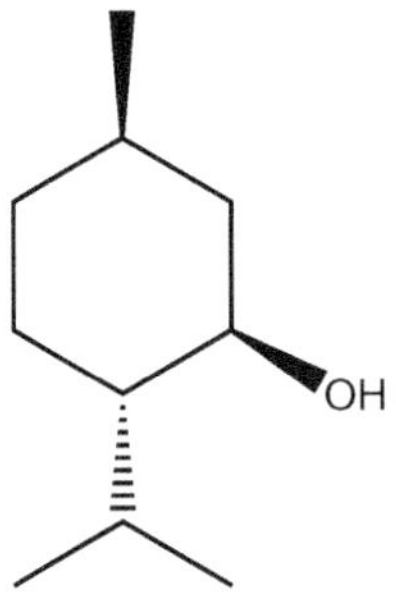

(-)-menthol

(24.1)

Menthol (24.1) is one of the earliest plant products to have been isolated as pure crystals, in 1771, by the German chemist Hieronymus David Gaubius, who was Professor of Medicine and Chemistry at the (Dutch) University of Leiden. You could call menthol an afterthought on the part of the mint plant. It's a secondary metabolite, which means that it is not essential to the structure of the plant or to the reactions which power its life or help it grow, but rather serves a 'secondary' purpose. Plants do not waste energy; menthol is there because of its antifungal and antibacterial properties.

The best place to find it is the corn mint plant, *Mentha arvensis*, which is cultivated largely in India, China and Brazil. Steam distillation gives a product which is dried, after which the oil (containing about 85% (-)-menthol) is cooled to crystallise the menthol. The essential oil of the peppermint plant, *Mentha piperita*, contains as much as 50–60% of menthol. The mint plants provide around 10,000 tonnes of menthol a year, not enough to satisfy demand, so the chemical industry makes at least another 3,000 tonnes a year. New synthetic routes have been devised, notably the Noyori synthesis of (-) menthol from myrcene; for this work, Ryoji Noyori shared in the 2001 Nobel Prize in Chemistry.

The biosynthesis of menthol in the plants involves nine enzyme-controlled reactions (Figure 24.1). Starting with isopentenyl diphosphate (IPP) and dimethylallyl diphosphate (DMAPP), they join together, catalysed by the enzyme geranyl diphosphate synthase (GPPS) to form geranyl diphosphate, which on cyclisation (with (-)-limonene synthase, LS) affords limonene. This, in turn, forms (+)-pulegone after four more stages. The enzyme (+)-pulegone reductase (PR) affords (-)-menthone, which is reduced by (-)-menthone reductase (MR) yielding the product (-)-menthol.

DOI: 10.1201/9781003783565-24

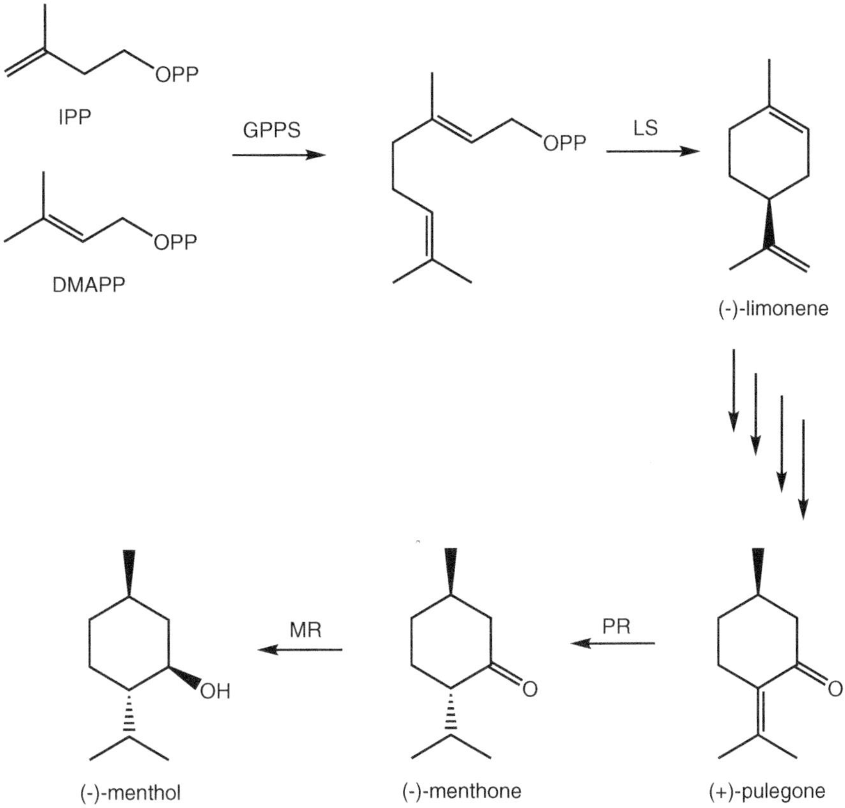

FIGURE 24.1 The biosynthesis of menthol.

The –OH group bound to a ring carbon makes menthol a secondary alcohol, capable of being oxidised to a ketone. This ketone, called menthone (24.2), is also found in peppermint oil, as is the ester menthyl acetate (24.3), formed by esterification of the –OH group. Both of these compounds also have minty smells.

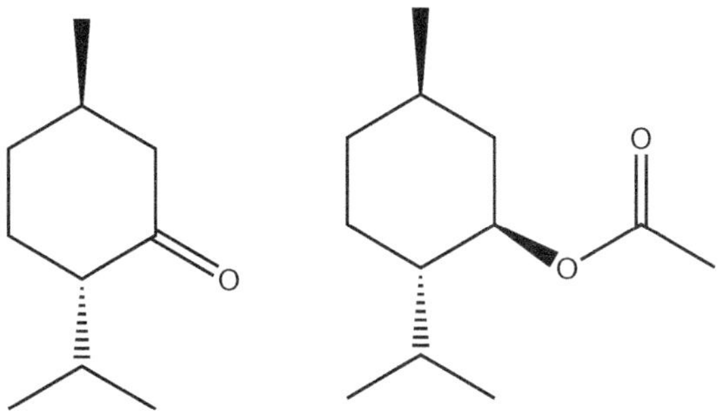

(24.2) (24.3)

The saturated ring system in menthol has six carbon atoms, each involved in bonds to four atoms, and unlike the unsaturated benzene ring it is not flat. There are three asymmetric (chiral) carbon atoms in a molecule of menthol; the molecule therefore occurs as four pairs of optical isomers. The other isomers are known as isomenthol, neomenthol and neoisomenthol, but menthol is the most stable of these molecules, as the three bulkier groups – OH, CH_3 and $CH(CH_3)_2$) – all occupy equatorial positions (24.4).

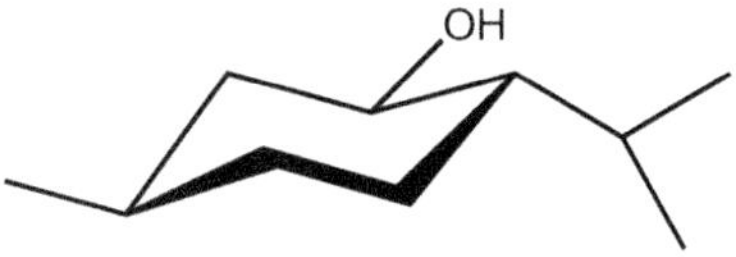

(24.4)

Natural menthol contains only one of its optical isomers, (-)-menthol, though (+)-menthol can be synthesised in the laboratory (Figure 24.2)

Of the two isomers of menthol, natural (+)-menthol has a more pleasant smell – 'fresh, sweet, minty, cooling, refreshing' – compared with the (+) isomer described as having 'musty, bitter, phenolic and herbaceous notes'. The (+)-menthol has about four times the cooling power of the (+) isomer.

When menthol binds to the TRPM8 receptor (transient receptor potential cation channel subfamily M member 8) on the surface of cells, it opens a channel which permits Ca^{2+} ions to enter the cell, changing the electric charge within the cell, resulting in a signal to the brain. This channel is also opened by cool temperatures, below about 26 °C. So, whenever the channel is opened, the brain gets the same signal, a 'cool' message. The temperature is not reduced, but it feels as if it has been. The opposite effect is produced by capsaicin in chilli peppers acting on the TRPV1

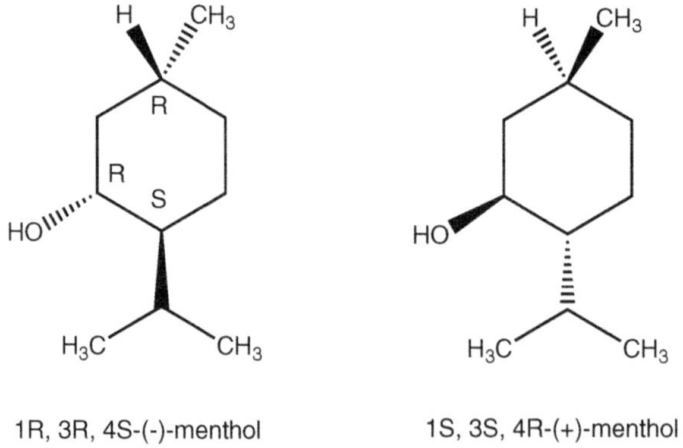

FIGURE 24.2 Optical isomers of menthol.

receptor, which is also sensitive to temperatures above 43 °C, which is why chillis and curries are 'hot', see pp. 23–24). Scientists have carried out molecular modelling studies on the interaction between menthol and the TRPM8 receptor; these indicate that menthol hydrogen-bonds via its –OH group to a particular amino acid in TRPM8 (as a 'hand'). It also uses the isopropyl group as 'legs' to 'stand' on two other amino acids.

So because menthol generates a cooling sensation when applied to the skin or to mucous membranes, it is used in pharmaceutical products, whether liniments such as *Tiger Balm* and decongestants such as Vicks VapoRub, or cough medicines, mouthwash and toothpaste. Its weak antibacterial properties help here.

It is widely used as a flavouring material in drinks, ranging from Crème de menthe and the American Mint Julep (vital for attendees at the Kentucky Derby) to peppermint tea. It is also used in sweets, notably mints, and in chewing gums, of course.

Menthol is generally regarded as a very safe substance. One case has been reported (2016) of fatal menthol intoxication, caused by accidental exposure in a worker at a peppermint factory, who got down into a tank he had been asked to clean and was overcome by the fumes of menthol. But there is a bigger downside. For around a century, menthol has been used as a cigarette flavouring, in quantities ranging from near-zero up to around 1.0%. The cooling effects of menthol, together with some anaesthetic properties, mean that smokers may take a deeper 'drag' and then hold the smoke in their lungs for longer, resulting in greater exposure to carcinogens in the smoke, as well as to nicotine, which feeds addiction. In the USA, African Americans comprise approximately 12% of the population, and it is among them that advertising has prominently featured mentholated cigarettes. Analysis implies that mentholated cigarettes are responsible for a disproportionately high number of premature deaths.

On a more positive note, inhaling peppermint has been linked with improved exercise performance. Moreover, a study of basketball players showed that they were able to run faster and perform more press-ups after inhaling it. Inhaling peppermint increases oxygen uptake, improves mood, and also enhances memory accuracy. It has been suggested that the alertness of drivers could be improved by using peppermint as an in-car fragrance.

25 Methamphetamine

It's a rare chemical that becomes the theme of a popular television programme, but methamphetamine (25.1) was the *leitmotif* of the series *Breaking Bad*, which featured the actor Bryan Cranston as a 50-year-old high-school chemistry teacher, Walter White. Diagnosed with terminal lung cancer, he turned to the synthesis and distribution of methamphetamine in order to provide for his family and cover his medical costs.

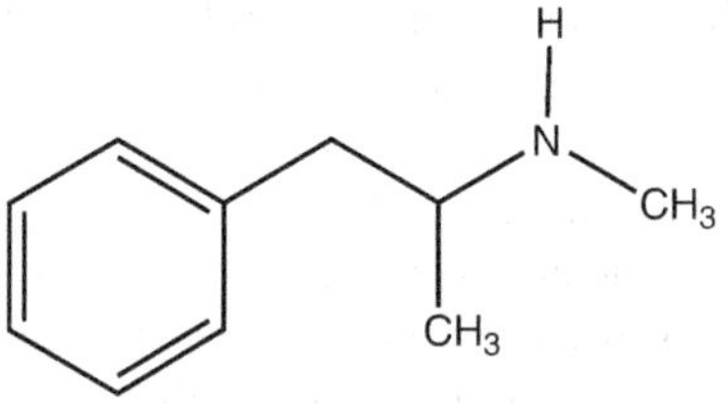

(25.1)

Many drugs, such as cocaine and nicotine, are chemicals that occur naturally. In 1994 it was claimed that amphetamine and methamphetamine had been found in the shrub *Acacia rigidula*, but subsequent research showed that the compound identified as amphetamine (25.2) was β-phenethylamine, an isomer of amphetamine. It is therefore likely that another compound was misidentified as methamphetamine. Thus, methamphetamine is synthetic in origin.

But, like amphetamine itself, it is structurally related to phenylethylamine, which does occur in some foods, such as certain cheeses and chocolate. Phenylethylamine does not have the psychoactive effects of amphetamines, not least because it is readily oxidised to phenylethanoic acid (phenylacetic acid) by monoamine oxidase B enzymes in the liver, preventing significant amounts of phenylethylamine from reaching the brain. Dopamine (25.3) is a natural neurotransmitter in the brain, used to transmit information. Sufferers with Parkinson's disease have low levels of dopamine; this can be treated by giving them the substance known as L-DOPA (3,4-dihydroxyphenylalanine), which the body converts into dopamine.

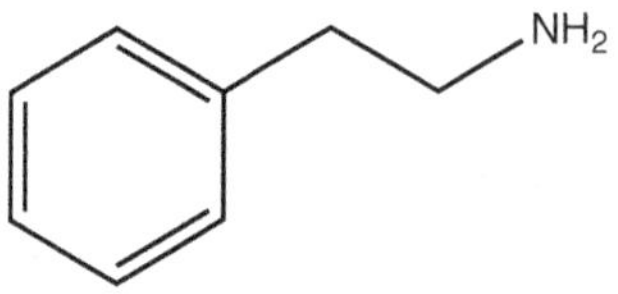

(25.2)

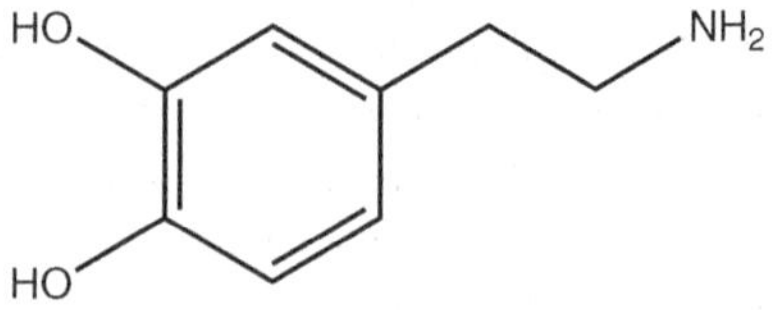

(25.3)

DOI: 10.1201/9781003783565-25

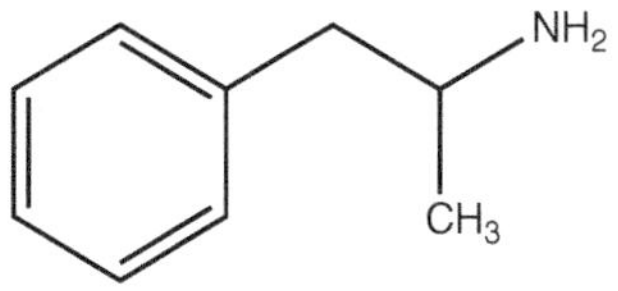

(25.4)

Alpha-methylphenylethylamine, whose name was rapidly shortened to 'amphetamine' (25.4), was first synthesised in 1887. Some 40 years later, it was found to shrink mucous membranes. Marketed from 1932 as *Benzedrine* by Smith, Kline and French, it entered clinical use to relieve nasal congestion. Mixed with menthol, it was absorbed into cotton and placed in 'Benzedrine inhalers'. Some people found that the amphetamine could be extracted from these inhalers and used to get 'high'. William Burroughs wrote about it, whilst Graham Greene used it to write his spy novel '*The Confidential Agent*' in six weeks flat, significantly improving a morning's output from five hundred to two thousand words. Similarly, Jack Kerouac hammered out '*On the Road*' in three weeks' work on his typewriter in April 1951.

Methamphetamine was first isolated as a pure chemical in 1919 by the Japanese chemist Akira Ogata. As with amphetamine, the presence of the methyl group in the side-chain of methamphetamine means that it does not fit the active site of the monoamine oxidase enzyme that oxidises phenylethylamine, thereby rendering it harmless. It was soon found that methamphetamine was a powerful stimulant, increasing the amounts of dopamine that the body releases.

In 1937, the German chemical company Temmler developed and patented a new synthesis of methamphetamine, which it marketed as Pervitin, a drug that was to become the 'people's drug' (*Volksdroge*). Initially available over the counter, Pervitin was actively promoted as a stimulant, both through advertising hoardings and also by targeting doctors and hospitals, not least with free samples. It was taken up across society, even added to chocolates, as something to help people get through their work, whether they were manual workers, lorry drivers, students studying for exams, or workers on night shifts.

From 1939, Pervitin pills were provided to the German armed forces, and it helped fuel the Blitzkrieg across Western Europe the following year, from Guderian and Rommel downwards. Luftwaffe pilots took Pervitin during the Battle of Britain, giving a whole new meaning to the expression 'flying high'. One person who did not seem to have taken the drug was Adolf Hitler. It appears that Dr Theodor Morell, Hitler's personal physician, usually gave opiates to 'Patient A' (Hitler).

Amphetamine use was not confined to Germany. The Japanese were heavy users – the tablets were not just given to Kamikaze pilots, or even the military, but methamphetamine was supplied to workers in the armaments industry, to boost their performance. At the end of World War II, there were still large stocks of methamphetamine left over in Japan, which many people found attractive, and it was only in 1954 that heavy legal punishments were imposed for methamphetamine use in Japan.

In Britain, the Royal Air Force had, by 1942, approved of the use of Benzedrine (amphetamine) tablets by bomber pilots on long night missions, and the drug was also taken by personnel in Montgomery's Eighth Army at the Battle of El Alamein. As far as American forces are concerned, it has been estimated that some 15 per cent of American servicemen took Benzedrine regularly. Aircrew were major users, with Benzedrine tablets being part of all emergency kits in American bombers by 1943.

Benzedrine helped Britain's Prime Minister Sir Anthony Eden through the Suez Crisis in 1956, though it may not have improved his decision-making. After 1951, amphetamines became prescription drugs in the USA; they were widely prescribed as pick-me-ups, commonly known as 'pep pills'. Long-distance lorry drivers in the 1950s used amphetamines to help them stay awake on their coast-to-coast trips, whilst college students knew that 'pep pills' containing amphetamines kept them alert when revising for exams and thereby improved their marks.

Methamphetamine promotes the release of high levels of dopamine, the neurotransmitter which stimulates brain regions linked with alertness and vigilance. A side-effect of amphetamines is appetite loss, so post-war they were widely used as prescription drugs to combat obesity. Amphetamines were controlled in the UK by the 1964 Drugs (Prevention of Misuse) Act, but continued to be abused, as in the USA, where because of their stimulant properties methamphetamine came to be known as 'speed'.

The 1967 'Summer of Love' in San Francisco is associated with psychedelics like LSD, but then things went badly wrong with an epidemic of intravenous methamphetamine injection in 1968, so the motto changed from 'peace and love' to 'speed kills'. Amphetamines contributed to the deaths of Judy Garland and Edie Sedgwick, Andy Warhol's muse, but Johnny Cash survived after he turned to God. In Britain, amphetamines fuelled youth culture, notably the Mods, hence the reference to 'speed' in the 1967 song 'Here Comes the Nice', by the group The Small Faces ('He's always there, if I need some speed'). Amphetamine abuse is a particular problem in the USA, with methamphetamine-related deaths increasing from 3,616 to 16,127 per year between 2013 and 2019. It especially affects the Midwestern states. Employees in manual industries under pressure, to increase their productivity (for example, meat processing), may resort to it, as described in the 2009 book *Methland*. It is often regarded as a forgotten epidemic, compared with opioid abuse, though sometimes users combine methamphetamine with opioids such as fentanyl.

It is reckoned that methamphetamine is the second most abused drug worldwide, after cannabis, with perhaps 35 million users. Methamphetamine has never caught on to the same extent in the United Kingdom as in the USA, probably because there is a ready supply of alternatives, such as cocaine and heroin.

25.1 SYNTHESIS

Although there is a licit but limited production of methamphetamine in the pharmaceutical industry, methamphetamine is largely manufactured illegally, despite the sale of the precursor chemicals being very tightly controlled. This production is carried out sometimes in very dangerous domestic settings, rather than in properly and

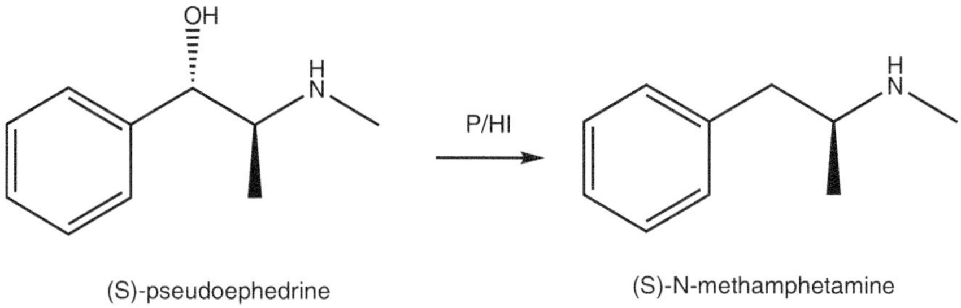

FIGURE 25.1 Synthesis of methamphetamine from pseudoephedrine.

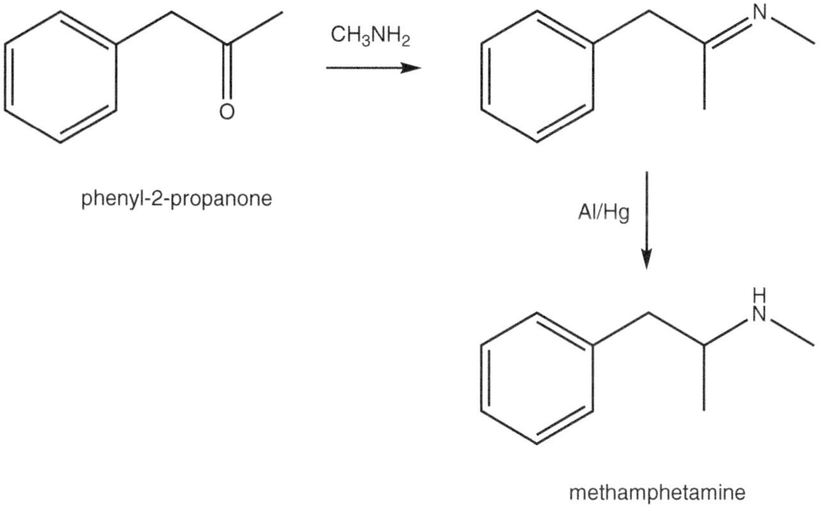

FIGURE 25.2 Synthesis of methamphetamine from phenyl-2-propanone.

safely equipped laboratories that are requisite for handling flammable solvents (such as acetone and toluene) and toxic chemicals (such as phosphorus).

Two routes are widely used. One involves the reduction of either ephedrine or pseudoephedrine (Figure 25.1), commonly obtained from medicinal products like decongestants. The reducing agent is usually a mixture of phosphorus and either hydrogen iodide or iodine/water. This method has the advantage of producing a single isomer of methamphetamine, namely the 'active' isomer.

The second method (Figure 25.2) begins with phenyl-2-propanone (P2P), also known as benzyl methyl ketone or phenylacetone. This undergoes a condensation reaction with methylamine; in a subsequent stage, the product is reduced to methamphetamine, often using aluminium amalgam as the reducing agent. This method has the disadvantage of producing a racemic mixture of the two optical isomers of methamphetamine, only one of which is a stimulant. A further step in the process thus involves the addition of tartaric acid to form a mixture of the two tartrate salts, which can be separated by selective recrystallisation.

Law enforcement forces are well aware of the chemicals used in these illegal syntheses, and the supply of the precursors, such as P2P, and solvents is carefully controlled. Often, they are smuggled in or imported under false declarations that they are industrial chemicals. Someone in the Far East in the 1980s found how to create large crystals of methamphetamine hydrochloride known as 'crystal meth', which does not have to be injected. In Thailand, it is called *Ya ba* ('madness drug'); its use spread from Hawaii to the United States.

25.2 THE ISOMERS OF METHAMPHETAMINE

The carbon atom next to the methyl group in the side-chain of the methamphetamine molecule has four different groups attached, so it is chiral. This means that methamphetamine exists as two optical isomers, or enantiomers, which can be isolated separately (Figure 25.3). Enantiomers often have different properties – thus, one enantiomer of the ketone carvone smells of spearmint, whereas the other enantiomer smells of caraway. The enantiomers of methamphetamine have significantly different properties in the body.

L-methamphetamine is a decongestant, used medicinally. It has no stimulant properties. However, its enantiomer, D-methamphetamine, is the stimulant variously known as meth, methedrine, crank, crystal meth and speed (amongst other names). Because the isomers have different three-dimensional structures, they fit into the body's protein receptors differently and therefore have different effects on the body. That the difference is not readily comprehended by non-scientists, it seems, is illustrated by the following story.

The Winter Olympic Games of 2002 took place at Salt Lake City in the United States. On 23 February, the Scotsman Alain Baxter finished in third place in the slalom event, the first Briton ever to gain an alpine skiing medal. Then his urine test revealed the presence of traces of methamphetamine. Baxter was stripped of his bronze medal.

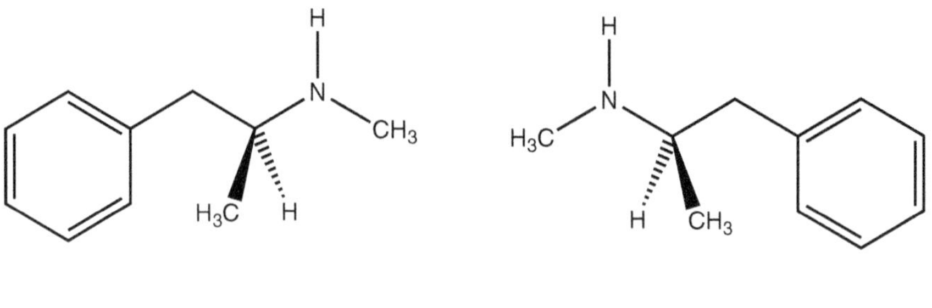

(*S*)-(+)-methamphetamine
D-methamphetamine

(*R*)-(-)-methamphetamine
L-methamphetamine

FIGURE 25.3 Optical isomers of methamphetamine.

Just another athlete doping, you say? Not so. At home, Alain Baxter suffered from nasal congestion, a longstanding medical condition, which he treated with OTC Vicks nasal inhalers. They contain a mixture of compounds, including menthol, camphor and methyl salicylate (present in oil of wintergreen), which shrink inflamed membranes. Faced with these recurrent symptoms in Salt Lake City, he consulted the team doctor. However, the doctor's suggestions did not work. Baxter spotted a familiar-looking Vicks inhaler in a shop, which looked the same as those he bought back in Scotland, and bought it. However, the American version of Vicks has a different formulation from the British version; for one thing, it includes L-methamphetamine, tiny amounts of which duly turned up in Baxter's urine sample.

The IOC having disqualified Baxter, he appealed. The Court of Arbitration for Sport supported the IOC but did accept that Baxter had acted honestly throughout. They did not follow up Baxter's request that his urine be tested to show which methamphetamine isomer was present – which was quite feasible, as the two isomers can be separated by chromatography using a chiral column – to show that he had obtained no advantage from taking the non-stimulant isomer. IOC regulations simply bar the use of 'methamphetamine', without distinguishing between the two isomers.

Sportsmen have been abusing amphetamines for over half a century. Back in 1952 at the Winter Olympics in Oslo, a number of speed skaters who had been taking amphetamines needed medical assistance; as others have remarked, this gave a new meaning to the expression 'speed skating'. At the Rome Olympic Games in 1960, the Danish cyclist Kurt Jensen collapsed and died of a heart attack, reportedly as a result of an amphetamine overdose. The most famous victim was Tom Simpson, the British cyclist who was the 1965 world champion and also the first Briton to wear the famous yellow jersey as leader of the Tour de France.

During the 1967 Tour de France, he was undertaking the famous ascent of Mont Ventoux in Provence on 13 July, a day of brutal heat, when he collapsed and died of heart failure. His death was brought on by a combination of the intense heat, some brandy he had drunk earlier, and Tonedron, a form of methamphetamine (tablets were found in the pocket of his jersey). People argue about the dangers of taking methamphetamine. In Tom Simpson's case, in the words of the 1960s slogan, 'Speed kills'.

26 Morphine and Heroin

Nature is a brilliant synthetic chemist, far more efficient than humans in its use of enzymes to carry out high-yield, multi-stage syntheses in plants, so that the opium poppy (*Papaver somniferum*, the second part of its name referring to its sleep-inducing properties) is much better at synthesising morphine than white-coated chemists in a laboratory. When the poppy's unripe seed capsules are pierced, they exude a liquid which on drying yields raw yellowish opium, whose pain-killing ability humans discovered thousands of years ago in the Eastern Mediterranean, where its use is documented over 5,000 years ago in Mesopotamia. The properties of opium, for example in poppy juice, appear to have been described over 2000 years ago by the Greek classical authors including Homer and Hippocrates. It was being used in Egypt by 1500 BC (as evidenced by the Ebers Papyrus); the production of opium in the city of Thebes is linked with an alkaloid in opium named thebaine. Opium was being used in Rome 2,000 years ago, with the Roman doctor Galen promoting the use of opium in the second century AD.

In the Middle Ages, the Swiss German physician Phillipus Aureolus Theophrastus Bombastus von Hohenheim (1493–1541), also known as Paracelsus, dissolved opium in an alcoholic solvent (such as brandy); this became known as laudanum, which became widely used as a medication. It contained a few other substances, including crushed gold and pearl, musk, saffron, nutmeg and amber. In the following century, a London physician named Thomas Sydenham popularised his version of laudanum, made by dissolving opium, cinnamon, cloves and saffron in sherry.

Laudanum became known as a general cure-all, especially useful for treating cholera and dysentery, and for the next two centuries was widely used by all social classes as an over-the-counter medicine, the aspirin of its age. Much of it was consumed in the English Fenland, where malaria – known as the 'ague' – was endemic until the Fens were drained. Laudanum was effective against fever and other problems, including diarrhoea.

At this time, opium was the equivalent of an over-the-counter drug, used by respected people, whether George Washington, who used it because of his painful teeth, or the reformer of slavery, William Wilberforce (1759–1833). Wilberforce provides a warning, having been prescribed opium in 1788 by his doctor, he was still taking it in 1818.

It assisted some famous writers, with Samuel Taylor Coleridge (1772–1834) penning the *Rubiyat of Kubla Khan* after a morphine-assisted dream, or Wilkie Collins (1824–1889), whose novel *The Moonstone* was influenced by laudanum. Likewise, the title of the autobiography of Thomas De Quincey (1785 –1859) – *Confessions of An English Opium-Eater* - gives the clue to his habits, up to 20 grams of opium a day. That is before we mention Lord Byron, Keats, Shelley, Southey, Lewis Carroll and George Crabbe.

Even Florence Nightingale took laudanum. The best known of proprietary medicines that depended upon opiates was Dr Collis Browne's Chlorodyne, which hit the

DOI: 10.1201/9781003783565-26

market in 1856, the year in which Dr Collis Browne left the British Army, partnering with John Davenport, a chemist. By this time, the chemistry of opium was better understood, so that this medicine combined morphine with tincture of cannabis, ether and chloroform, as well as flavourings including liquorice and peppermint oil. Opiates were gradually removed, and today's Chlorodyne is opioid-free.

Opium fuelled trade, with the two Opium Wars resulting in part from the British Empire's promotion of opium sales to China – the reverse trade brought tea to Britain. Chinese sailors came to London with the East India Company, bringing opium 'dens' with them – likewise Chinese labourers who came to the USA as miners or as builders of railways. Grocers sold opium-containing products until they were regulated by the Pharmacy Acts of 1868 and 1908 so they could only be obtained from pharmacists (or 'chemists').

Chemical understanding improved rapidly in the nineteenth century, and this affected opium. It gradually became clear that it contained a mixture of basic molecules known as alkaloids, of which the best-known ones are morphine, codeine and thebaine. The first two are painkillers, whilst thebaine is useful in the synthesis of substances like oxycodone (p. 57). A young pharmacist's assistant, Friedrich Wilhelm Adam Sertürner, obtained morphine (26.1) in 1804 by extracting opium with water and then precipitating the morphine by the addition of ammonia. He found that it turned red litmus blue, confirming its basicity by reacting it with several acids and isolating the sulphate, nitrate, acetate, tartrate and bicarbonate salts.

Morphine was soon discovered to be a very effective painkiller. Although in the eighteenth century opium was occasionally used to treat postoperative pain, morphine revolutionised the treatment of chronic pain, once it became available in significant quantities in the 1820s and 1830s. It is recorded that in 1656, Sir Christopher Wren and Robert Boyle once injected opium, dissolved in fortified wine, into a dog, resulting in anaesthesia followed by a full recovery, but this observation was not followed up. In the American Civil War, the Crimean War and the Prussian Wars, morphine was often given by oral and subcutaneous routes, but the invention of the hypodermic syringe changed the means of administration. It is said that after the American Civil War, some 100,000 soldiers suffered from addiction to morphine.

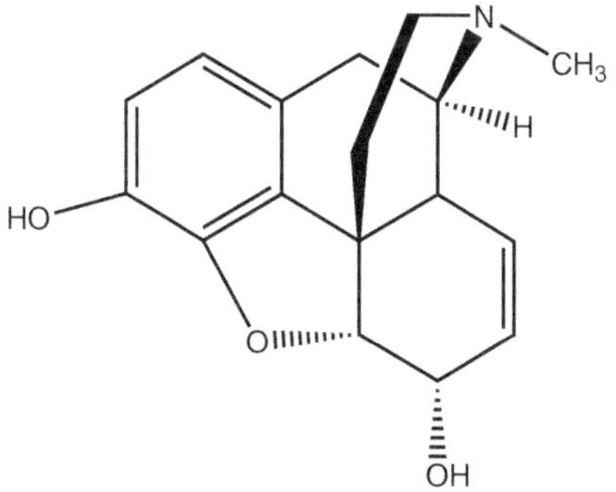

(26.1)

In the human body, receptors for opioids are found in the central nervous system, the brain and spinal cord. Endorphins and enkephalins are natural molecules that stimulate them. Morphine binds to the mu (μ) receptors, which are responsible for analgesia and sedation, as well as respiratory depression, amongst others. Fentanyl also binds to this same receptor, as does codeine, but more weakly.

Part of the development of organic chemistry in the nineteenth century consisted of developing new synthetic pathways, and acetylation was one such reaction. This was used to synthesise acetylsalicylic acid, commonly known as aspirin (p. 13) and in 1874 Charles Romley Alder Wright of St. Mary's Hospital Medical School in London applied it to morphine, heating it with ethanoic anhydride, obtaining diacetylmorphine (26.2), in which morphine's two –OH groups have been esterified. Animal testing on dogs showed that it produced rapid anaesthetic effects, with faster pain relief than morphine itself.

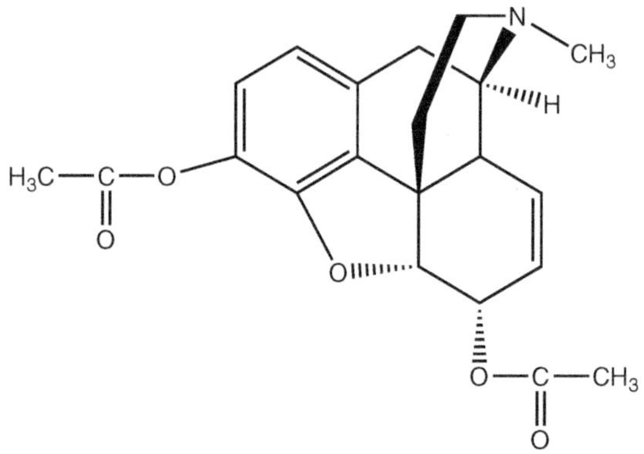

(26.2)

Then, in 1897, Heinrich Dreser and Felix Hoffmann of the German pharmaceutical firm Bayer reinvestigated diacetylmorphine. Believing that it was an addiction-free version of morphine, Bayer promoted it, enthusiastically, as an improved drug with '*heroisch*' properties, and so gave it the name Heroin. In the light of history, this optimism was misplaced. Its faster action than morphine – and quicker pain relief – is likely due to the absence of morphine's hydrophilic –OH groups, so that the more lipid-soluble diacetylmorphine crosses the blood-brain barrier faster. Moreover, by 1900, people were starting to become addicted to heroin, an addiction stronger than dependence on alcohol, nicotine or cocaine.

In the early twentieth century, legislation was passed that made use of opium (for example, in opium dens) illegal, while also restricting cocaine. Along with the arrival of Prohibition (of alcohol), people turned to heroin. Its quicker 'kick' than morphine added to its attractions. It was associated with urban areas, spreading out from New York, and was increasingly associated with gangs. Addicts became 'junk men', sorting through waste to find metal scrap to sell in order to support their habit – hence the word 'junkie'.

The banning of heroin use by the US Congress in 1924 had no effect on heroin addiction, but gangsters, who had grown rich from illicit alcohol under Prohibition, saw another way to make money, and moved into heroin. Originally the heroin trade was run by Jewish gangs, sourcing their heroin from China, France and the Middle East, but in the 1930s Mafia gangs took over, notably that of Salvatore C. Lucania ('Lucky Luciano').

Of course, World War II meant that smuggling in heroin from countries like China and France was impossible, so that the number of heroin addicts in the USA fell to perhaps 20,000. A drive against organised crime by the FBI saw Luciano convicted. In 1946, he was paroled and deported to Italy. There, he set up a heroin smuggling network, under the cover of a business exporting fruit and vegetables, sending heroin to the USA hidden in shipments of his produce. Marseilles became the place where heroin was refined and eventually became the heroin capital of France. Now you know where '*The French Connection*' originated.

When shipments to New York became more difficult, destinations were changed to places such as Cuba and Florida (before Castro came to power). Around 1970, these networks were disrupted by police on both sides of the Atlantic, and Turkey, hitherto the source of the majority of the opium, outlawed poppy cultivation in 1972 under pressure from the USA. The Turkish ban lasted just two years, long enough to disrupt supplies to France, whereupon the Mafia started to source it in the 'Golden Crescent' of Pakistan, Afghanistan and Iran, and more importantly the 'Golden Triangle' of Laos, Thailand and Burma. At present it is believed that most of the heroin supplied to the United States comes from Colombia and Mexico.

Heroin addiction is often associated with popular entertainers, whether jazz musicians (for example, Billie Holliday, Chet Baker and Charlie Parker) or figures in the world of pop music (for example Janis Joplin and Jim Morrison). A variant is the heroin-cocaine 'speedball', which claimed the lives of John Belushi, River Phoenix and Philip Seymour Hofmann. Such victims attract publicity, but many thousands of others are lost in statistics, with the number of American addicts rising from 100,000 in the late 1950s to an estimated 600,000 in 1985, with the 1960s counterculture and the Vietnam War, when many GIs became addicted, being important contributing factors. The twenty-first century has seen four 'waves' of opiate mortality in the USA. The first wave, starting around 2000, was due to natural and semisynthetic alkaloids, notable oxycodone (see pp 57), with a second wave of deaths primarily due to heroin starting around 2010, with a third wave caused by due to fentanyl and other synthetic opioids starting around 2013, followed by a fourth wave particularly associated with fentanyls, methamphetamine and cocaine use.

Heroin is classed as a prodrug, since its effects are caused by metabolites. Once in the human body, it is quickly broken down by hydrolysis of the two ester linkages, ultimately resulting in morphine. The first stage of the hydrolysis is especially rapid (estimates suggest that heroin in plasma has a half-life of around five minutes), yielding an intermediate compound, 6-monoacetylmorphine (6-MAM) (26.3).

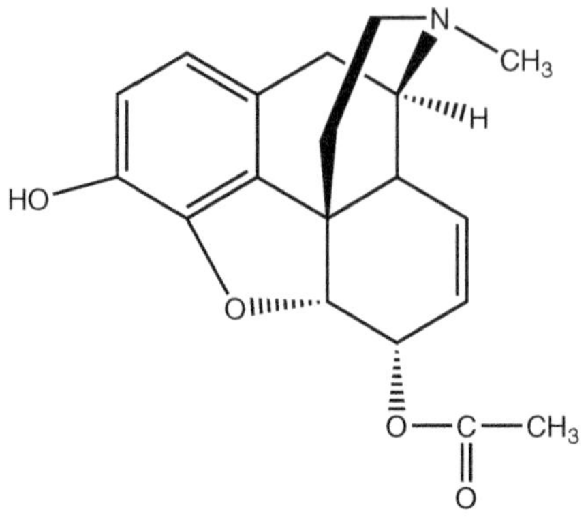

(26.3)

In rats, 6-monoacetylmorphine was found to be the principal opioid present in both the blood and the brain for the first 30 minutes after heroin was injected, and its concentration was found to correlate with behavioural effects noted. This suggests that 6-MAM may be responsible for many of the effects of heroin intake. In other words, heroin appears to be a prodrug, a molecule with little or no pharmacological activity, which the body's metabolism converts into the active drug, in this case 6-MAM. Further hydrolysis, converting 6-monoacetylmorphine into morphine, removes 6-MAM from the body quite rapidly, so that it cannot be detected in urine after a day, whereas morphine can be detected for several days. The presence of 6-MAM in urine indicates that the person concerned has used heroin. In contrast, the presence of morphine in urine is ambiguous, as the person could have ingested either heroin, morphine or codeine.

27 Muscone and Synthetic Musks

'Musk' has been used in fragrances for over a thousand years, as it possesses an animalic, sensory, even erotic smell. For most of that time, it was synonymous with a secretion from the anal gland of the male musk deer (*Moschus moschiferus*), which the deer uses both to attract females and mark its territory. The word 'musk' is derived from the Sanskrit word for testicle, *muská*. Perfumes commonly contain a 'musky' compound, both to act as a fragrance material and, because of its low volatility, to retain (fix) the odour of the other molecules.

Musk deer have been hunted almost to extinction by poachers who believe that musk is an aphrodisiac, and they are also at risk due to habitat loss, notably in China and the Himalayas. The musk odorant can be obtained without killing the deer, but hunters do not worry about that – one dead musk deer supplies no more than 25 g of dried musk glands ('musk pods'). *Moschus moschiferus* became a protected species in 1979. The secretion from the deer is a mixture of molecules, but the key odorant is a chemical called muscone (27.1). A similar musky substance, civetone (27.2), is obtained from the anal glands of the African civet cat (*Civettictis civetta*), in this case without killing the animal.

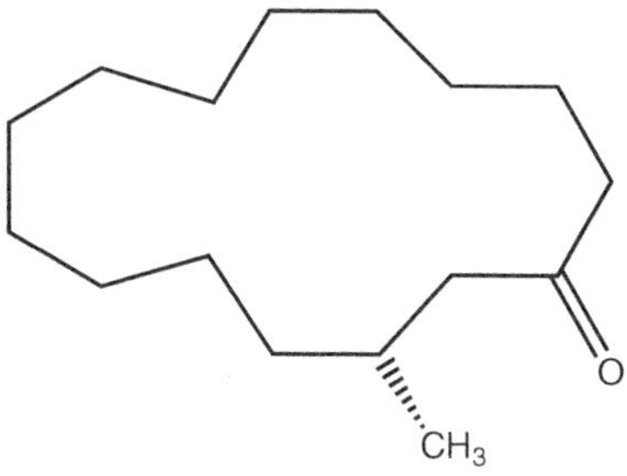

(27.1)

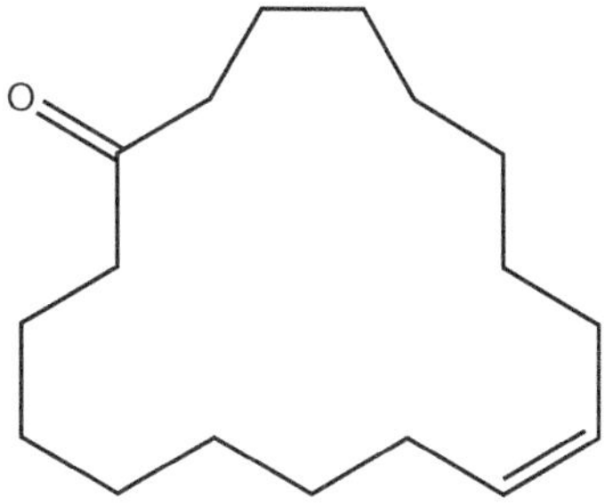

(27.2)

By the early twentieth century, chemists knew that both these odorants were ketones, muscone being $C_{16}H_{30}O$ and civetone being $C_{17}H_{30}O$, but they had no idea of their structures. Ring structures were thought unlikely, as in 1885, Adolf von Baeyer predicted that rings containing more than eight atoms were too strained to be stable, assuming that they were flat (Baeyer's strain theory). It was the great Croatian-born chemist Lavoslav (Leopold) Ružicka (1887–1976) who did the key research that established their structures (he would go on to share the 1939 Nobel

DOI: 10.1201/9781003783565-27

Prize in Chemistry), showing that muscone was 3-methylcyclopentadecanone (subsequently shown to be the (-)-(*R*)-isomer) and that civetone was (9*Z*)-1-cycloheptadec-9-enone. Ružička knew that if heavy-metal salts (for example, Ca , Ce , Y , Th) of carboxylic acids were heated, a ketone was produced:

$$(RCOO)_2Ca \rightarrow CaCO_3 + RCOR$$

He reasoned that if this reaction was carried out with the salts of dicarboxylic acids, where the two R groups were linked, a cyclic ketone would result, and he prepared several macrocyclic musks in this way (Figure 27.1), by what became known as the Ružička 'large-ring synthesis' (in small yields, 5% or less), showing that muscone and civetone were in no way exceptional. He found that the $C_9 - C_{12}$ compounds had camphor-like smell, C_{13} was woody and that the $C_{14} - C_{20}$ compounds were musk-like.

Ružička successfully made cyclopentadecanone, known commercially as Exaltone, by this method (Figure 27.1). This is a component of the musk produced by the Louisiana muskrat (*Ondatra zibethicus*), along with cycloheptadecanone. Major advances in musk chemistry were indeed made in the 1920s, as musks were also identified in plant sources.

Synthetic musks had been produced long before Ružička. The most notable examples, nitromusks, developed through the pioneering work of Albert Baur around 1890. However, Olivier David has tracked down an experiment performed in 1758–1759 by the German chemist Andreas Sigismund Marggraf (1709–1782). In this experiment, Marggraf nitrated the 'rectified oil of yellow amber' and obtained a product with a very strong musk odour.

Nitration became a popular reaction among chemists in the nineteenth century, particularly in the manufacture of explosives. Pioneer compounds include nitrobenzene (Eilhard Mitscherlich, 1834), nitroglycerine (Ascanio Sobrero, 1847), and 2,4,6-trinitrotoluene (TNT; Joseph Wilbrand, 1863). Nitro compounds were, of course, key precursors to the amines needed to make azo-dyes.

Baur's key experiment involved a Friedel-Crafts alkylation of methylbenzene with isobutylbromide (which is isomerised to give a tert-butyl group during the reaction),

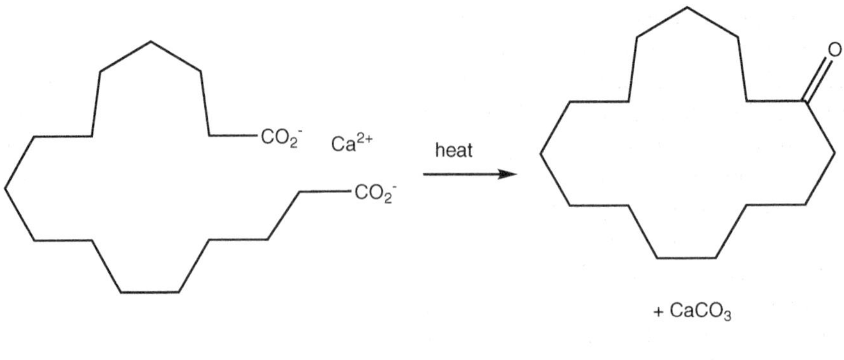

FIGURE 27.1 Synthesis of cyclopentadecanone by pyrolysis.

followed by nitrating the initial product. The final product, 2-*tert*-butyl-4-methyl-1,3,5-trinitrobenzene, was not suitable for explosives, but it had a splendid musk smell. In the ultimate compliment to a chemist, this *künstlicher Moschus* ('artificial musk') came to be known as 'Musk Baur'(27.3). Baur patented this synthesis in 1888.

He went on to make even better nitromusks, in the forms of Musk Xylene, Musk Ketone and Musk Ambrette (27.4); the syntheses of musk xylene and musk ketone are shown in Figure 27.2. These and other nitromusks (for example, Musk Moskene (27.5) and Musk Tibetine (27.6)) became essential to the perfume industry. For example, Ernest Beaux is claimed to have used over 10% nitromusks, especially Musk Ketone, in *Chanel No. 5* (1921).

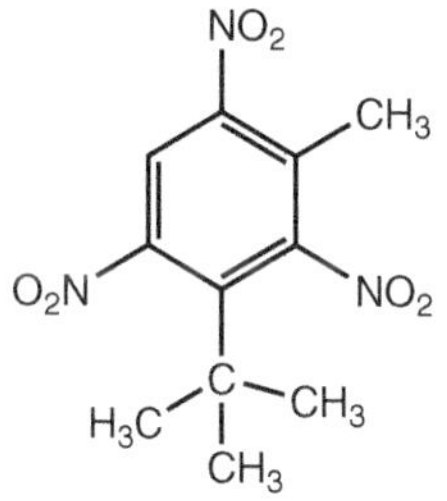

(27.3)

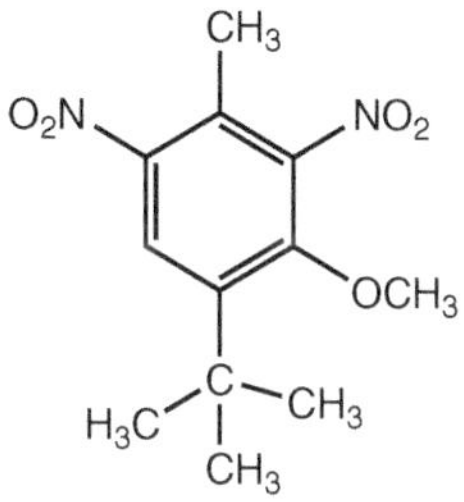

(27.4)

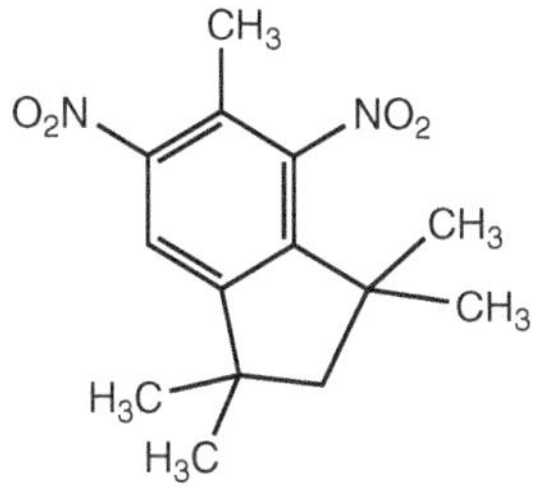

(27.5)

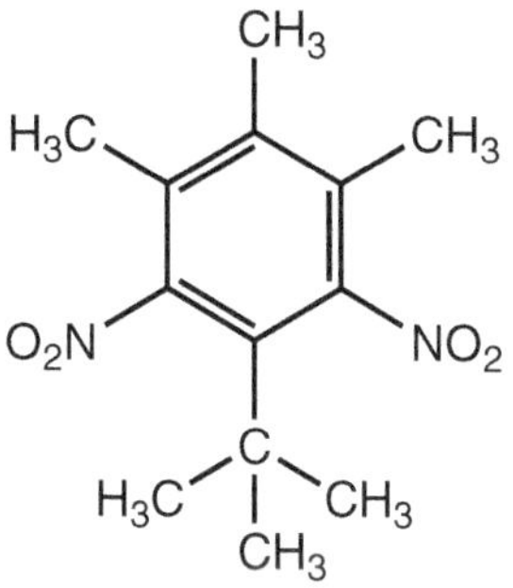

(27.6)

Nitromusks were industry leaders until World War II, after which their place was usurped by the cheaper polycyclic musks, which came to be widely used in applications such as detergents. The decline in nitromusks was hastened by the concerns about their environmental impact. Musk Ambrette was found to be a photosensitiser of human skin, whilst metabolites, such as those formed from musk xylene, have been detected both in sewage plants and as metabolic products of mammals. Nitromusks are not easily degraded and therefore bioaccumulate, leading to their detection in human milk and fat tissue, as well as in fish.

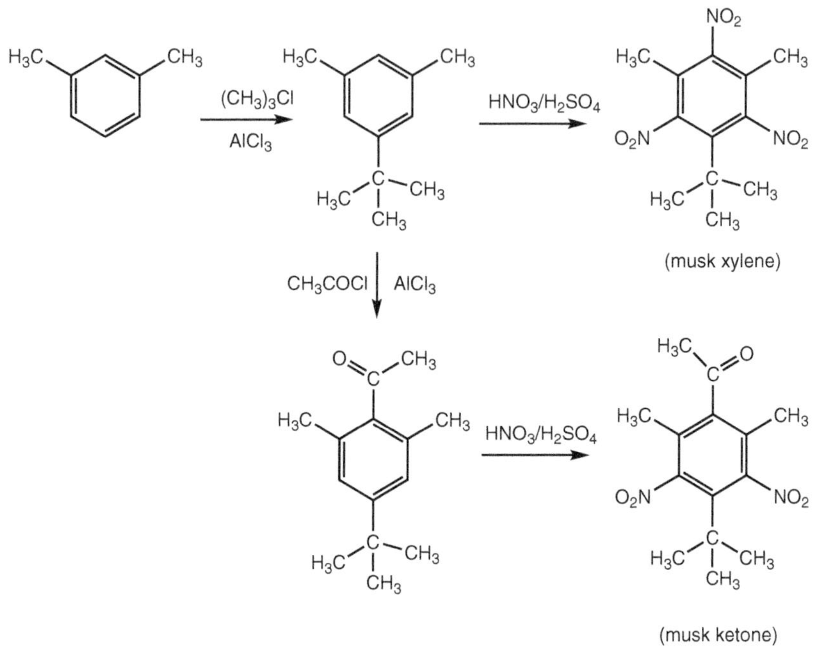

FIGURE 27.2 Syntheses of musk xylene and musk ketone.

Chemists developed polycyclic musks that could be used in detergents on account of their stability under alkaline conditions (unlike nitromusks). These hydrophobic molecules tend to adhere to clothes, instead of washing off, so that the fragrance stays on the garment. The original polycyclic musk was Phantolide ((27.7); 1951), so called because its structure was unknown at first. The most successful polycyclic musk is Galaxolide ((27.8); 1965), used on a very large scale in detergents and fabric softeners, whilst its clean, musky smell also led to its use in perfumes, such as in the original formulation of *White Linen* (Estée Lauder, 1978). Galaxolide and Tonalide (27.9); have been the two most widely used polycyclic musks in both Europe and the United States, comprising over 90% of those markets, and were used together in *Le Mâle* (Jean Paul Gaultier, 1995), by the perfumer Francis Kurkdjian.

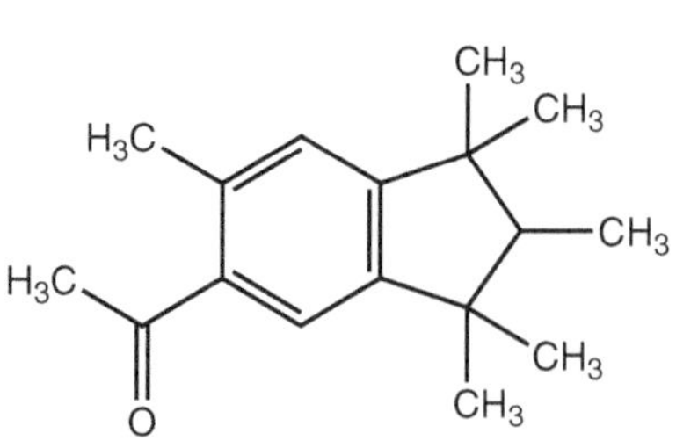

(27.7)

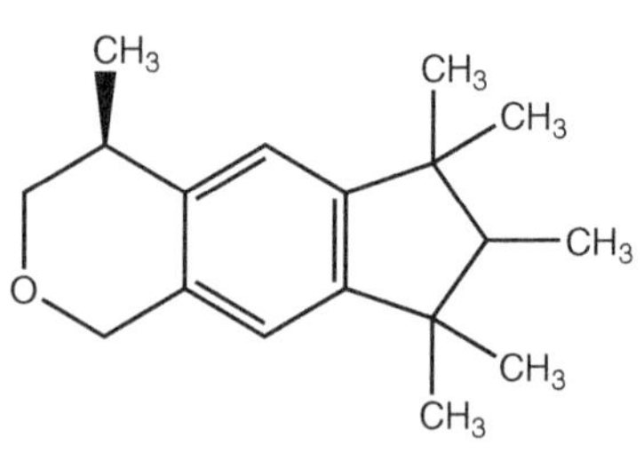

(27.8)

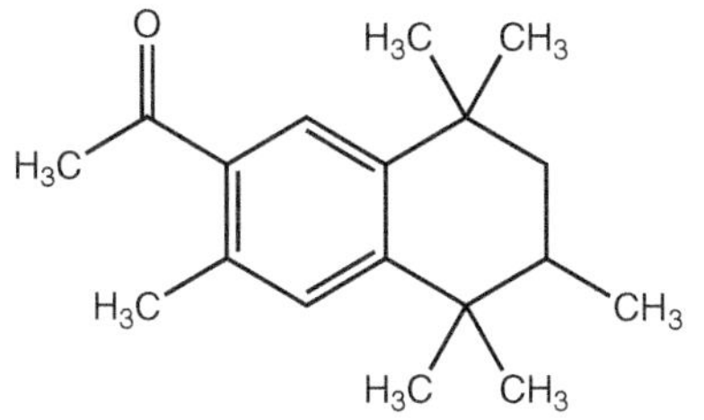

(27.9)

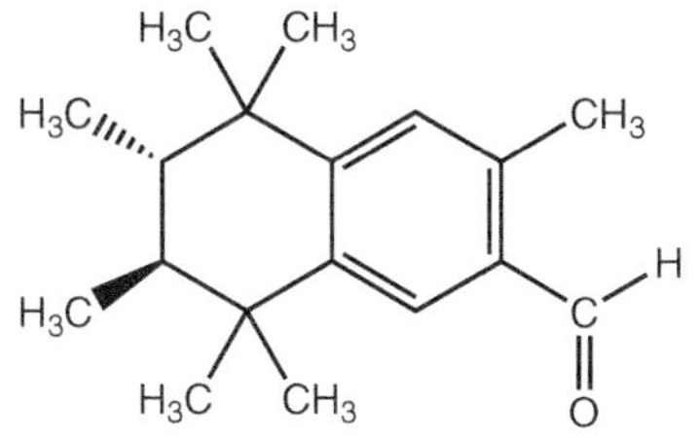

(27.10)

The polycyclic Vulcanolide (27.10); is an excellent example of the effect that chirality can have on odour, as (*S,S*)-(-)-Vulcanolide (2002) has a powerful musky, earthy and slightly amber smell, whereas the isomeric (*R,R*)-(+)-Vulcanolide has a very weak smell which is less earthy but slightly woody. The unreactivity of polycyclic musks makes them non-biodegradable, whilst their lipophilicity means that (like nitromusks) they accumulate in the environment – detected in marine mammals, aquatic organisms like fish, wastewater treatment plant effluents and in adipose tissue and human milk. They are therefore used less extensively in most consumer applications but are still being used in perfumes.

Attention thus turned to macrocyclic musks, some of which like muscone and civetone had been known for many years. Many of these are ketones, often obtained from animal sources, but plants have yielded musk-smelling lactones.

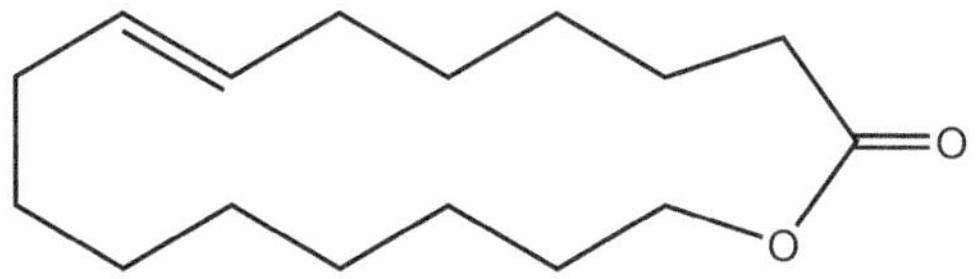

(27.11)

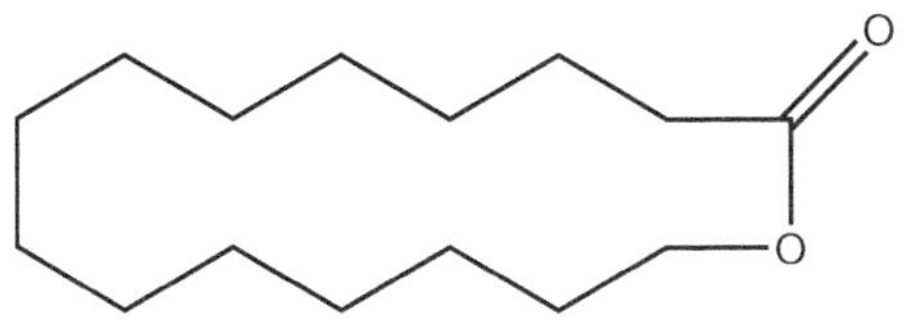

(27.12)

Thus Ambrettolide (27.11) was discovered by Kerschbaum (1927) in the seed oil of the ambrette plant (*Abelmoschus moschatus*); in that same year, cyclopentadecanolide, known as Exaltolide (27.12) or Thibetolide, was found in Angelica root oil (*Angelica archangelica*). Globalide (Habanolide) is a mixture of the isomeric

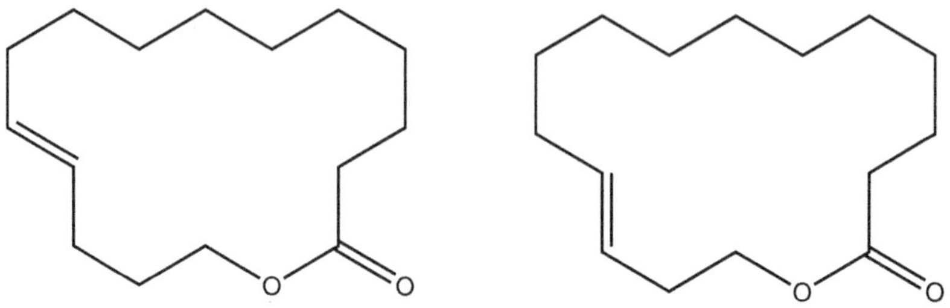

FIGURE 27.3 Isomeric pentadecen-15-olides found in Globalide.

11- and 12-pentadecen-15-olides (Figure 27.3) and has been used in perfumes such as *Armani White for Her* (Giorgio Armani, 2001) and *Glow by J.Lo* (2002).

(12*R*, 9*Z*)-Nirvanolide (27.13) has an intense musky, fruity, powdery odour, in contrast to the (12*S*, 9*Z*)-isomer, which is odourless. The smell of Nirvanolide is said to closely resemble Baur's Musk Ketone, and it has been used in perfumes such as *Forever Elizabeth* (David Apel for Givaudan, 2002). Ethylene brassylate ((27.14), Astratone or Musk T) is the most widely used macrocyclic musk. A diester, it is synthesised from ethylene glycol and brassylic acid (tridecane-1,13-dicarboxylic acid), which is obtained from oilseed rape, the cheapness of the raw materials being a major reason for its use in a wide range of products.

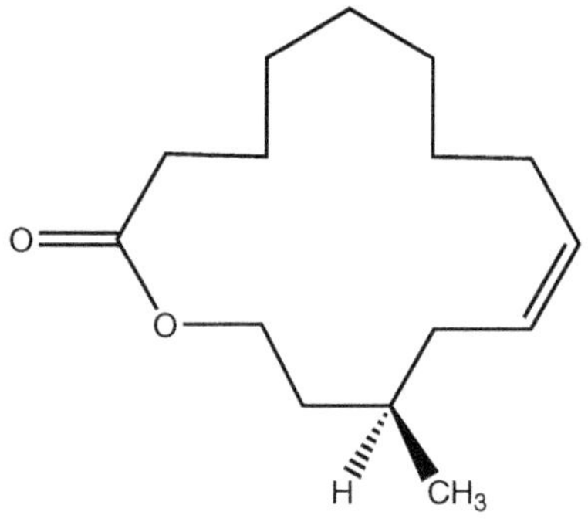

(27.13)

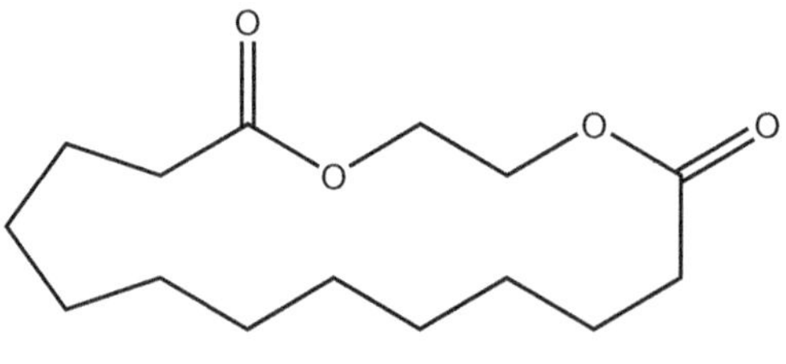

(27.14)

The ring size affects the odour. Molecules with about 14–18 atoms in their rings tend to have a musky smell. With fewer atoms than that, the molecules tend to have cedarwood smells; with more than 18 atoms, they lose their smell as they become involatile $(Mr \geq 300)$. The ester linkage in the lactones makes them more reactive and susceptible to biodegradation than the carbonyl groups usually found in polycyclic musks, making them environmentally more desirable.

Alicyclic musks are the fourth and newest type of musk, featuring alkyl ester linkages. The most important of these is Helvetolide (27.15), discovered in 1990, whose musky smell has nuances of fruit (particularly blackberry) and ambrette, very suitable for use in the fruity perfumes fashionable in recent years. Alberto Morillas incorporated it into *Armani White for Her* (2001). Romandolide ((27.16), 2000) is reported to have a smell similar to that of Galaxolide, with a touch of ambrette.

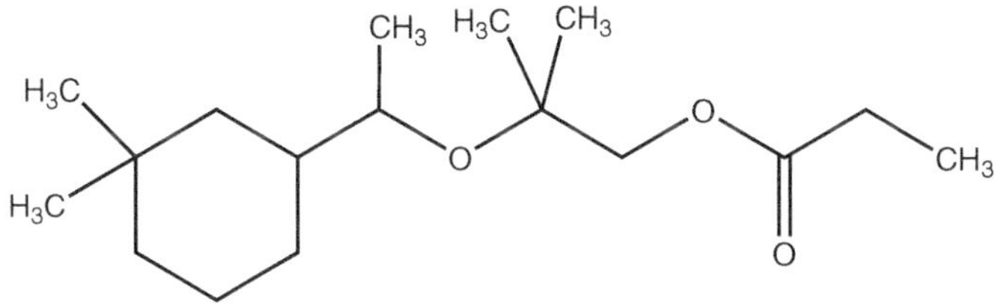

(27.15)

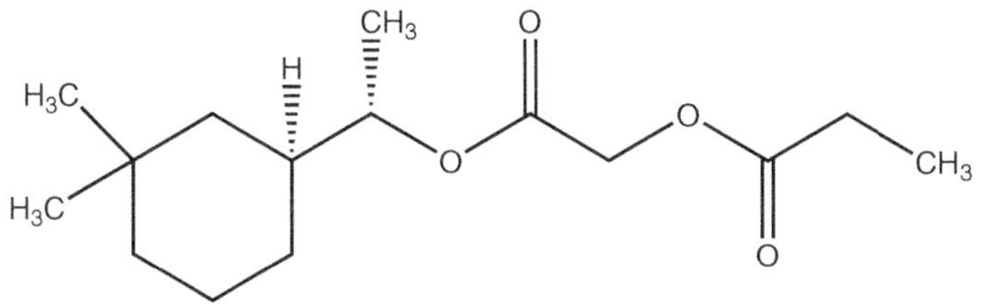

((27.16)

The importance of musks in consumer products is not confined to perfume odourants and fixatives; it also includes laundry detergents, after-shave lotions, domestic cleaning agents and soaps.

Perhaps the most remarkable thing about musks is how compounds with such a range of structures can all produce a 'musk' smell. Within recent years real progress has been made in identifying the human olfactory receptors that respond to the different types of musks. The first receptor found to respond to musk odorants was the OR5AN1 receptor – this responds to macrocyclic musks and also some nitromusks, responding to musk ketone, musk xylene and musk Tibetine, but not to musk Ambrette, nor to non-musk nitro compounds such as 1,3,5-trinitrotoluene. It responds only weakly to certain macrocyclic lactones and not to any polycyclic musks.

The OR5A2 receptor is responsive to a wider range of musks, including macrocyclic ketones and most macrocyclic lactones; OR5A2 responds to both Musk Ketone and Musk Xylene (again) but, unlike OR5AN1, it also responds to Musk Tibetine, though not to Musk Ambrette. It also responds to polycyclic and linear musks.

In contrast, the OR1N2 receptor is responsive to macrocyclic lactones, but not to many other musks – for example it does not sense any nitromusks. Genetic mutations in these receptors also affect their sensitivity to some musks.

28 Nylon

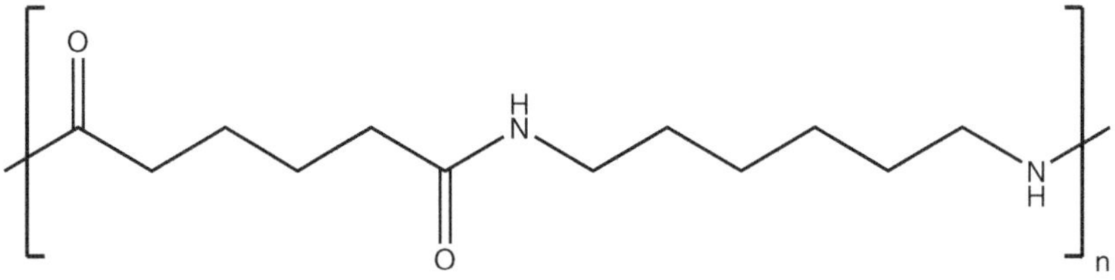

(28.1)

Early in the twentieth century, ideas about the structure and bonding of molecules were limited. Chemists generally believed that there was an upper limit to the size of molecules. From his studies of the chemistry of rubber, during which he obtained some very high molecular masses, the German chemist Hermann Staudinger (1881–1965) was convinced that giant molecules existed, though he was pretty much in a minority of one. In 1920, Staudinger proposed the idea of small molecules joining together and forming very long chains – in other words, the idea of polymerisation. As time went on, more data supported Staudinger's viewpoint, and eventually, in 1953, he was awarded the Nobel Prize in Chemistry 'for his discoveries in the field of macromolecular chemistry'.

In the intervening years, nylon (28.1) was developed. The American chemical company, Du Pont, set up a new laboratory for 'fundamental research' in Wilmington, Delaware. They turned to Harvard University to recruit the man who would be the lead chemist for organic chemistry, Dr Wallace Carothers (1896–1937). He had incentives, his salary was doubled, and he was told that 'as for funds, the sky is the limit'.

Carothers' insight was this. When a molecule has a functional group at one end, it can only react at that end. Thus, when a monocarboxylic acid reacts with a simple alcohol to form a molecule of an ester, that is the end of the process (Figure 28.1). No further reaction can occur.

Likewise, when a monocarboxylic acid reacts with a primary amine, it forms just one amide linkage (Figure 28.2).

Carothers' insight lay in using reactants with two functional groups, one at each end of the molecule. Thus, when both ends of the molecule reacted, it could become part of a chain. As a result, lots of 'monomers' could join to form a 'polymer'. Initially, the researchers focused mainly on polyesters, which they were successfully synthesising, but did not have the physical properties for a commercial fibre. Hence, in 1934, they concentrated on polyamides.

The key reaction, the first synthesis of nylon 6,6, was carried out on 28 February 1935 by Dr Gerard Berchet, who led one of the research groups at DuPont under

DOI: 10.1201/9781003783565-28

FIGURE 28.1 Synthesis of an ester from a carboxylic acid and an alcohol.

FIGURE 28.2 Synthesis of an amide from a carboxylic acid and an amine.

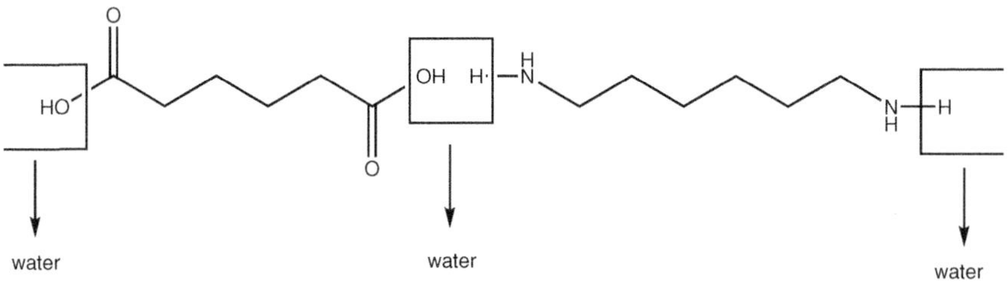

FIGURE 28.3 Synthesis of Nylon 6,6.

Carothers' direction (Figure 28.3). Berchet mixed adipic acid (hexane-1,6-dioic acid, $HOOC-(CH_2)_4-COOH$) and hexamethylene diamine (1,6-diaminohexane, $H_2N-(CH_2)_6-NH_2$), then heated the mixture with some *meta*-cresol at 215°C for three hours before distilling off the cresol at 255–260 °C.

Both the acid and the amine molecules reacted at both ends. An OH group is lost from the acid molecule and a hydrogen atom from the amine molecule, so a molecule of water is eliminated at each end, with the formation of an amide link. This kind of polymerisation, in which a small molecule such as water is eliminated, is known as condensation polymerisation. The resulting poly(amide) is made from two monomers, each containing six carbon atoms, and is therefore known as nylon 6,6 (or polyamide 6,6). Berchet obtained 12.5 grams of the polymer, with a rather high melting point of 252–254 °C.

Overall, then:

$$(n+1)\ H_2N\text{-}(CH_2)_6\text{-}NH_2 + (n+1)\ HOOC\text{-}(CH_2)_4\text{-}COOH \rightarrow$$
$$H_2N\text{-}(CH_2)_6\text{-}NH[CO(CH_2)_4\text{-}COHN\text{-}(CH_2)_6\text{-}NH]_nOC\text{-}(CH_2)_4\text{-}COOH +$$
$$(n+1)H_2O$$

The team at DuPont explored a range of combinations of dicarboxylic acid and diamines. Within a year, they had come up with two candidates for commercially suitable materials: (a) the already mentioned polyamide 6,6 and (b) the polyamide 5,10, made from $H_2N-(CH_2)_5-NH_2$ and $HOOC-(CH_2)_8-COOH$. Carothers favoured the polyamide 5,10, but the chemical director argued for the polyamide 6,6 on the grounds that the chemicals needed were readily obtainable from benzene, which was commercially a reasonable idea. Thus, the researchers concentrated on what we now know as nylon 6,6. It was found that if molten polymer was forced through a hole in a steel plate, filaments were produced; twisting a number of these filaments together yielded a thread. These threads could be knitted together to form a nylon fabric.

Carothers had achieved all that he could have hoped for. More than that, he was elected to the National Academy of Sciences on 30 April 1936, the first industrial organic chemist to be so honoured. But Wallace Carothers could not shake off the depression and mental instability that had plagued him for years. The death of his favourite sister may have been the last straw – Isabel Carothers contracted a streptococcal infection; in the days before penicillin, this was a death sentence. Shortly after applying for a patent on nylon, Carothers committed suicide by drinking potassium cyanide dissolved in lemon juice on 29 April 1937. The US Patent 2,130,948, entitled 'Synthetic Fibers' was issued on 20 September 1938.

The original name that DuPont intended for the new material was Nuron, which both suggested novelty and was also 'No run' backwards, just the thing for ladder-proof hosiery. However, Nuron resembled existing trademarked names, so Nylon was the next idea. In the end, DuPont did not patent the name, hoping that it would become a generic term, like glass.

A different approach to nylon utilised a single monomer, which contained two different functional groups – a carboxylic acid and a primary amine group – at opposite ends.

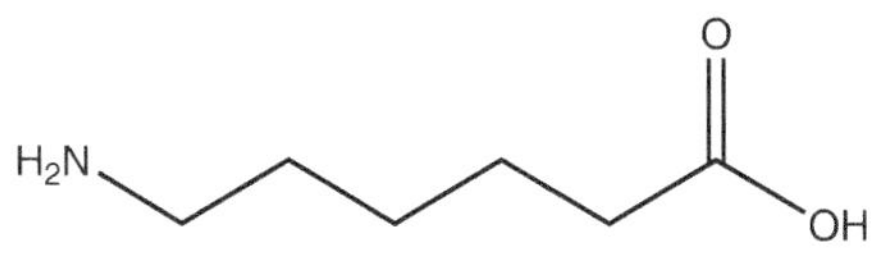

(28.2)

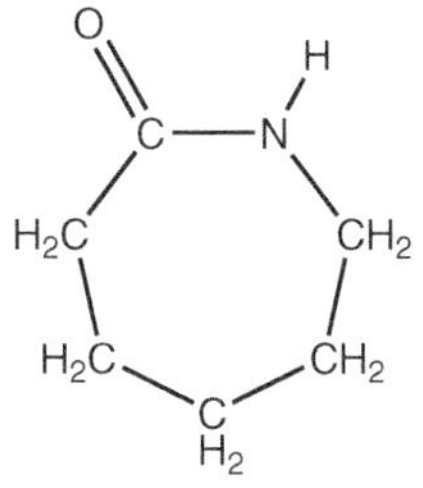

(28.3)

Formally, this is 6-aminohexanoic acid (28.2), but in practice this undergoes an intramolecular condensation forming a cyclic lactam (caprolactam, 28.3). However, when this molecule polymerises, the ring opens, and the monomers join together to form the nylon-6 chain (28.4), the name indicating that it is made from a single monomer containing 6 carbon atoms. This polymer, first synthesised on 29 January 1938, was largely the work of Paul Schlack at IG Farben in Germany. It was marketed under the name Perlon, though there are other trade names. Nylon 6 has a slightly lower melting point at 218 °C than nylon 66. In 1930, Carothers had briefly examined caprolactam but did not pursue it further.

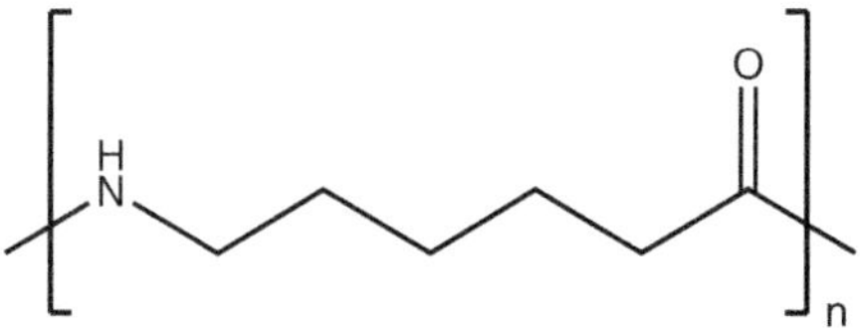

(28.4)

Carothers' team had given DuPont a new and quite remarkable material. It did not corrode, was flexible, had a low density, unlike metals, and was easy to colour, unlike bakelite and many natural materials. It was also non-biodegradable, though today we may view that as a mixed blessing. Nylon was chemically very inert, though acids can hydrolyse it. It was DuPont vice-president Charles Stine who announced nylon publicly on 27 October 1938 at a meeting of women's club members in New York. This was appropriate, as the company would be targeting women for sales of the product that would be associated with the name of their new polymer. But the first application of nylon was in toothbrushes, replacing boar's bristles, which had been hard to obtain since the Japanese invasion of Manchuria in 1937. And then 'Dr West's Miracle-Tuft Brushes' were advertised as toothbrushes that could not shed their bristles and would not become soggy.

DuPont opened its first commercial plant for the manufacture of nylon at Seaford, Delaware, on 15 December 1939. It was manufacturing a pioneer synthetic fibre, ready to be made into products for the consumer and offering properties superior to the natural materials like silk, which it replaced. Initially, nylon stockings were offered just to DuPont's employees at the factory, from 24 October 1939.

National advertising heralded the launch of nylon stockings across America on 'N-Day', 15 May 1940. On that day alone, five million pairs were sold at a price of \$1.15 a pair, much cheaper than silk stockings. In the first year of their manufacture, 64 million pairs of nylon stockings were sold.

Following the Japanese attack on Pearl Harbour on 7 December 1941, the United States entered World War II, and nylon became indispensable to the war effort. For one thing, it replaced silk, no longer obtainable from Japan, as the material used for making parachutes. Nylon was also used to manufacture a wide range of products, including ropes, tyre cords, life rafts, weather balloons, mosquito nets and machinery parts such as gears.

Nylon stockings were not available again to civilians until the end of the War, in September 1945, so in the interim black-market prices skyrocketed. To begin with, supplies were limited to just a few stores-with immense queues. Thus, at Pittsburgh in June 1946, 40,000 people formed a queue over a mile long to get their hands on 13,000 pairs of stockings. When the nylons sold out, many customers were disappointed and some rioted, in what came to be known as 'Nylon Riots'. Although the supply situation improved, demand continued to exceed supply, and in 1951 DuPont licensed nylon to other manufacturers.

By the 1950s, all sorts of clothing were being made from nylon, including men's drip-dry suits. It became an everyday material and ceased to be so novel. The 1951 film, *The Man in the White Suit*, starring Alec Guinness, celebrated this new age. Increasingly, however, the position of nylon was usurped by polyesters such as Dacron (US) and Terylene (UK), though even today much of the world's plastic production is still nylon, around five million tonnes a year.

The classic experiment used to demonstrate the synthesis of nylon is known as the 'nylon rope trick'. It was devised by Stephanie Kwolek of DuPont, who was also responsible for Kevlar, the material famously used in bullet-proof vests. This is a famous demonstration experiment, widely used in schools and colleges. In its standard form (Figure 28.4), it employs an aqueous solution of 1,6-diaminohexane and a solution of sebacoyl chloride (1,10-decanedioyl dichloride) in a non-polar solvent, such as heptane (or even tetrachloromethane).

Using the acyl chloride means that the reaction occurs at room temperature, rather than having to be heated to around 200°C, as would be the case if an acid were used. The denser solution, containing amine, is poured into a beaker, and the acyl chloride solution is then carefully poured on top so that the solutions do not mix (traditionally poured slowly down a glass stirring rod). Because the two solutions do not mix, they only react at the interface between them, forming a thin film. The film can be pulled carefully away as a thread, then wrapped round a windlass and steadily wound up. As the nylon product is removed from the interface, fresh reactant molecules are exposed and can meet, allowing the reaction to continue. It will only stop when one of the reactants is exhausted.

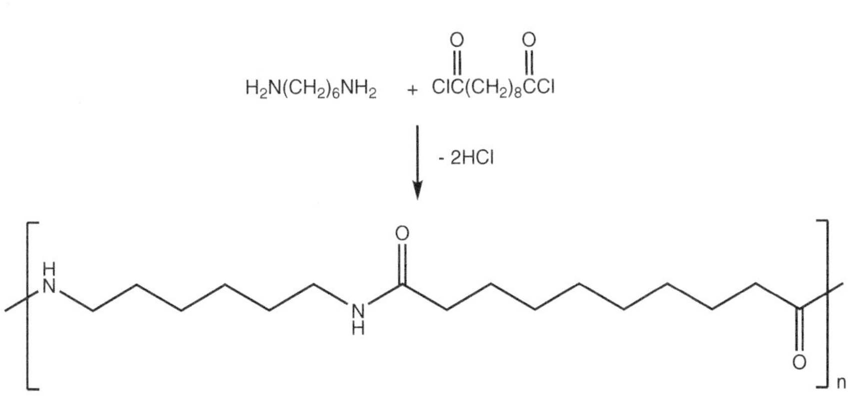

FIGURE 28.4 Synthesis of nylon 6,10 in the 'nylon rope trick'.

29 Octanal

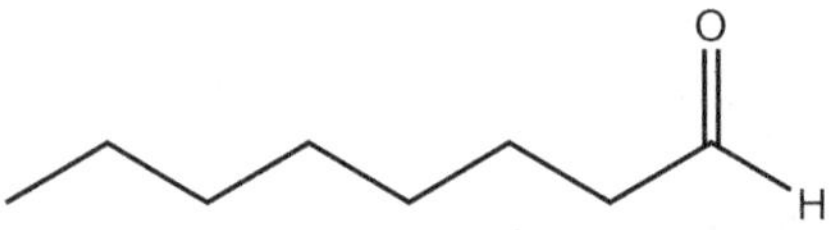

(29.1)

Aldehydes are a kind of molecular halfway house, as they can be oxidised to carboxylic acids or reduced to alcohols. As its name suggests, octanal (29.1) is an aldehyde containing eight carbon atoms. Thus, it can be oxidised to octanoic acid or reduced to octan-1-ol. It is much easier to remember the name 'octanal' than the older name, caprylaldehyde. This name derives from the related carboxylic acid caprylic acid (now known as octanoic acid), which in turn takes its name from the Latin noun *capra*, meaning she-goat, because this acid was first identified in the lipids found in goats' milk.

Octanal smells nothing like goats. Not all aldehydes have the same smell; hexanal has a 'green grass' smell (see p. 77) – and one might detect a green note in octanal, whilst smaller molecules such as butanal and ethanal have rather unpleasant odours. Pure, undiluted, octanal has an intense and pungent citrus smell, but this becomes more moderate upon dilution.

Octanal is found in both orange oil and orange juice, though the amount varies from one variety to another, and fresh juice is unstable. The principal component of orange oil is (*R*)-(+)-limonene (29.2). Analysis of orange essence oil revealed that limonene was by far the most abundant compound (94.5%), followed by other terpenoids – myrcene (1%), valencene (0.8%) and linalool (0.7%) – and then by the aldehydes octanal and decanal, together with the ester ethyl butanoate ((29.3) – 0.3%). Similar compositions have been reported for oils from a variety of sources.

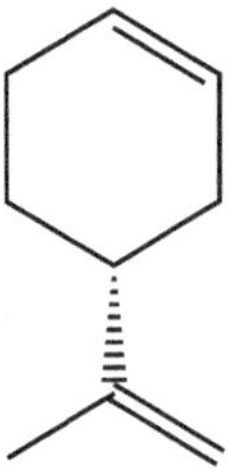

(29.2)

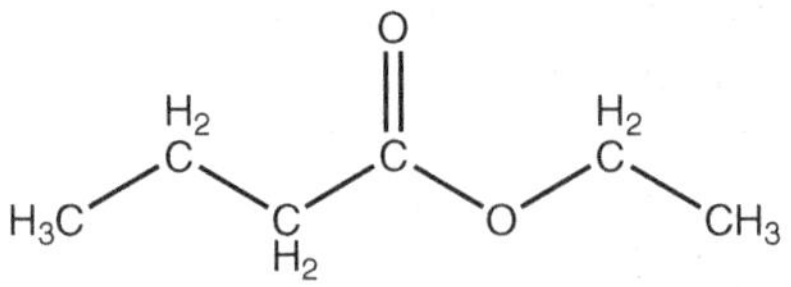

(29.3)

DOI: 10.1201/9781003783565-29

Octanal and decanal are considered to be the major constituents of aroma-active substances in orange juice. In one study of freshly squeezed orange juice, octanal, hexen-2-al and decanal were the three aldehydes detected. Octanal concentrations decrease on storage of orange juice, eventually disappearing. A study of orange juice freshly reconstituted from concentrate revealed 40 odour-active constituents and measured their odour activity values (OAVs), the ratio of concentration to odour threshold – the higher the OAV, the greater the smell impact. This study concluded that the most important odorants in the overall aroma were (*R*/*S*)-linalool, (*R*)-limonene and (*S*)-ethyl 2-methylbutanoate, with the highest OAVs (greater than 1,000), followed by octanal, then some others. By itself, limonene has a rather weak smell, but if you buy limonene from commercial sources, it contains significant amounts of smelly aldehydes and other impurities, as it is very difficult to purify, and therefore it will probably have a citrus smell.

The authors noted that no single one of the key odorants, by itself, smelled like orange juice, and that the overall aroma of the orange juice arises from a combination of these key odorants in their actual concentrations. A study of fresh, 'hand-squeezed' Valencia orange juice showed the importance of ethyl butanoate, (*Z*)-3-hexenal (see pp. 76–80) and decanal ('bigger brother' of octanal) as well as the esters ethyl 2-methylpropanoate, (*S*)-ethyl 2-methylbutanoate and wine lactone, though octanal did not get a mention. Thus, the 'orange' smell of fresh orange juice and of oil squeezed from orange peel seems to arise from octanal and the similar C_9-C_{11} aldehydes from nonanal to undecanal, along with unsaturated aldehydes such as neral. Ethyl butanoate and certain other esters, such as ethyl octanoate and ethyl decanoate, also contribute.

Chemists synthesised and studied aldehydes such as octanal in the mid-nineteenth century but were content to use natural oils in perfumery. It was only around 1900 that they started to use 'synthetic aldehydes', as it was not known that these simple aldehydes were natural ingredients. Aldehydes started to be used regularly in perfumes from the first decade of the twentieth century, most famously in *Chanel No. 5* (see pp. 34–37). Today, octanal is used in low concentrations in perfumery, in eau de cologne, and in artificial citrus oils.

Why octanal has the smell it does is not known. Humans have around 350 different smell receptors that detect odorants; these are linked to the olfactory bulb in the forebrain, which processes the signals and sends messages to the brain; each receptor responds to some molecules, but not others. A molecule will usually produce a response from several receptors, and the 'smell' is the brain's response to the pattern of signals arriving from all the receptors activated by that molecule.

An MRI study was carried out on the olfactory bulb of a mouse when the mouse was exposed separately to each one of the C_4-C_8 aldehydes (from butanal to octanal inclusive); each one caused different globular activity patterns, corresponding to different 'smells'. The mouse I7 odorant receptor is very similar to the rat I7 receptor, differing by a single valine-to-isoleucine substitution; this receptor gives its strongest response to heptanal, the seven-carbon aldehyde.

The rat I7 odorant receptor gives a strong response to straight chain C_7-C_{10} aldehydes, most strongly to octanal, but not to aldehydes with less than seven or more than ten carbons. Short-chain aldehydes appear to bind to this receptor but not to activate it; aldehydes above C_{11} are too big to fit into the binding pocket. Some unsaturated aldehydes activate it too, but few other odorants – for example, neither ketones nor esters, nor alcohols like octan-1-ol. The shape of the carbon chain ('conformation') in the aldehyde is important – aldehydes with unbranched chains ('straight chains') bind best.

Octanal does one thing that is utterly unexpected. In their breeding season, crested auklets (*Aethia cristatella*) – small planktivorous seabirds which are cousins of the puffin, found in the Bering Sea off Alaska and the Okhotsk Sea off Siberia – give out a strong tangerine-like perfume. This is a mixture of aldehydes with an even number of carbon atoms – notably hexanal, octanal and decanal, as well as the unsaturated aldehydes (Z)-4 decenal (29.4), (Z)-4-dodecenal (29.5) and (Z)-6-dodecenal. Octanal is present at more than twice the concentration of the next most abundant aldehyde.

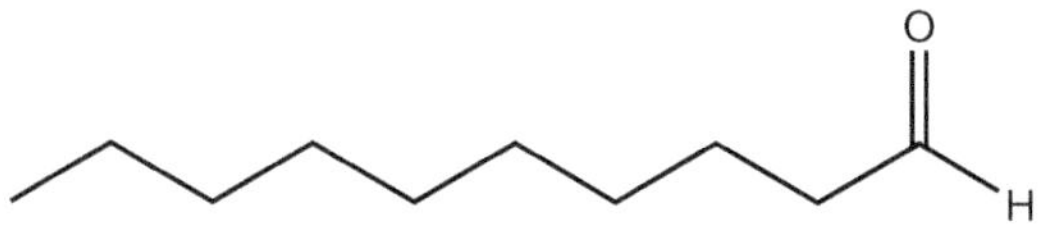

(29.4)

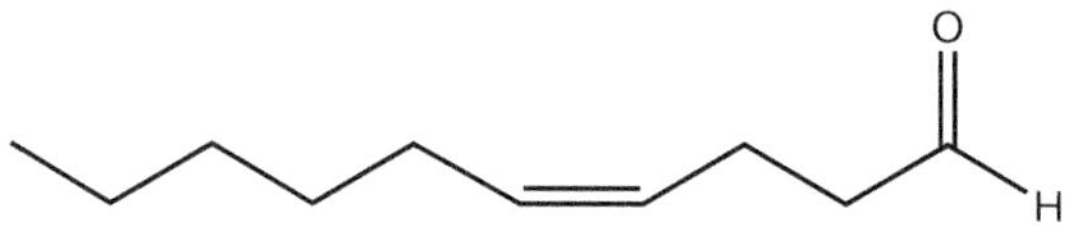

(29.5)

Pairs of these very gregarious birds engage in 'alloanointing' – they rub their bills, breasts, heads and necks over the wick feathers of their partner, thereby spreading odorant molecules over the head, neck and face areas that their partners cannot reach themselves. The birds probably use this mixture to repel pests such as arthropods and ectoparasites. This mixture is known to repel female *Aedes aegypti* mosquitoes, and octanal by itself has the same effect (as do nonanal and decanal).

30 Octane

Octane (30.1) is one of many hundreds of different molecules that occur in crude oil, also known as petroleum (which is not the same as petrol). Crude oil originated from marine organic matter like plankton, which was present millions of years ago in warm seas. Organic deposits formed on the seabed, and these deposits gradually got buried. Over time, these deposits were forced deeper and deeper, so that the pressure and temperature increased, causing the molecules to break down into a mixture of smaller molecules. This mixture is crude oil; this impregnates porous rock, and gets trapped between layers of impermeable rock. This 'source rock' is what oil companies search for before drilling for the crude oil, which is then forced to the surface by pressure.

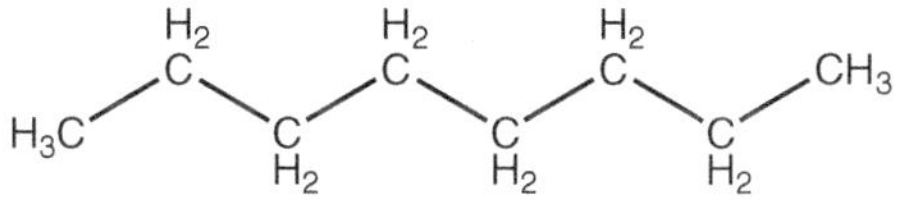

(30.1)

A chemist who is also a football supporter might be tempted to chant, 'There's only one octane'. True as far as it goes, but this requires elaboration. Octane is an eight-carbon alkane (30.1), with the molecular formula C_8H_{18} ; its structure can also be represented as $H_3C(CH_2)_6CH_3$. It is not, however, the only compound with the formula C_8H_{18}. Only the three smallest alkanes – methane, ethane and propane – have just one isomer. Alkanes with four or more carbon atoms have at least two isomers.

There are 18 isomers with the molecular formula C_8H_{18} (some of which additionally have optical isomers). Aside from the one isomer with eight carbons in the chain – the only one properly known as octane - there are several possibilities for compounds with a seven-carbon chain; these must also contain one methyl group as a side chain. This side chain can be positioned at carbon 2; carbon 3; or carbon 4. So there is a 2-methylheptane; a 3-methylheptane; and a 4-methylheptane. There is not a 5-methylheptane, as the methyl group is nearer the other end of the carbon chain, so what you thought was 5-methylheptane is actually 3-methylheptane.

If the main chain has six carbon atoms, the possibility arises of having two methyl substituents or one ethyl substituent; again, the position(s) in the chain lead to further isomers. And so on, with some isomers based on four- or five carbon backbones. The isomers are depicted in Figure 30.1.

DOI: 10.1201/9781003783565-30

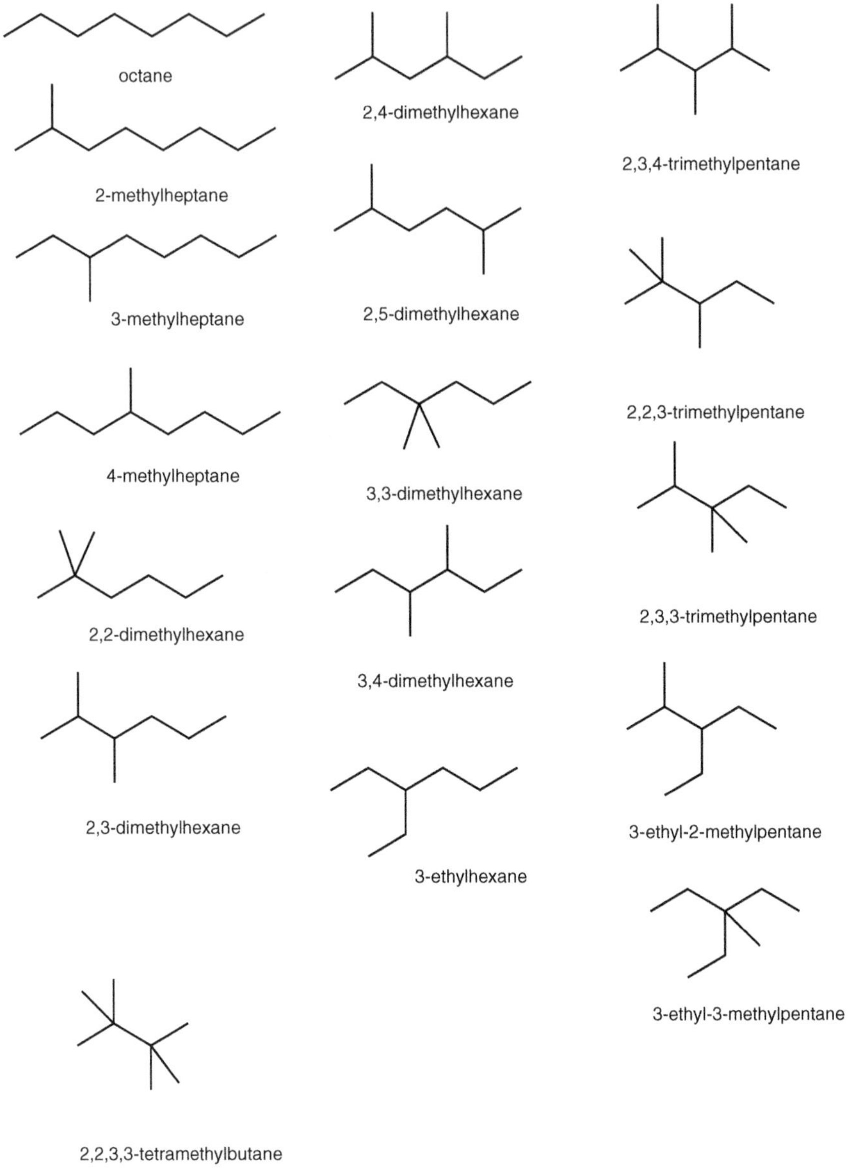

FIGURE 30.1 Structures of the isomers of octane.

Why is this important? The answer is that it impinges upon the use of the isomers of octane as fuels. These molecules have different shapes and that affects the weak attractive forces between them. Although they contain exactly the same atoms, they are arranged in different ways, so that the intermolecular forces between them will vary slightly. When you boil a liquid and convert it into a gas, the molecules are separated from each other and move much further apart. To make this happen, you have to overcome the attractive forces between the molecules. These 'intermolecular forces' are weaker than the chemical bonding forces that hold molecules together; think of them as making molecules 'tacky' - 'Molecular Velcro' if you like. The bigger the molecule, the stronger the intermolecular forces. Consequently, molecules of roughly similar size tend to have similar – but not identical – boiling points.

Oil companies purify the petroleum by refining – in a refinery. This is done by boiling the crude oil and sending the vapour up a fractionating column, a tower that is hot at the bottom and cold at the top. Because the different-sized molecules have different boiling points, they condense out from the vapour at different points. There are trays at different levels of the tower that collect the fractions, the name given to the portions of the crude oil that condense at different points in the column. Each fraction contains molecules with similar boiling points.

The boiling points of the isomers of C_8H_{18} are similar but not identical – the values for octane and three of its isomers are as follows: octane, 126 °C; 2-methylheptane, 99.3 °C; 2,2,4-trimethylpentane (isooctane), 99.2 °C; and 2,2,3,3-tetramethylbutane, 106 °C. This is a reflection of the fact that although these four substances contain exactly the same atoms, these molecules have different shapes, because the atoms are arranged in different ways, so that the intermolecular forces will vary slightly. Molecules of roughly similar size are collected in the same fraction, though a fraction still contains many different molecules. Each fraction can be separated further. Some of the isomers of octane are more valuable than others, and another job that refineries do is to convert a straight chain octane into an isomer with more branches. These are valuable on account of their higher octane ratings, the numbers seen on petrol pumps in garages. Octane ratings have been used for around a hundred years.

Petrol engines operate by the combustion of a gaseous mixture of air and hydrocarbon fuel, which is injected into the cylinder as the piston moves downwards and is then compressed as the piston moves up. At a particular stage, a spark is generated, igniting the mixture. It burns to form a very hot mixture of carbon dioxide and steam, which forces the piston down. This drives the engine, sending power to the wheels. More advanced engines compress the fuel-air mixture to a greater extent before ignition, a feature called higher compression ratio, leading to greater power. However, it also means that it is more likely that the fuel-air mixture would pre-ignite, before being sparked. This misfiring is known as 'knocking', which would result in reduced power. An American chemist called Russell Marker, working for the Ethyl Corporation, found that putting branched chain hydrocarbons into the fuel mixture meant that knocking was less likely to occur. He studied the knocking characteristics of mixtures with varying amounts of heptane and isooctane. A fuel described as 95-octane has the same knocking characteristics as a mixture containing 95% isooctane (2,2,4-trimethylpentane) and 5% heptane.

TABLE 30.1
Octane Ratings (Research Octane Numbers) of Individual Chemicals

Compound	RON value
hexane	24.8
heptane	0
octane	-19
nonane	-17
2-methylheptane	~ 25
2,4-dimethylhexane	65.2
2,2,3-trimethylheptane	106
2,2,4-trimethylheptane	100
benzene	101
toluene	~ 120
ethanol	~ 109

The higher the octane rating, the more the gas-air mixture can be compressed before it is ignited, and the more power you get when it burns. Research Octane Numbers (RON, commonly referred to as Octane Ratings) can be measured for individual compounds by burning the compound in a test engine, the results being compared with those obtained for heptane-octane mixtures. Table 30.1 lists some RON values for certain straight chain alkanes, including octane, and branched-chain isomers of octane, including 2,2,4-trimethylpentane (isooctane), as well as benzene, toluene (methylbenzene) and ethanol, three common petrol additives. You can see that the 'branched' alkanes have higher octane ratings than the straight chain alkanes.

Octane ratings helped shape history. Between 1927 and 1931, British seaplane aircraft won the high-speed Schneider trophy competition for Britain three times, aided by high-octane fuels that were developed by Air Commodore Rod Banks and his colleagues. The seaplanes were designed by R. J. Mitchell, who later went on to produce the Supermarine Spitfire fighter aircraft. Until early 1940, the RAF's Hurricanes and Spitfires largely used 87-octane fuel. During that summer, however, they increasingly switched to 100-octane fuel supplied from a variety of refineries, including those in Aruba, off the Venezuelan coast, as well as British refineries and some in the USA and elsewhere. Along with the introduction of the newer Spitfires with the variable-pitch propeller (a very important improvement), this was a factor – together with the early warning system made possible by observer spotting and the British development of radar – that enabled the Royal Air Force's fighters to compete with the Luftwaffe's Me (Bf) 109 aircraft, running on 87-octane fuel, notably shortening the time taken in reaching combat altitude at around 20,000 feet, enabling better interceptions.

31 Paracetamol Acetaminophen, Tylenol and Other Coal-Tar-Derived Painkillers

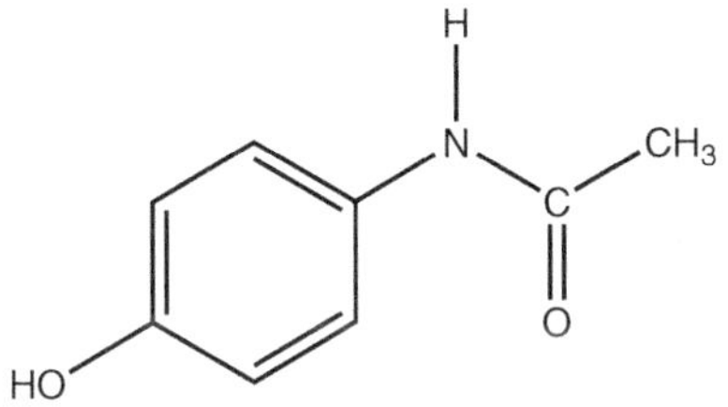

(31.1)

Today paracetamol (31.1) is a ubiquitous medicine; few, if any, other medicines are found in as many homes, apart from aspirin. Yet, unlike aspirin, it does not have its roots in nature. Moreover, it only came into routine use in the second half of the twentieth century, and it would surprise most of its users to learn that it was 'only the third cab off the rank'.

The Industrial Revolution was fuelled (literally) by coal. In innumerable towns and cities, gas works grew up to provide the fuel that lit and heated homes, whilst their by-products were also important. Coke was a fuel as well as being essential for the manufacture of iron in blast furnaces, and the coal tar distillate turned out to be the source of many useful organic compounds. One of these was aniline, which became the base of a number of dyes ('aniline dyes') and also of some medicines.

The early nineteenth century was the period during which organic chemistry began to develop rapidly, with German chemists taking a leading role. Aniline was first isolated from coal tar in 1834, but an important development occurred in 1854, when Antoine Béchamp discovered how to make it. Nitrobenzene can be synthesised by nitration of benzene itself; the resulting nitro group can then be reduced – in Béchamp's case using iron and hydrochloric acid – to afford aniline (Figure 31.1). This became known as the Béchamp reaction (a sign that

DOI: 10.1201/9781003783565-31

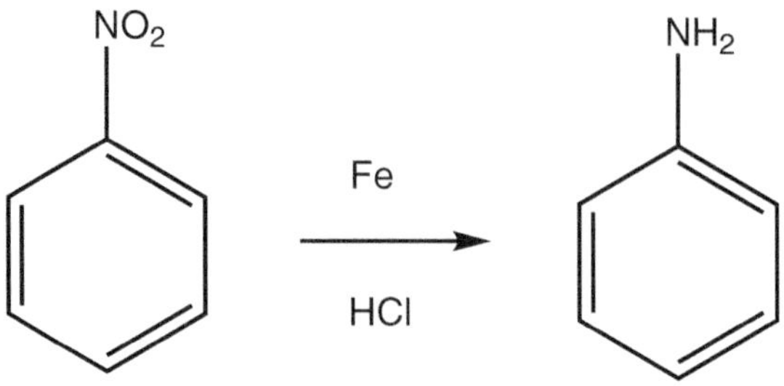

FIGURE 31.1 Synthesis of aniline (phenylamine) from nitrobenzene.

you have discovered something important is when a compound or reaction is named after you).

Apart from dyestuffs, medicines became an essential adjunct to the chemical industry. The first of the aniline-related drugs was acetanilide (31.2).

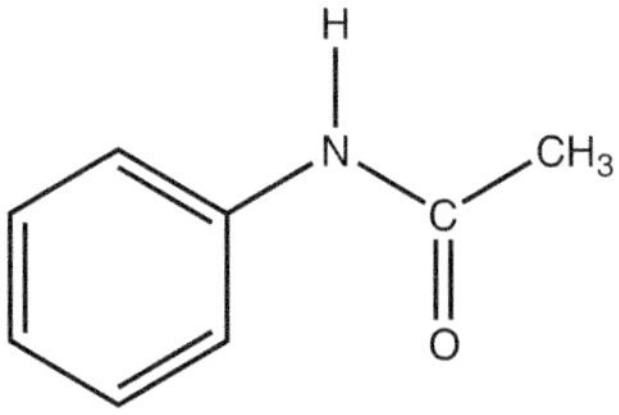

(31.2)

Acetanilide is easily prepared by the acetylation (ethanoylation) of aniline, either with ethanoic anhydride or ethanoyl chloride.

Acylation was a key synthetic reaction discovered by the nineteenth century chemists. Subsequently, it was used to prepare acetylsalicylic acid (aspirin) from salicylic acid (p. 14), and to make diacetylmorphine (heroin) from morphine (p. 110).

Medicinal use of acetanilide (N-phenylethanamide) was a classic case of serendipity. Arnold Chan and Paul Heppa, two researchers at the University of

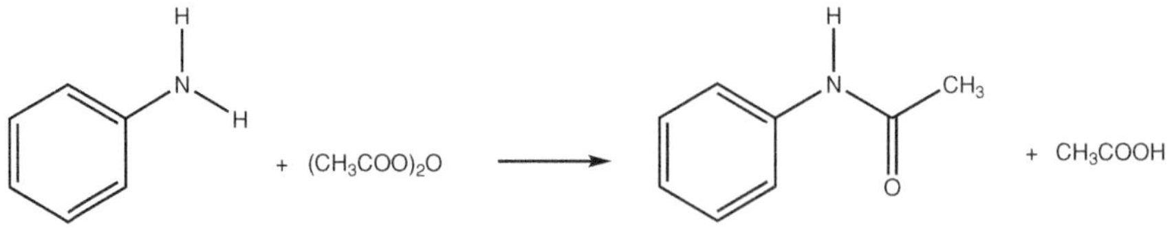

FIGURE 31.2 Synthesis of acetanilide.

Strasbourg, were investigating intestinal worms in patients. The standard treatment for the worms was the hydrocarbon naphthalene, so they ordered some. A worker at the wholesalers made an error and sent acetanilide instead. This did nothing to the worms, but lowered the temperature of some patients with fever, so the researchers investigated further. It so happened that Heppa's brother was a chemist at Kalle and Co., and he realised that this cheap and easy-to-make substance had real potential. As a known chemical, acetanilide could not be patented. So, in 1886, they marketed acetanilide as 'Antifebrin', a medicine with analgesic (pain-killing) and antipyretic (temperature-lowering) properties. Acetanilide's use was short-lived, since it was found that it oxidised haemoglobin and thus reduced the oxygen-carrying capacity of the blood (methaemoglobia). It also caused liver and kidney damage.

Another painkiller soon replaced acetanilide. The company Farbenfabriken Bayer needed to do something with a leftover chemical, para-nitrophenol (4-nitrophenol), a waste material from dye manufacture. They used Béchamp's method to turn it into 4-aminophenol, which could be acetylated to form 4-hydroxyacetanilide. Treatment with iodoethane under basic conditions (a reaction known as the Williamson synthesis for ethers) produced 4-ethoxyacetanilide.

4-hydroxyacetanilide is converted to the aryloxide ion (shown below as RO^-) under basic conditions so that the aryloxide can carry out a nucleophilic attack upon the iodoethane.

4-ethoxyacetanilide (31.3) was also found to be an analgesic and antipyretic, and was therefore marketed from 1887–1888 under the name Phenacetin. It remained a widely used painkiller until the 1960s. In the UK, it was well known as Phensic. It was often used in combination with codeine (and other medicines, including caffeine). Much like paracetamol today.

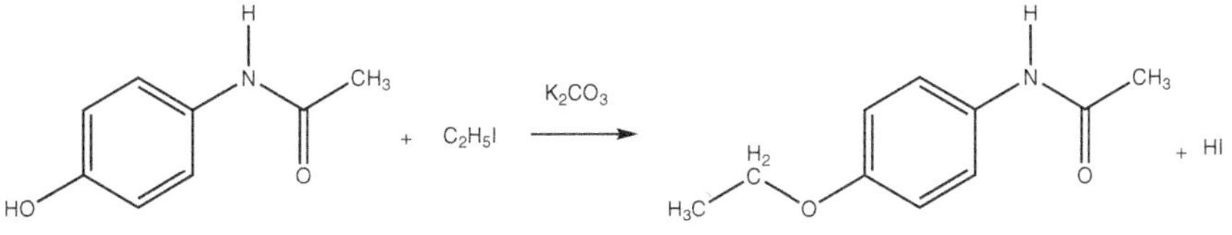

FIGURE 31.3 Synthesis of 4-ethoxyacetanilide.

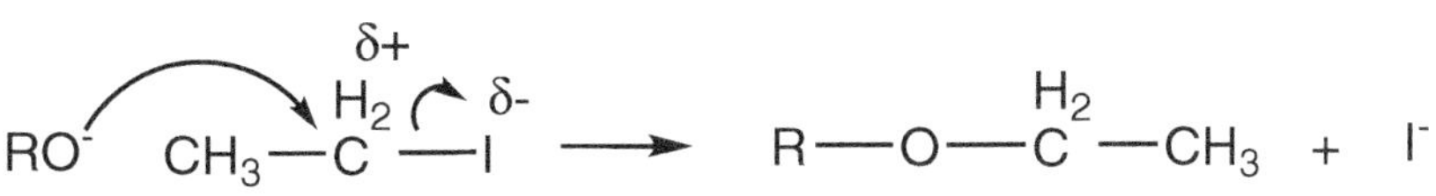

FIGURE 31.4 Mechanism of the synthesis of 4-ethoxyacetanilide.

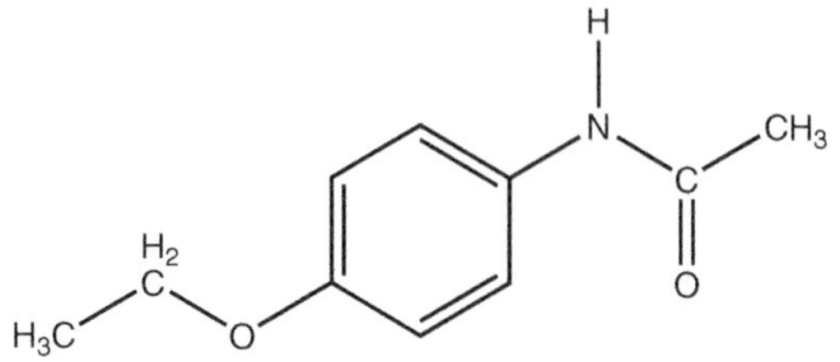

(31.3)

However, eventually phenacetin was found to cause methaemoglobia, like acetanilide. It was also found to cause kidney damage and to be a carcinogen, so that it gradually fell out of use from around 1970.

The most famous victim of phenacetin was Howard Hughes (1905–1976). Hughes was a remarkable man with an incredibly diverse range of interests. He was an entrepreneur involved in manufacture of chemicals, including the fledgling plastics industry, while he also had interests in the entertainment industry, notably filmmaking. Famously, he designed a bra for the actress Jane Russell, famous for films such as *The Outlaw* and *Gentlemen Prefer Blondes* (spoiler: Jane Russell did not wear the bra, but instead wore one of her own). Hughes also designed aircraft – and flew them, as a test pilot. The 'Spruce Goose' flying boat was one of his creations.

On 7 July 1946, Hughes was testing an experimental reconnaissance plane, the XF-11, which crashed shortly after take-off from Santa Monica, California. He somehow survived, but with life-threatening injuries, including widespread third-degree burns; fractures of his chin, jaw, left knee, left elbow and 13 ribs, along with multiple haemorrhages, with blood loss requiring transfusions. Hughes required four skin grafts and three chest drainings, and was on morphine for five weeks, throughout his hospitalisation. Hughes had intractable (chronic) pain for the rest of his life. He took a daily dose of 1,200–2,700 mg codeine for 30 years, often as 'oral compounded codeine' – codeine combined with phenacetin or aspirin. It is thought that long-term phenacetin use caused the kidney failure that killed him.

Paracetamol (31.1) had first been made in 1877. At the time – and for many years – it was believed that phenacetin was safer than paracetamol (which was suspected of causing methemoglobinemia), so paracetamol was not investigated further.

In the late 1940s, investigations confirmed that acetaminophen (Paracetamol) did not cause methemoglobinemia. They also showed that, in the body both phenacetin and acetanilide were metabolised to acetaminophen, and this is what caused their beneficial effects. Thus, both phenacetin and acetanilide were prodrugs. It was also established that acetaminophen had none of the side-effects of phenacetin and acetanilide.

Within the next few years, McNeil laboratories (an American firm) began to market acetaminophen under the name of Tylenol, the name being universally recognised in the USA to this day. It was first marketed in the UK as Panadol, but its generic form is known as Paracetamol. Paracetamol is sold on its own, as an inexpensive OTC medicine, but is also used in combination with codeine as a painkiller ('Co-codamol').

Paracetamol is made in a similar way to acetanilide (Figure 31.5), starting from phenol, nitration of which results in the two isomeric products, the 2- and 4-nitrophenols, which can be separated, for example, by steam distillation. Proceeding with 4-nitrophenol, it can be reduced to 4-aminophenol, readily acetylated to form acetaminophen, paracetamol. The chemicals used to produce it are inexpensive, which is why the product is a relatively inexpensive medicine.

How paracetamol works is not wholly understood. Some researchers think that a proportion of it is converted by the body's metabolism into 4-aminophenol, which combines with the fatty acid, arachidonic acid, to generate *N*-arachidonoylphenolamine ((31.4): AM404). This is a naturally occurring cannabinoid that may bind to the cannabinoid CB1 receptors, notably found in the brain and spinal cord, thereby blocking pain signals.

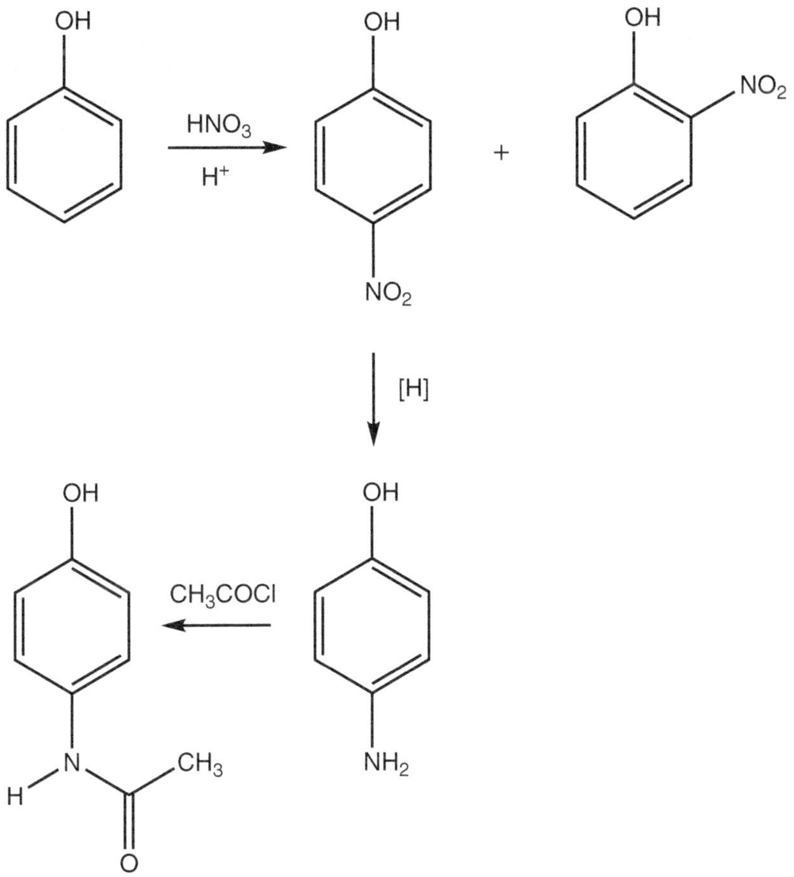

FIGURE 31.5 Synthesis of paracetamol.

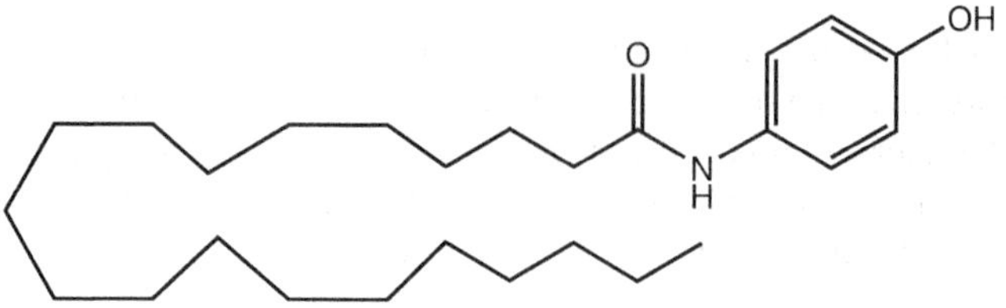

(31.4)

Paracetamol is a safe drug, provided it is taken in the prescribed dose. Overdoses, however, can cause serious problems and may be fatal. The body mainly metabolises it to harmless products, except for around 5% of it, converted into *N*-acetyl-*p*-benzoquinone imine (NAPQI). The human body contains small amounts of a compound called glutathione (GSH), which mops up small amounts of NAPQI. However, an overdose of paracetamol generates an excess of NAPQI, which overwhelms the body's GSH content, leaving the excess of NAPQI free to attack the liver. The result is often fatal liver failure, along with kidney failure. Some 200 people are killed by paracetamol overdoses in the UK each year.

Antidotes to paracetamol poisoning exist, provided they are taken in time. Either L-methionine or *N*-acetyl-L-cysteine can help the body overcome an overdose; they probably promote the synthesis of GSH. Thus, one safeguard would be to add some methionine to every paracetamol tablet, but few manufacturers do this. One problem with paracetamol overdoses is that the antidotes ideally need to be administered within two hours, and certainly within eight to ten hours. Symptoms of an overdose can take a day or more to emerge, so someone having a change of mind after a suicide attempt, for example, may think that they are out of danger until it is too late.

Some 40 years ago, America was shaken by the so-called 'Tylenol murders' in the Chicago area – though they were not caused by Tylenol itself. When 12-year-old Mary Kellerman complained to her parents of a runny nose and sore throat on 29 September 1982, they opened a pack of Tylenol and gave her a capsule. She died – the capsule had been contaminated with cyanide. Within days, six other people who, like Mary, lived around Chicago, died in the same way. Tylenol was still manufactured by McNeil Laboratories, by then a subsidiary of Johnson and Johnson, a large America-based pharmaceutical and medical company, which acted very promptly. The company withdrew over 31 million bottles of Tylenol from circulation (and offered replacements to consumers). Three more bottles of cyanide-laced capsules were found in shops. It was found that the contamination had happened in shops, but the perpetrator was never caught. Most importantly, the manufacturers quickly introduced tamper-proof packaging.

The most unusual use of paracetamol has been on the small Pacific island of Guam. During World War II, brown tree snakes (*Boiga irregularis*) arrived on this island, probably having been accidentally transported there on a cargo ship or aircraft. Within a few years, they had flourished, in the process eliminating much of the island's wildlife, notably birds. As a countermeasure, paracetamol-baited dead mice

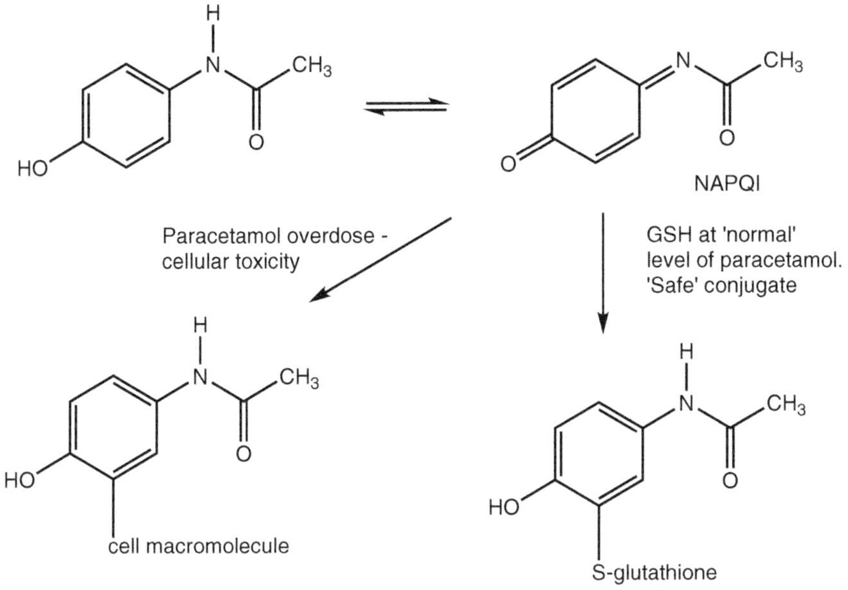

FIGURE 31.6 Formation of *N*-acetyl-*p*-benzoquinone imine (NAPQI) and its consequences.

have been dropped onto Guam by parachute, landing in trees. Laboratory tests found that the cause of the snakes' death is acute methemoglobinemia and respiratory failure; this is believed to be a humane method of killing the snakes.

One last thing. Possible links between the use of paracetamol (Tylenol) during pregnancy and autism in the child have attracted attention. A 2025 umbrella review of the literature, taking into account factors such as the quality of the evidence, could not find evidence of such a link.

32 Penicillin

War kills, not always as one might think. Of the ten million soldiers who died during World War I (1914–1918), just two-thirds died on the battlefield. The other third died of disease, many of these due to bacterial infection of their wounds. As early as October 1914, the Director-General of the British Army Medical Service said: 'We have in this war gone straight back to all the septic infections of the Middle Ages'. Many doctors and scientists who survived the battlefields remembered this. One of them was the German physician Gerhard Domagk, who served in field hospitals and saw the casualties of gangrene; later he was responsible for the sulfanilamide family of drugs (p. 168). Another was the Scot Alexander Fleming, who served in the British Army's Royal Army Medical Corps. After World War I, Fleming made bacteria and antibacterial agents his field of study, at St. Mary's Hospital in London, and it was a decade after the end of the war that he made his great discovery, whilst examining the bacterium *Staphylococcus aureus* on nutrient agar in culture dishes. While he was away on holiday, one of the dishes became adventitiously contaminated with a *Penicillium* mould, which had come in from a neighbouring laboratory possibly via an open door.

On 3 September 1928, Fleming observed that the mould destroyed the colonies of staphylococci which surrounded it, causing an area free of the bacteria between the mould and the staphylococci. Evidently, the mould contained a substance that prevented bacterial growth. The mould was eventually identified as *Penicillium notatum*. Fleming continued to study penicillin, but little progress was made for a few years, as he could not purify and isolate this intriguing chemical.

The story moves to the University of Oxford. The Australian pathologist Howard Florey, who was appointed Professor of Pathology in 1935, recruited the German-born biochemist Ernst Boris Chain – who had left Germany in 1933 when Hitler came to power – to his team, and it was Chain who suggested the problem of penicillin separation to Florey. Working on antibacterial agents, they started studying penicillin in 1938. Another of the team, Norman Heatley, was responsible for building a pilot plant for the manufacture of penicillin, famously assembling bedpans, milk churns and baths. Heatley later devised special culture vessels. The team improved the yield and were the first to isolate solid penicillin, a form which became known as penicillin F.

Although they only had tiny quantities of penicillin F, they carried out trials on mice, finding out in May 1940 that they were cured of lethal streptococcal infections. There were obvious risks in testing penicillin on humans, so they chose a patient who had a deadly illness, with nothing to lose.

Albert Alexander was a 43-year-old police constable and a patient in the Radcliffe Infirmary in Oxford. His police station in Southampton was damaged by German bombs on 30 Nov. 1940, when flying shrapnel caused wounds in Alexander's face that subsequently got infected, leading to staphylococcal and streptococcal septicaemia

 DOI: 10.1201/9781003783565-32

(blood poisoning). At the time, for reasons of wartime morale, the story was circulated that Alexander had contracted the infection from a prick from a rose thorn in his garden. By the time his treatment began, Alexander had lost one eye and had abscesses all over his head and also in his lungs. In dreadful pain, he was on a path towards death. On 12 February 1941, 200 mg of penicillin F was administered intravenously, with further doses of 300 mg every three hours. Penicillin was in desperate short supply, so that all Alexander's urine was collected. Every morning, Charles Fletcher, the doctor in charge, took the urine by bicycle to the laboratory, where any excreted penicillin could be extracted and reused.

There was an immediate improvement in the patient's condition, which continued. Five days later, Alexander was vastly better and eating well, whilst the abscesses were healing. However, the penicillin supply had run out and ten days after the initial treatment Alexander began to relapse, and he died on 15 March. A few more patients were successfully treated successfully.

This shortage of penicillin was a serious problem; the production needed to be improved and scaled up. However, with Britain on the receiving end of German bombs, it was no place for such expansion. On 27 June 1941, Florey and Heatley travelled to the USA to study their practices, taking penicillin mould from Oxford with them. Heatley stayed at the Northern Regional Research Laboratory in Peoria, Illinois, for several months, working with the American scientist Andrew Moyer. It was there that several improvements were introduced, notably the use of corn steep liquor in place of yeast and lactose in place of glucose. A mouldy cantaloupe melon provided a strain of *Penicillium chrysogenum* that greatly improved yields. It was also discovered that adding different chemicals to the corn steep liquor produced different penicillins. For example, adding phenylacetic acid to the brew proved an excellent source of what became known as penicillin G, benzylpenicillin, in contrast to the penicillin F that had been produced in England.

Heatley returned to England in July 1943. Moyer published the results of their research, but omitted Heatley's name from the paper. By taking the credit as sole author, Moyer was able to apply for a patent naming himself as the inventor. Florey had earlier been persuaded that it would be unethical to patent a medical discovery.

Mass production of penicillin became possible, thanks to the American adoption of deep fermentation plants in place of the UK practice of growing the moulds on the surface of nutrients. Whilst UK and US production levels of penicillin were comparable in 1943, the United States moved ahead in 1944 by a factor of at least 40, largely due to Pfizer's new deep-fermentation plant.

Almost a year after Albert Alexander's death, doctors at a hospital in New Haven, Connecticut, administered penicillin to Anne Miller, a 33-year-old patient who was close to death after developing streptococcal septicaemia following a miscarriage. Her temperature was hovering around 106 °F. Another patient of her doctor was a friend of Florey, who managed to get 5 ½ g of penicillin sent from the pharmaceutical company Merck in New Jersey, which had begun producing the drug. Her temperature fell overnight, after the first administration of the drug, which continued to be given over a month's convalescence. Anne Miller lived to the age of 90, dying in 1999.

Later that year, on the night of 28 November 1942, a dreadful fire broke out at the 'Cocoanut Grove' nightclub in Boston, USA, with 492 fatalities and many more injuries. Following the fire, 444 burn victims were hospitalised, of whom just 130 survived. Penicillin, still in short supply, was given to a number of survivors of the fire; their recovery and freedom from infection influenced the subsequent use of penicillin by the US armed forces.

Back in England, in August 1942, 52-year-old Harry Lambert – whose employer was Robert Fleming, Alexander Fleming's brother – developed streptococcal meningitis, which proved resistant to sulfonamides (pp. 168–174). Alexander Fleming tested the dying patient's spinal fluid and found that it was penicillin-sensitive, so he asked Florey for penicillin. Fleming injected penicillin F into Lambert's spinal canal on 13 August, following this with intramuscular injections of penicillin every two hours and a daily spinal injection. Lambert immediately began to recover, and was discharged from hospital, completely recovered, a month after treatment began.

Mass production in the USA began in 1943, so that penicillin was available to British and US troops in time for the invasion of Sicily in July 1943 and the ensuing Italian campaign. Penicillin production expanded greatly in 1944, so that penicillin was available in large quantities by mid-1944 to treat the wounded following D-Day.

It was not generally available to civilians – unless you were a celebrity – thus an exception was made to save the life of the German-American singer and actress Marlene Dietrich, who contracted pneumonia while entertaining US troops in Bari, Italy, in 1944.

By the end of the war, American factories were making large amounts of penicillin. It prevented infection of the wounds of many Allied soldiers, saving their lives – as well as those who had contracted venereal disease. Penicillin became generally available to civilians from American pharmacies at the end of the war, in August 1945, whilst doctors in the UK were able to prescribe it from June 1946. Within a decade, it was saving millions of lives across the world.

Until 1945, little was known about the structure of penicillin. Analytical evidence showed that in addition to carbon, hydrogen and oxygen, it contained nitrogen and sulphur, but for all that time, no one knew its molecular structure. Chemists considered that two possible structures: one containing an oxazolone-thiazolidine system (32.1), the other a beta-lactam.

Finally, in 1945, the great Oxford crystallographer Dorothy Hodgkin determined the structure of penicillin G, or benzyl penicillin (32.2). To the surprise of most scientists involved, it contained a 4-membered beta-lactam ring, which gave the clue to how penicillin works. As any UK A level chemistry student can tell you, a four-membered ring, with 90° bond angles (compared to the normal angle at carbon of around 109 ½°) is under strain and therefore very reactive. The beta lactam ring binds to the enzyme peptoglycan transpeptidase, which the bacterium needs to make cell walls, so bacteria grow bigger but cannot divide, and eventually they burst. Animal cells do not have cell walls, so humans do not have peptoglycan transpeptidase. Hence, penicillin is not usually toxic to humans.

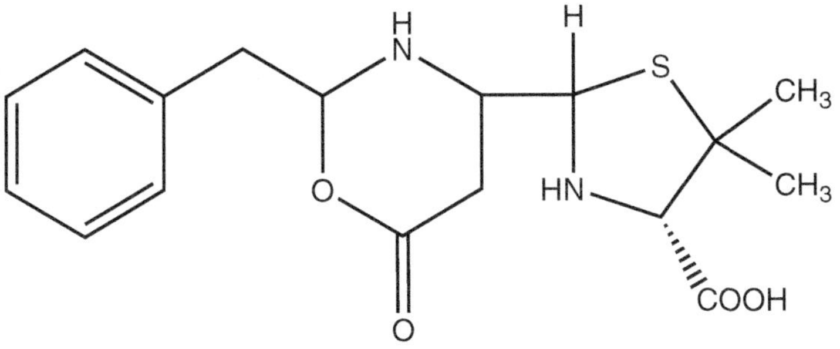

(32.1)

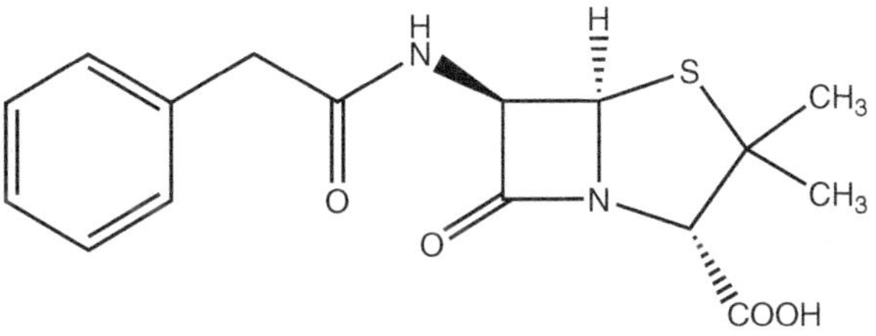

(32.2)

By this time, scientists had discovered more than one type of penicillin, differing in the side chain attached to the beta-lactam ring. When Hodgkin solved the structure of penicillin, she had the choice of studying crystals containing either penicillin F or penicillin G, and chose the latter as the crystals were simpler. Penicillin F (32.3) has an aliphatic 2-pentenyl side chain.

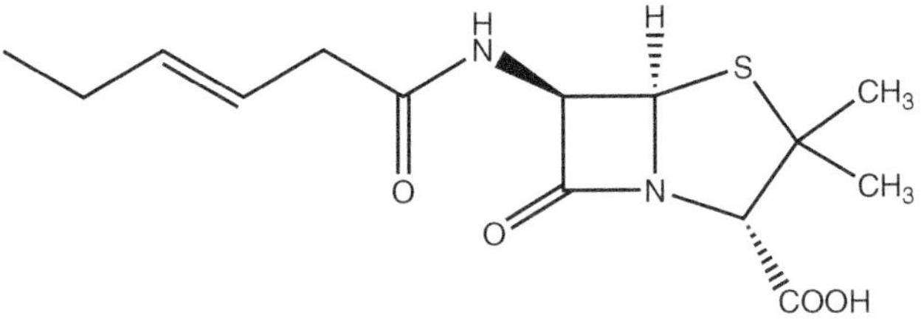

(32.3)

Many earlier penicillins, including benzylpenicillin (Penicillin G), are decomposed by acid. The acid in our stomachs means that they have to be injected and cannot be taken orally. However, it was later discovered that if phenoxyacetic acid ($PhOCH_2COOH$) is added to the fermentation broth, the resulting phenoxymethylpenicillin, commonly known as penicillin V (32.4), is quite acid-stable to be taken orally. Since then, many more penicillins suitable for oral administration have been developed.

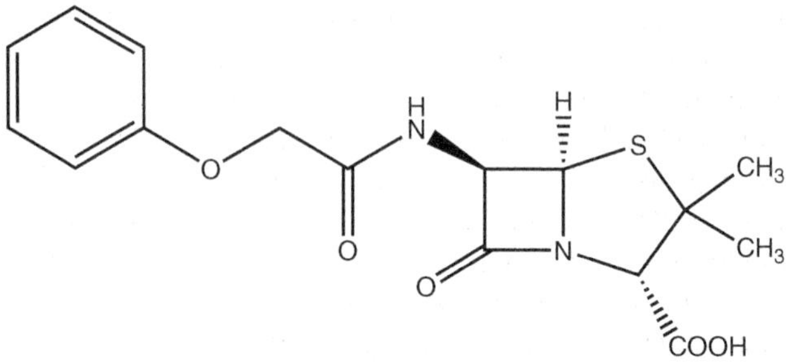

(32.4)

The key to producing the newer penicillins was a discovery made by Beechams in 1959. Researchers found that 6-aminopenicillanic acid (6-APA; (32.5)), which already contained the β-lactam group, could be isolated from the culture broth used to make penicillin, if it was deficient in the carboxylic acid that supplied the R group sidechain. This enabled the addition of the appropriate 'R group' to make the penicillin required, a classic example of 'semisynthesis'. It is now known that 6-APA can also be made from penicillin G.

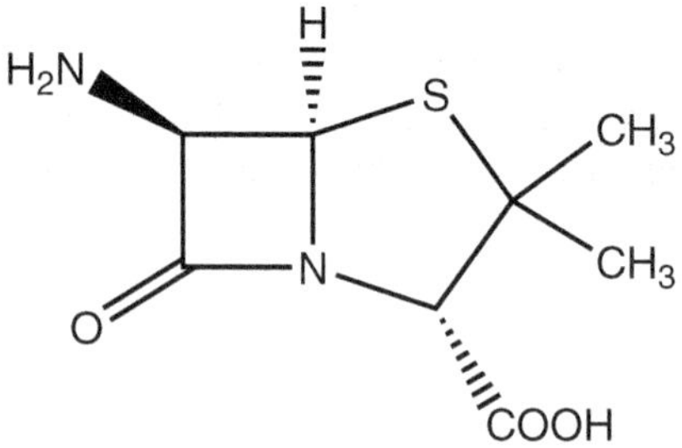

(32.5)

Fleming predicted in his Nobel lecture of 1945 that bacteria would acquire resistance to penicillin. By 1946, very soon after the widespread use of penicillin G, some strains of *Staphylococcus aureus* had already become resistant to it. Because bacteria adapt quickly, mutations in chromosomes can alter receptor sites so that penicillin cannot bind effectively. Other bacteria can make the enzyme penicillinase, a β-lactamase enzyme which catalyses the hydrolysis of the β-lactam ring by cleaving the amide linkage, thereby destroying the activity of the penicillin. The causes of antibiotic resistance vary from the 'individual level' where people may not complete their course of antibiotics, to larger scale misuse of antibiotics as growth promoters in animals.

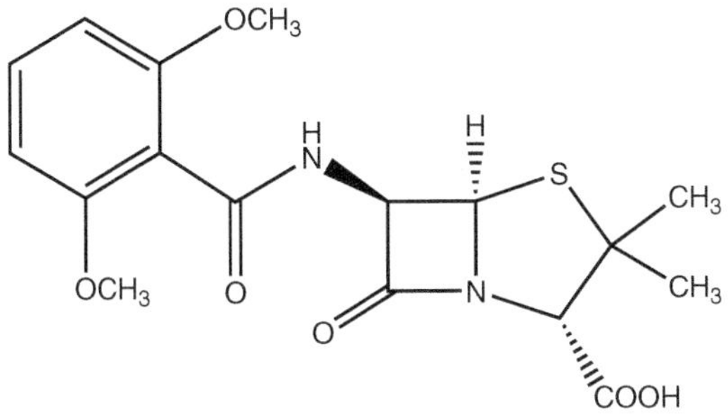

(32.6)

Although only a weak antibacterial agent, methicillin (32.6) has methoxy groups in the 2, 6-positions of the benzene ring that make it resistant to β-lactamase enzymes. Methicillin was introduced in 1959 for the treatment of infections due to penicillin-resistant *Staphylococcus aureus*. It hit the headlines in 1961 when the actress Elizabeth Taylor was at death's door with pneumonia while filming *Cleopatra* – methicillin cured her. But that year, there were reports of *S. aureus* that were resistant to methicillin (methicillin-resistant *S.aureus*, MRSA), owing to a methicillin-resistant penicillin-binding protein. MRSA is now a worldwide problem.

One way to counteract penicillinase is to add clavulanic acid (32.7) to the penicillin. Clavulanic acid, isolated by Beechams in 1976, has a structure that is quite similar to that of penicillin, though it has no significant bacterial activity. It binds to the β-lactamase via a serine residue and thereby permanently inhibits it (irreversible inhibition), thus protecting the penicillin.

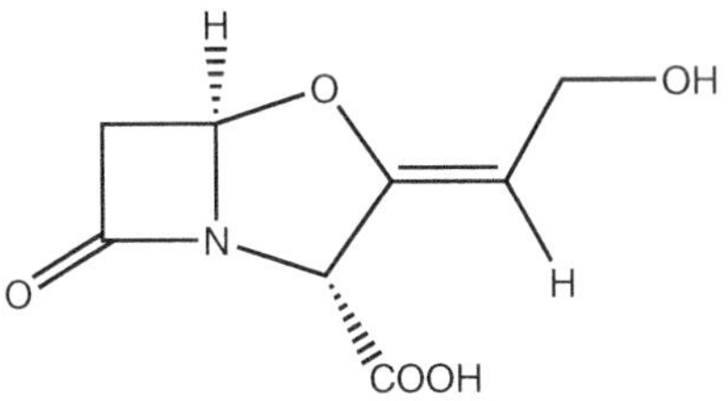

(32.7)

One widely used combination pairs clavulanic acid with amoxycillin (32.8), marketed by Glaxo as Augmentin. Amoxycillin (introduced in 1972), like ampicillin (32.9), is a semisynthetic penicillin. Traditional penicillins are only active against Gram-positive bacteria, but amoxicillin and ampicillin are active against some Gram-negative bacteria and are therefore referred to as broad-spectrum antibiotics. The reason for this is that the $-NH_2$ side chain enables the molecules to penetrate the outer membrane of Gram-negative bacteria.

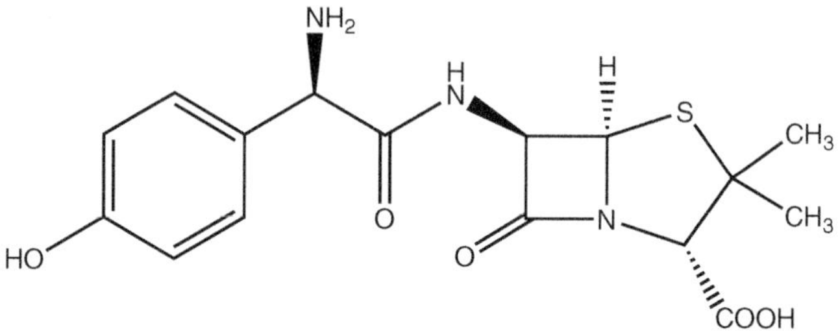

(32.8)

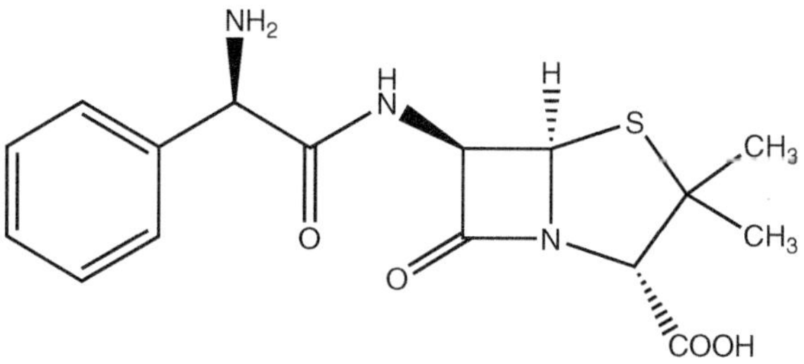

(32.9)

Thus, an adventitious discovery led to the penicillin family, and the antibiotic era that we know today. However, it took acumen to recognise the significance of Fleming's initial observation in that Petri dish.

33 Polythene

Even the Beatles' greatest fans would not claim that *Polythene Pam* merits a high place in the Beatles' canon of music. Written in 1968 by John Lennon while in India (possibly 'chemically fuelled'), allegedly about a fan who ate polythene, it was recorded for posterity on the *Abbey Road* album the following year. Yet it is a significant historical document, showing how polymers like polythene were part of people's consciousness then. This could not have been said a decade earlier, a time when paper bags, together with carrier bags, were the norm when shopping.

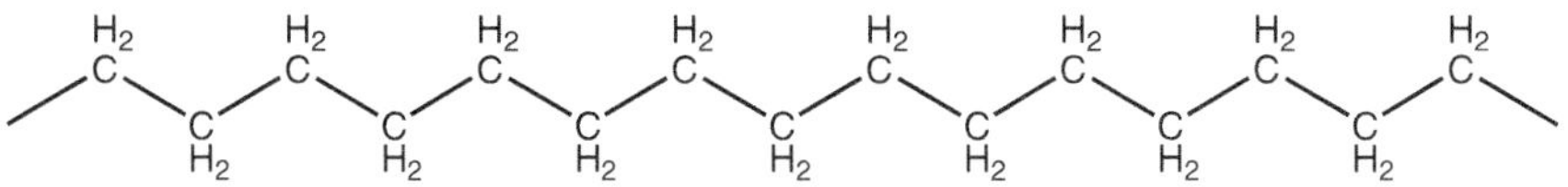

(33.1)

Polythene (33.1) was first made in 1898 by the German chemist Hans von Pechmann. He was a brave man, as he manufactured polythene by heating diazomethane. Of course, he did not know what this waxy solid really was, but he recognised that it had the empirical formula CH_2, and was not a simple small molecule, terming it *polymethylene*. Then, on Monday 27 March 1933, polythene was again produced. This time it was two chemists working for Imperial Chemical Industries (ICI) at Northwich in Cheshire. They attempted to react ethene (at that time known as ethylene) with benzaldehyde, attempting to insert ethene into the C-H bond of the aldehyde group and make ethyl phenyl ketone, $CH_3CH_2COC_6H_5$ (propiophenone, as we would now term it). They carried out the reaction at 2000 times normal atmospheric pressure and 180°C in an open laboratory (this was the days before risk assessments) and – what you may think was fortuitously – they obtained a white waxy solid (resembling paraffin wax), analysing as CH_2. Subsequent attempts sometimes gave just a black deposit of carbon; sometimes they resulted in an explosion and a fire, so ICI suspended this line of investigation for two years. When another ICI team, led by Michael Perrin, had another try, it was done in an isolated laboratory with one-foot-thick walls. Again, they got a white powder, 8.5 grams of it. As had been found two years earlier, the experiment was irreproducible, and eventually it was realised that a trace of oxygen was needed as a catalyst to start polymerisation. No oxygen – no reaction; too much oxygen – explosion. This is a classic example of a great scientific discovery made by accident.

The electron-rich C = C double bond in ethene makes it more reactive than an alkane, so that, unlike alkanes, it is seldom found in nature, but – on the other hand – it is a valuable petrochemical feedstock, not least in making polythene. ICI now had

DOI: 10.1201/9781003783565-33

a reliable synthesis of what we call polythene. Although unreactive, involatile, insoluble and a non-conductor, it was a material searching for applications. Initially, ICI suggested it as an insulating material for submarine cables, deep under the oceans. In the event, this application did not work out (though a mile of cable was laid between the Isle of Wight and the mainland), but the plant for the process had been constructed. It came on stream in September 1939, on the fateful day when the Germans invaded Poland. Polythene was the material chosen for insulating the coaxial cables of radar sets and came into use for this early in 1940.

Sir Robert Watson Watt, inventor of radar, said:

> The availability of polythene (polyethylene) transformed the design, production, installation and maintenance of airborne radar from the almost insoluble to the comfortably manageable. . . . And so polythene played an indispensable part on the long series of victories in the air, on sea and land, which were made possible by radar.

During World War II, commercial sales of polythene were suspended, as the material was required for the war effort. When peace came, the market opened up for plastics, including polythene. This was embodied in the empire built by the Massachusetts entrepreneur Earl S. Tupper. He got acquainted with polythene early in the 1940s, when working with DuPont, which was making polythene under licence from ICI. Towards the end of the war, he got hold of some polythene waste and investigated ways of commercialising it that he could put into practice after August 1945, when polythene came onto the open market. He changed the face of kitchenware, which had previously been based on easily broken glass and ceramic products. Moulded polythene containers were marketed as Tupperware, combining toughness with softness. Suddenly, Tupperware bowls, tumblers, containers and other artefacts were everywhere in the kitchen. Around 1950, marketing of Tupperware was turned over to the home, in the form of Tupperware parties.

New polythene applications mushroomed, such as the first 'plastic macs'. The first polythene washing-up bowls emerged in 1948. Although initially they were more expensive than the traditional enamelled metal-based bowls, advantages of the plastic bowls included resistance to corrosion and chipping, as well as their range of colours. These were probably the first plastic objects to really impact upon the domestic market. Plastic buckets were close behind. Transparent polythene films for wrapping food had obvious advantages. Milk bottles were traditionally made of glass, of course, but the use of polythene film coatings enabled the use of card cartons for milk as well as soft drinks. With the spread of domestic refrigerators, plastic trays to make ice cubes have become universal, as have the sacks used to put out domestic waste for collection. And, of course, the use of polythene and other polymers transformed the world of children's toys.

The ICI route to polythene – high temperature and high pressure – gives what is now known as low-density polythene (LDPE). In this form, there is quite a bit of branching of the chains, which means that they do not pack together neatly in the solid state, causing the material to be somewhat soft and weak, which in turn leads to applications that do not require strength, like making wrappings and bags. Then, in 1953, Karl Ziegler, working at the Max Planck Institute, found that using a mixture

FIGURE 33.1 Regular carbon-based chains are capable of packing well.

of $Al(C_2H_5)_3$ and $TiCl_4$ provided a catalyst that would polymerise ethene at low temperature and pressure. Shortly afterwards, the Italian Giulio Natta developed a similar catalyst for polymerising alkenes. Ziegler and Natta shared the 1963 Nobel Prize in Chemistry for developing what has come to be known as the Ziegler-Natta process.

The product obtained by this route has less chain branching, helping the chains pack better (Figure 33.1) and producing a stronger material with a higher density, which can be used for applications like washing-up bowls and milk bottles. It is known as high-density polyethylene (HDPE). Other polymerisation catalysts exist and various versions of polythene are known, with different chain lengths – between 2,000 and 20,000 ethene units – and different amounts of chain branching, leading to different mechanical properties.

More than 100 million tonnes of polythene are manufactured each year, with most of it used in packaging. In turn, a sizeable fraction of the 350 million tonnes of plastic waste discarded each year is the downside of polythene – this waste has low biodegradability. This is a reflection of the structure of polythene. When ethene is polymerised, its reactive double bond disappears, replaced by a single bond. Polythene is essentially a giant alkane, so – unlike nylon or terylene – it has no functional group that is susceptible to hydrolysis or to enzyme-catalysed breakdown. Plastics are cheap, which discourages recycling.

34 Quinine

In the 2020s, when we were introduced to deadly diseases and pandemics, our thoughts naturally turned to COVID-19. After a moment's reflection, we also remembered the Black Death and AIDS. The Spanish flu pandemic of 1918–1920 came to mind as well, as did viral haemorrhagic fevers emanating from deep within African forests. But deadly disease means far more than that. Each year, over 300 million people suffer from malaria and nearly 600,000 people die from it. Nearly all (close to 90%) of these deaths are in sub-Saharan Africa – mainly young children. Malaria is caused by a protozoan parasite of the genus *Plasmodium*, which attacks the bloodstream; it seems to have originated in sub-Saharan Africa. Of the four strains affecting humans – *P. falcipanum*, *P. malariae, P. ovale* and *P. vivax* – *P. falciparum* and *P. vivax* are the most common and most dangerous. An infected female *Anopheles* mosquito transmits the parasite via its bite. Inside its human host, the parasite multiplies in the liver and in red blood cells. The victim gets fevers and chills, but if the *P. falciparum* parasites accumulate in the brain, the result is lethal cerebral malaria, with around 1% of *P. falciparum* infections proving fatal. This continuous pandemic has claimed countless victims, including, long ago, the pharaoh Tutankhamun (c. 1323 BC), Alexander the Great, Alaric the Goth, the Roman emperor Titus and, later, the poet Dante, as well as George Gordon, Lord Byron (1824) and the explorer David Livingstone (1873). At least four Holy Roman Emperors are believed to have been victims of malaria – Otto III (1002); Henry VI (1197); Henry VII (1313); and Charles V (1558). As late as 1960, the famous Italian cyclist Fausto Coppi died of malaria.

Malaria reached Europe well over two thousand years ago, most likely from Africa, and within a couple of hundred years it became endemic in Rome, by around 300 BC. Richer citizens, who could afford it, spent their summer months in the cooler hills well away from the city. In his *Satires*, the poet Horace (65–68 BC) wrote that, on his highland estate, he did not fear an unhealthy autumn. The custom was later adopted by Popes, who spent some of their summer in the Alban Hills at Castel Gandolfo, starting with Pope Urban VIII. Rome became the malaria centre of Europe. In 1588, Pope Sixtus V established a Congregation of Waters and Streets with the responsibility for draining the Roman marshes, a breeding ground for mosquitoes. However, this was not completely achieved until the Pontine Marshes to the south of Rome were finally drained by Mussolini's government in the 1930s. Historians have named 15 Popes who died of malaria between the years 999 and 1590, a period of less than six centuries. The Papal Conclave of 1623 elected Pope Urban VIII; within a fortnight, eight cardinals died of malaria and the new Pope also contracted it, though he survived to enter upon a 20-year pontificate, during which he issued a bill excommunicating users of tobacco, and also to force his former friend Galileo Galilei to abjure his support for the Copernican system. Around 1630, Jesuits brought cinchona bark from Peru to Rome, whereupon it helped improve the cardinals' chances of surviving consistories and malaria (the 1655 Papal Conclave was the

DOI: 10.1201/9781003783565-34

first to not see a malarial death). Henceforth, it was familiarly known as Jesuits' bark, or Jesuits' powder when ground up.

Malaria expanded across Europe, including England. From the sixteenth to the nineteenth centuries malaria, known as the '*ague*', was endemic in the Fens and coastal marshlands of East Anglia and south-east England, largely due to *P. vivax* and *P. malariae*, with the inhabitants using laudanum as a popular remedy. It was only diminished by factors like improved drainage, eliminating the mosquitoes' breeding grounds. That stern Protestant, Oliver Cromwell, the Lord Protector of England, Wales, Scotland, and Ireland, died of malaria in 1658, reportedly refusing 'Jesuit's bark' for his malarial fever as he lay dying. Which brings us to quinine (34.1).

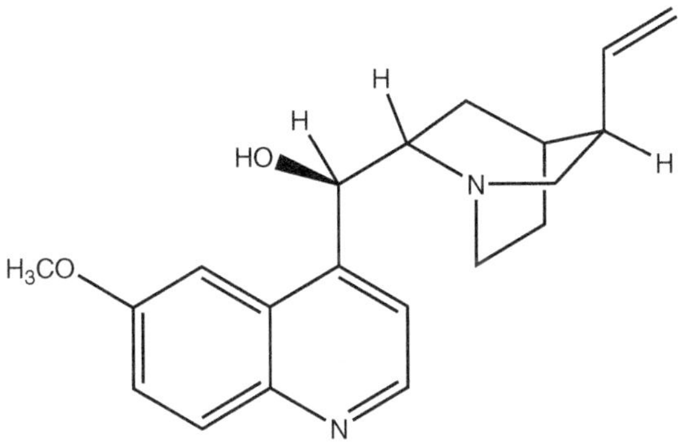

(34.1)

Europeans who colonised South America, notably the Spanish, found that the inhabitants knew that the ground bark of the Cinchona trees, native to the Andes, was effective in the treatment for fevers, among other complaints. Jesuit missionaries returning to Europe brought this bark from the 'fever tree' with them in the 1630s, and it duly became known as 'Jesuit's powder' and a recognised treatment for malaria (Italian: *mala aria*). By 1650, Jesuit's powder was an accepted cure for malaria in Rome and the surrounding region, and its use spread across Europe, though (as already noted) a Puritan England in the 1650s was less accepting. Following the Restoration of the monarchy to England, things changed, and the Puritan physician Thomas Sydenham became an advocate for its use.

In 1672, Robert Talbor (1642–1681), a young English apothecary, published a book *Pyretologia: A Rational Account of the Cause and Cures of Agues,* in which he said: 'Beware of all palliative cures, and especially of that known as Jesuits' Powder'. But things were not all that they seemed. In 1672, Talbor cured one of King Charles II's officers of a fever; the King was so impressed that he made Talbor one of his physicians in ordinary. That was not all. In 1678, he knighted Talbor, who in the following year used his remedy to cure the King of a serious fever. In November 1680, Talbor cured the Dauphin, son of Louis XIV of France, in the same way. It was only following the death of Talbor in 1681 that the true nature of his 'secret potion' was revealed: it was based on 'Jesuit's powder' dissolved in white wine with added ingredients such as roses and lemon juice. The popularity of this Cinchona-based remedy increased – as did the price – and the trees began to disappear in the areas where they were cultivated. Cinchona cultivation was a Spanish monopoly but in the mid-nineteenth century cinchona seeds were smuggled out of Peru to start the cultivation of cinchona in the colonies of other European powers, notably Dutch Java. Finally, in 1820, two young French pharmacists, Pierre Joseph Pelletier and Joseph Caventou, isolated quinine from cinchona bark, which hereafter became the treatment for malaria. Without the increasing availability of inexpensive quinine, major tropical enterprises such as the Panama Canal, not to mention the colonisation of Africa, would not have been possible.

Queen Victoria's Consort, Prince Albert, was influential in establishing the Royal College of Chemistry in London (now Imperial College), and he also encouraged August Wilhelm von Hofmann, a protégé of Justus von Liebig, to become its Director (1845). By 1849 Hofmann was announcing his intention of finding a synthesis of quinine, for which the French Pharmaceutical Association offered a prize of 4,000 francs to the first to achieve the target. This was a difficult problem, as it was not until 1854 that Strecker determined that the chemical formula of quinine was $C_{20}H_{24}N_2O_2$ (Hofmann had suggested $C_{20}H_{22}N_2O_2$) – and no one had any idea of its structure. Armed with this information, Hoffmann's student, the young William Henry Perkin, was assigned the problem. Perkin thought that since *N*-allyl toluidine had the molecular formula $C_{10}H_{13}N$, roughly half that of quinine, if he carried out an oxidative coupling of two molecules of N-allyl toluidine, he might obtain quinine. He did not, he got a brown-red solid. Perkin then tried the reaction again, with aniline (phenylamine) as the starting material. This time he obtained a black precipitate, which gave a purple ethanolic solution that stained his mopping-up cloth. The dye proved to be very good at colouring silk. Perkin had just made the dye which proved especially popular with the French, who called it 'mauve', and in doing so he helped kick-start the aniline dye industry. However, he, of course, had not developed quinine.

The structure of quinine was not known to Perkin; indeed, it would not be known for another half-century, until it was established by the German chemist Paul Rabe in 1907; he and Karl Kindler reported the synthesis of quinine in three steps from d-quinotoxine in 1918. The next step towards establishing quinine's identity came through the work of Woodward and Doering, who in 1944 synthesized quinotoxine from 7-hydroxyisoquinoline, in 15 steps, so that taking into account the earlier paper had completed 'The Total Synthesis of Quinine', as the title of their paper says. Quinine has four chiral carbon atoms, making its stereoselective synthesis challenging, finally achieved by Gilbert Stork's research group, and reported in 2001.

That being said, these syntheses have not achieved the synthesis of quinine for the marketplace – cinchona trees remain the only economic source of quinine.

Artemisinin is now the preferred medication against malaria, but quinine remains the best alternative, though it does have side-effects, such as dizziness, tinnitus and vertigo. Quinine is toxic to the malaria parasite *Plasmodium falciparum*, as it affects its metabolism of haemoglobin, possibly by inhibiting haem polymerase, the enzyme used by the parasite to polymerise ferriprotoporphyrin IX (haem). This is believed to generate species that are toxic to the parasite.

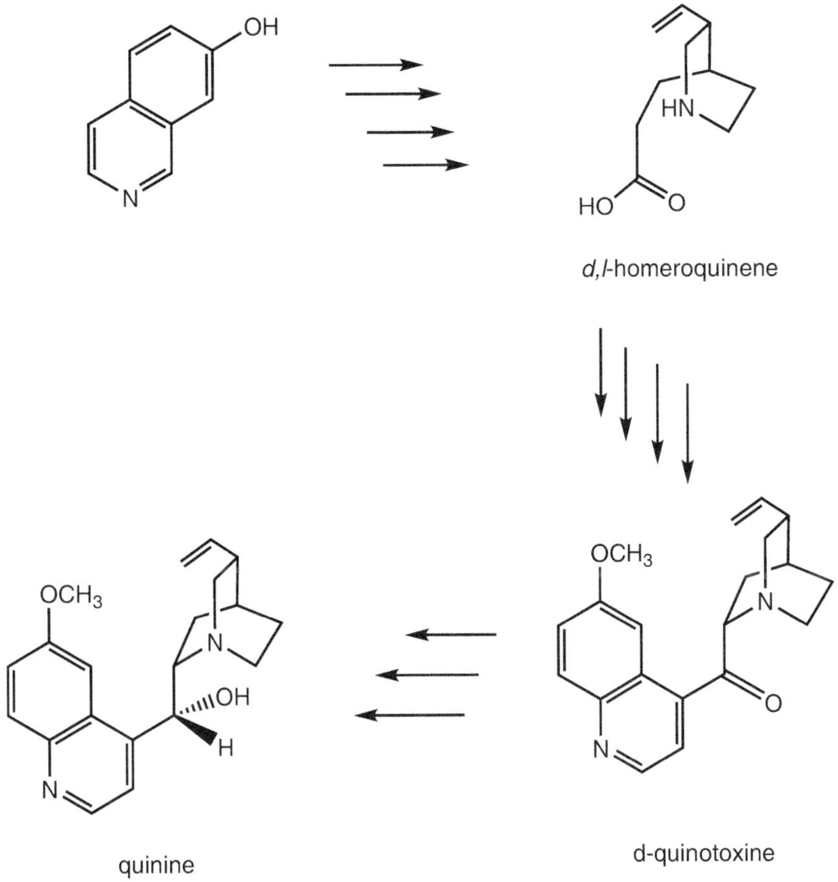

FIGURE 34.1 The 'total synthesis' of quinine.

35 Scopolamine (aka Hyoscine)

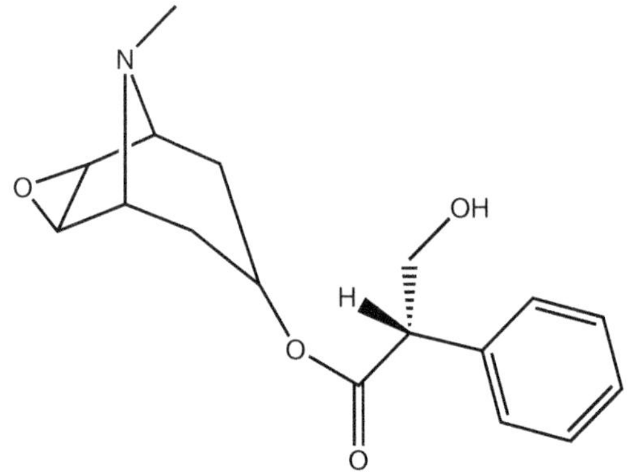

(35.1)

Like atropine (page 18), which has a very similar structure, scopolamine is a tropane alkaloid. As with atropine, its biosynthesis begins with the amino acid ornithine. It is also known as hyoscine. Scopolamine is found in various plants, especially in members of the nightshade family, such as deadly nightshade (*Solanum dulcamara*), mandrake (*Mandragora officinarum*), henbane (*Hyoscyamus niger*), thorn-apple or jimsonweed (*Datura stramonium*), scopolia (*scopolia carniolica*) and angel's trumpets (*Brugmansia* species, such as *B. suaveolens*). You may have encountered mandrakes in *Harry Potter*. Hundreds of years ago, mandrake was said to emit screams fatal to anyone within earshot if it was dug up. Thus, Professor Sprout showed the students how to pot mandrakes, while they were wearing earmuffs.

Scopolamine has some medical uses, often as a drug to counter seasickness. It was also once used as an anaesthetic-analgesic during childbirth, administered along with morphine. It can induce what was called 'twilight sleep', a state of consciousness lacking in memory. For some years, particularly during the 1920s, it was used in American and Canadian courts as a 'truth serum', as it was believed that people could say only the truth when under its influence. However, it was soon replaced by barbituric acid derivatives such as sodium pentothal.

One peculiar effect of scopolamine is that, as an acetylcholine blocker, it affects memory. This was first exploited by South American shamans who gave it to their

DOI: 10.1201/9781003783565-35

clients to induce hallucinations. It is possible that during the later medieval period, witches used medications containing scopolamine to create out-of-body experiences like the sensation of flying. Today, this short-term effect of scopolamine is exploited by criminals to rob or to sexually assault their victims, who may enter a state of submissive anaesthetic amnesia. In particular, it has been associated with the state of Colombia; here it is known as *Burundanga*, owing to its extraction from the brugmansia (angel's trumpet) plant. Criminals may introduce it into their victim's food or drink, though it is also said to be blown into peoples' faces. Once it takes effect, the victim is parted from their possessions or led compliantly to the nearest cash machine.

Scopolamine was the instrument of murder in the famous case in 1910 featuring 'Dr Crippen'. Hawley Harvey Crippen trained in homeopathic medicine in the USA, his home country, then came to England in 1898 with his second wife Cora, a music-hall entertainer under the name Belle Elmore. She was last seen alive at a dinner party on 31 January 1910. Her husband told people that she had gone back to the USA; by the end of March, his story was that she had died. Meanwhile, he had been giving some of her clothes and jewellery to his secretary, a young woman called Ethel le Neve, who was also his mistress. Crippen's lies started to catch up with him when Cora's American friends said that she was not in the USA. They became suspicious and involved Scotland Yard. On 8 July, Inspector Walter Dew called at Crippen's home, whereupon Crippen decided to disappear. The next day, he and Ethel – disguised as a young boy– boarded the boat train from Liverpool Street station and travelled to Brussels. After a few days, they went to the port of Antwerp and caught the SS *Montrose* on 20 July, heading for Quebec, under the identity of a Mr John Robinson and his son. By this time, Inspector Dew realised that his bird had flown and searched Crippen's house. A police sergeant saw some loose bricks in the floor of the cellar floor, which were concealing bodily remains. A warrant was issued for Crippen's arrest on 16 July.

The story became hot news in the British press. Captain Kendall of the *Montrose* was in radio contact with land and therefore knew the story. He realised that he had Crippen on board, a man who was holding hands with his effeminate 'son'. Kendall contacted Inspector Dew in England. The intrepid Inspector made haste to Liverpool, where he boarded a fast liner, the SS *Laurentic*, which docked in Canada two days before *SS Montrose*. This enabled Dew to board the *Montrose*, greet two unsuspecting travellers and arrest Crippen and Le Neve. Crippen's trial began on Tuesday, 18 October 1910. Dr Willcox, the Home Office pathologist who examined the remains discovered in the cellar, found that the stomach, kidneys, liver and intestines all contained an alkaloid, which was eventually narrowed down to scopolamine, with the help of the great pathologist Sir Bernard Spilsbury. Crippen was known to have purchased scopolamine from a pharmacist on 19 January. The jury were sent out on Saturday, 22 October, and were back with a guilty verdict in 27 minutes flat. The judge sentenced Crippen to death and, after an unsuccessful appeal, Crippen was hanged on 23 November. In a separate trial, Ethel le Neve was found 'not guilty'; later she remarried, living on in anonymity until 1967.

Subsequently, results were published much later (2010) of the DNA tests carried out on samples of scar tissue said to be from the body of Mrs Crippen. Comparison with DNA samples from maternal descendants of Cora Crippen shows that the tissues could not have belonged to her.

36 'Spice'

'Spice' is a drug that has recently come on the scene – starting around 2004 – but its story is not as simple as its name indicates. It has links with cannabis, but it is a synthetic chemical, which does not occur in nature. More than that, there are quite a number of molecules that go by the name of 'Spice'. Smokeable herbal mixtures, not just going by the name of 'Spice' but also other names including K2 and 'Black Mamba', were being sold by the mid-2000s. They consisted of dried leaves or other plant material, the vegetable matter having the synthetic chemical added. They were usually smoked (either using a pipe or rolled in cigarette paper). They were called 'legal highs', as they were claimed to be natural plant products, while being nothing of the sort and were also being marketed as herbal incense, bath salts, air fresheners or the like.

By 2008, chemists across the world were analysing these mixtures to find out which chemicals they contained. Groups in Germany and Japan almost simultaneously identified a known substance, JWH-018 (36.1).

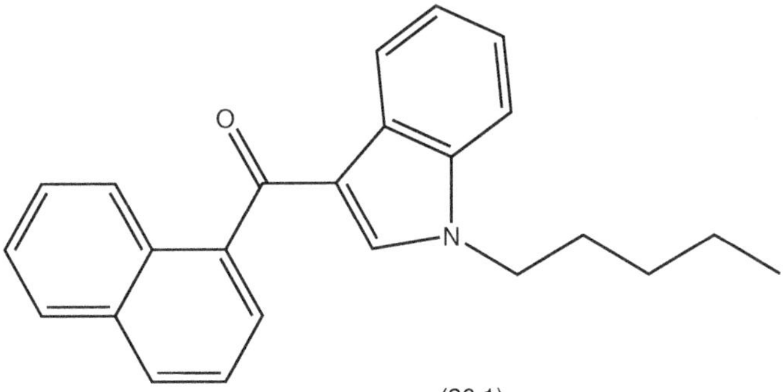

(36.1)

JWH-018 had first been reported by an American chemist, John W. Huffman, Professor of Chemistry at Clemson University, South Carolina. He was studying cannabinoid receptors, receptors found in various parts of the body. These receptors are 'switched on' by three kinds of molecules – (a) molecules produced by the body, known as 'endocannabinoids' (b) plant-derived cannabinoids (such as those found in marijuana) and (c) synthetic or 'man-made' cannabinoids. Stimulating these receptor sites affects mood, thoughts, feelings and senses, as well as causing physical effects. These receptors also bind to some molecules made by plants and also to synthetic ('man-made') cannabinoids. The 'cannabinoid receptors' got their name because they interact with Δ-9-tetrahydrocannabinol ((36.2), delta-9-THC or Δ^9-THC), the psychoactive molecule in cannabis; the best known of these receptors are CB1 (which stands for Cannabinoid receptor type 1) and CB2, discovered in the early 1990s. Huffman was making synthetic cannabinoids, molecules that are not found

DOI: 10.1201/9781003783565-36

in nature but which also bind to the CB receptor, and which he believed could help explain how the cannabinoid receptors work.

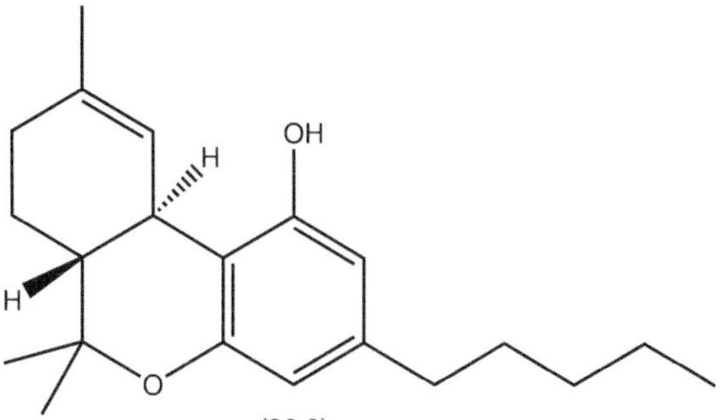

(36.2)

During his research, Huffman developed a lot of these synthetic cannabinoids, sometimes known as synthetic cannabinoid receptor agonists (SCRAs). He gave them labels based on his initials (JWH), which were easier to use than the chemical names of the molecules. The 'official' chemical name of JWH-018 is 1-pentyl-3–(1-naphthoyl)indole or naphthalen-1-yl-(1-pentylindol-3-yl) methanone, so you see his point. JWH-018 is reported to be a very potent agonist for the CB1 receptor with four times the potency of THC.

If people smoke JWH-018, they are said to get a 'buzz' like the one obtained from marijuana. So clearly someone had been reading the research paper published by Huffman's group in 1994 and had decided that there was money to be made from this chemical – and from other synthetic cannabinoids. Remember, the term 'synthetic cannabinoid' means that the compound interacts with the cannabinoid receptors in the body, not that they are found in marijuana. These include JWH-073 ((36.3): naphthalen-1-yl-(1-butylindol-3-yl)methanone), which was sold with claims that it was a 'fertiliser', 'Forest Humus'; JWH-200 ((1–(2-morpholin-4-ylethyl)indol-3-yl)-naphthalen-1-ylmethanone) and JWH-250 ((36.4): 2–(2-methoxyphenyl)-1–(1-pentylindol-3-yl)ethanone), both of which appeared in 'herbal' smoking blends.

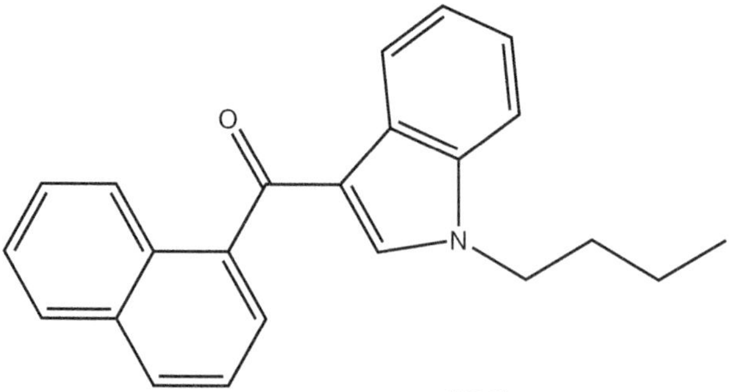

(36.3)

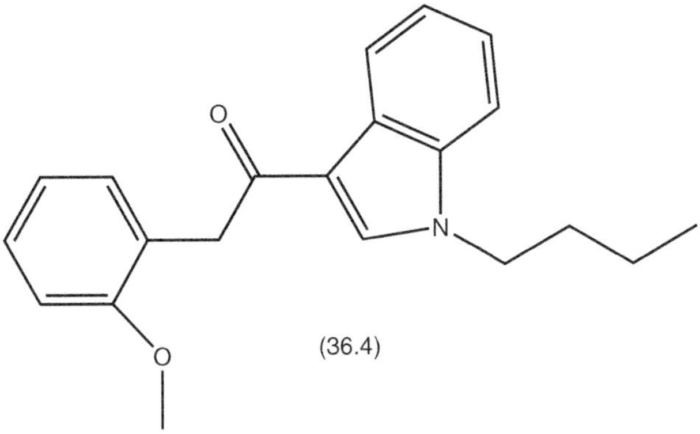

(36.4)

These and other molecules of this kind are collectively known as 'new psychoactive substances' (NPS), though 'designer drugs' or 'legal highs' are terms more often employed. When these compounds are first marketed, they sidestep existing legislation on illicit drugs. The methods for making them are available in the chemical literature and are simple for trained chemists. Such molecules are made by chemists across Europe, America and the Far East. And when legislation is introduced to declare a particular drug illegal (and such molecules are banned across much of Europe), illicit chemists produce a slightly different molecule, maybe with a methyl group in a different position, for example – something that is not covered by the legislation and therefore is not 'illegal'.

Another family of street drugs emerged at much the same time as 'Spice'. These molecules, often going by names including 'bath salts', 'pond cleaner' and 'plant food', have very different structures to the JWH family, as they have backbones like the amphetamines. Most notably, this family included mephedrone, a member of the cathinone family which includes the active ingredient of 'khat' (see page 25). These are stimulants and do not bind to the cannabinoid receptors.

A further development in the 'Spice' story occurred around 2013, when a new ingredient made its appearance. A molecule known as ADB-FUBINACA (36.5) had been synthesised and patented by Pfizer in 2009, but it was never brought to market. Along with the similar AMB-FUBINACA ((36.6): also known as MMB-FUBINACA) and other molecules in this indole carboxamide family, it had been discovered to be an active CB1 receptor agonist.

Within less than five years after its announcement, these molecules had been pirated, and they have spread across the world, as you can judge by the initial detection of ADB-FUBINACA in Japan in 2013, followed by AMB-FUBINACA detected in Sweden in 2015. Post-mortem analysis revealed that ADB-FUBINACA was present in a synthetic cannabinoid product ('Mojo') fatally smoked by a 41-year-old Louisiana woman, that year. In the following year, 2016, the New York Emergency Medical Services had to attend 33 zombie-like casualties in Brooklyn on 12 July 2016, of whom 18 were hospitalised. One victim had been smoking an herbal 'incense' product known as 'AK-47 24 Karat Gold', found to contain AMB-FUBINACA. Furthermore, a metabolite of AMB-FUBINACA turned up in blood and urine samples from eight of the patients, rather than any other abused drugs. Though none of the New York victims died, around 20 deaths were attributed to AMB-FUBINACA during 2017 in New Zealand.

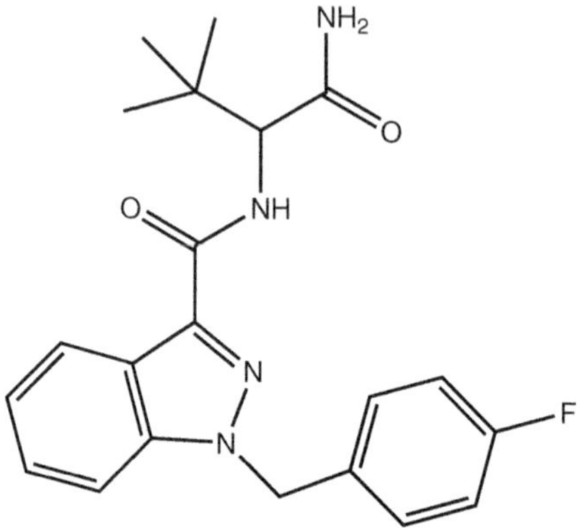

(36.5)

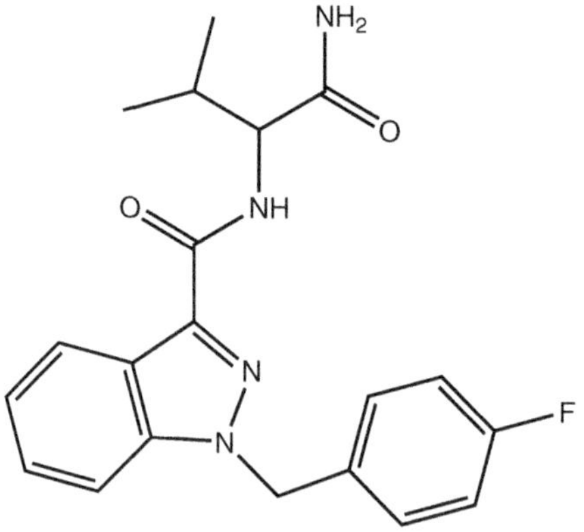

(36.6)

The United Kingdom has been part of this pattern, with AMB-FUBINACA and related molecules arriving in the UK. In some cases, drugs made abroad have been imported; in other cases, the precursor chemicals were imported from elsewhere, then the actual syntheses carried out in Britain.

In 2016, a gang of four received prison sentences of 19 years in total, with the leader getting nine years. Without any prior chemical training, they had been preparing AMB-FUBINACA in the kitchen of a flat on the sixteenth floor of a high-rise building in Oldham, Greater Manchester. They used the highly flammable diethyl ether (ethoxyethane) and corrosive sulfuric acid, plus other chemicals which had been purchased from sellers in Hong Kong, Shanghai and Poland. Fellow residents of flats were, of course, facing real danger.

Beyond the streets, many of the synthetic cannabinoid users are prison inmates, apparently a higher proportion of users than in the world outside prisons. The cannabinoids are smuggled into the gaols in various ways, with one method involving the dissolution of the drug in a solvent, the solution then being absorbed by writing paper. The paper is then used for writing a letter – either from a friend, a prison visitor, or even disguised as a lawyer's letter. Once inside the jail, the drug can be extracted out of the paper and consumed.

These 'cannabinoids' are much more potent than 'ordinary' cannabis. Users can be seen (and photographed) in a zombie-like state, slumped in public places in a state of semi-consciousness, creating a significant public health concern. Quite apart from the dangers to consumers of drugs sold by pushers for human consumption, these substances have never been tested on animals, let alone humans; there is the further issue of more potent drugs, such as fentanyls, getting cut with synthetic cannabinoids; or cutting ordinary cannabis with Spice. The effects of 'Spice' include a tight chest, racing and irregular pulse, breathing problems, dizziness and collapse.

37 Streptomycin

And other Drugs for the Treatment of Tuberculosis

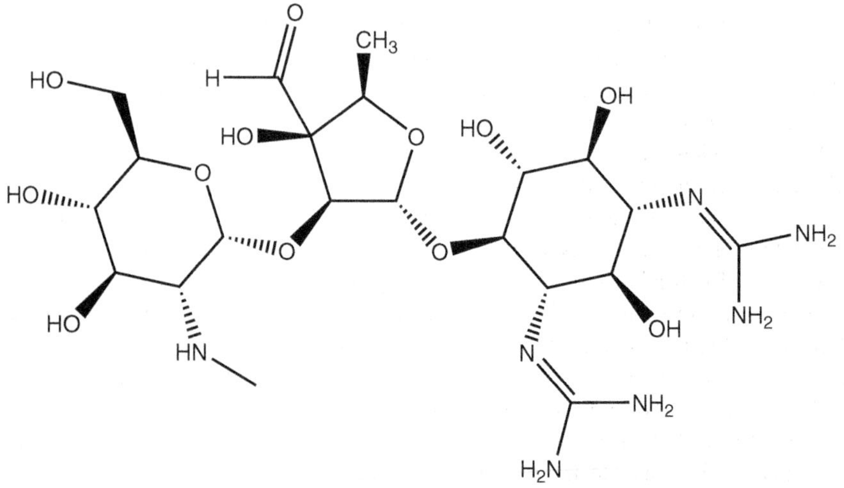

(37.1)

The global COVID-19 pandemic served as a reminder of past epidemics that have swept across the world. Many people in the UK think first of the Black Death, caused by the bacterium *Yersinia pestis*, which was also responsible for the 'Plague of Justinian' (named after the Byzantine Emperor at that time), which swept across Europe, the Mediterranean region and the Near East during the decade AD 540–550, and it even reached Britain.

The Black Death killed around one third of the population of Europe, and the earlier plague was similarly deadly. However, people in Europe have largely forgotten about the White Death (better known as tuberculosis, TB), which acquired this nickname because of the pallor often associated with it. Especially during the nineteenth century, tuberculosis was generally referred to as 'consumption', owing to the wasting and weight loss associated with it.

Worldwide, tuberculosis is estimated to have killed around a thousand million people in the nineteenth and twentieth centuries. Although the first cure was discovered mid-way through the twentieth century, it is still responsible for many deaths, notably in the Third World nations. Around two million people die annually from TB, and there are some ten million new cases each year. It has been

DOI: 10.1201/9781003783565-37

traced back at least six thousand years, including an Egyptian mummy in the fourth millennium BC.

Tuberculosis, which mainly affects the lungs, has always been more of a threat to the poor, who are more likely to live in crowded and insanitary conditions, but it is no respecter of class. Although you may not know the names of the vast majority of its victims, but there is a list of famous sufferers and victims as long as your arm.

You have heard of the writers Elizabeth Barrett Browning, Franz Kafka, John Keats, D. H. Lawrence, Aubrey Beardsley, Anton Chekhov and George Orwell. It also killed Elizabeth, Maria, Emily and Anne Bronte, their brother Bramwell, and probably Charlotte too. Several composers died of it, notably Frédéric Chopin, Luigi Boccherini and Giovanni Battista Pergolesi. Then there was the Australian Test cricketer Archie Jackson (1933); Simon Bolivar (the liberator of Colombia, Venezuela, Ecuador, Peru and Bolivia); politicians including George Canning, Prime Minister of the UK (1829) and James Monroe, President of the USA (1831); King Edward VI of England (1547–1553) as well as his ancestor King Henry VII (d. 1509), victor of the battle of Bosworth Field. The king whom Henry defeated, Richard III, lost his young wife Anne Neville to TB in 1485, five months before Bosworth. King Louis XIII of France is believed to have died of TB in 1643 at the age of 42.

Gavrilo Princip, the Bosnian Serb who lit the fuse leading to World War I in 1914 by assassinating Archduke Franz Ferdinand and his wife Sophie at Sarajevo, died of tuberculosis in 1918. Eleanor Roosevelt, widow of FDR, succumbed to it in 1962. Several saints, including Bernadette Soubirous, Theresa of Lisieux, Gemma Galgani, Casimir of Poland and Dominic Savio, were also victims of tuberculosis.

Many of the victims were young, so that several operas featured fictional heroines as victims of tuberculosis – Mimi in *La Bohème*, Violetta in *La Traviata*, Marguerite in *The Lady of the Camellias* and Antonia in *The Tales of Hoffmann*, not to mention Fantine in *Les Misérables*.

Towards the end of the nineteenth century, there were important developments in treatment, with sanatoria providing rest, controlled exercise, fresh air, and a respite from conditions under which TB flourished. It was in 1882 that the *tubercule bacillus* responsible for TB was discovered by Robert Koch (who also identified the causes of anthrax and cholera); he won the Nobel Prize in Physiology or Medicine for this discovery in 1905, but the search for a cure was to go on for over half a century.

Selman Waksman, who came with his family from Ukraine to the USA when he was young, began this research at Rutgers University in New York, where he had begun his studies before carrying out his PhD research in California, then returning to Rutgers. Waksman started his programme in 1939 – this was the decade that saw the discovery of sulfanilamide, the first antibiotic (page 168) and penicillin (page 140). Waksman was particularly interested in micro-organisms in the soil – which were to lead to antibiotics like the tetracycline family in the 1940s (page 183) – and began studying soil micro-organisms that might have antibiotic properties against pathogenic bacteria.

Albert Schatz was the member of Waksman's team carrying out the research that led to streptomycin. He was working with a highly virulent form of *Mycobacterium tuberculosis*, supplied by William Feldman of the Mayo Clinic in Rochester, Minnesota. Schatz, therefore, worked in a basement laboratory separate from

Waksman's other laboratories and was easily isolated in case the bacterium escaped. In 1944, within two years of starting his work, Schatz had found two strains of a bacterium, which made an antibiotic that he named *streptomycin*. This was active against *Mycobacterium tuberculosis* and was subsequently found to be active against many other pathogens. Schatz and Waksman published their results on streptomycin's activity against tuberculosis in test organisms in 1944, then Feldman and Corwin Hinshaw, his co-worker at the Mayo Clinic, showed that streptomycin would successfully treat TB in humans. Schatz was to receive his PhD in 1945, and he and Waksman were named on the application to patent streptomycin, with the patent being granted in 1948.

Unfortunately, there was a controversy. Schatz was persuaded by Waksman to make over his royalty rights to Rutgers University, but later he claimed that Waksman had an agreement with Rutgers to receive 20% of the royalties. Schatz later sued Rutgers and obtained a substantial settlement. But in 1952, when Waksman was awarded the 1952 Nobel Prize in Physiology or Medicine 'for his discovery of streptomycin, the first antibiotic effective against tuberculosis', no one else shared in the prize – not even Schatz, who did the work at the bench, nor Feldman and Hinshaw, who showed that streptomycin worked against tuberculosis in humans.

Unlike penicillin, which is only active against Gram-positive bacteria, streptomycin is active against a range of both Gram-positive and Gram-negative bacteria, including *Y. pestis*. It stops the tuberculosis bacterium from synthesising protein, by binding to the rRNA [ribosomal RNA] in the ribosome, distorting the ribosomal decoding site and inhibiting the binding of tRNA [transfer RNA]. It has saved millions of lives.

One TB victim for whom it was unsuccessful was the writer George Orwell (1903–1950), who had lung problems, including bronchitis for most of his life. His tuberculosis was diagnosed in 1938, during pre-streptomycin days. In early 1948, he bought some streptomycin from the USA (using royalties from his new book *Animal Farm*) but suffered an allergic reaction, so treatment was discontinued. He managed to finish his last book, *1984*, before taking permanently to his bed. Attempts at treatment with another new TB drug, p-aminosalicylic acid, were also unsuccessful, as was a second trial of streptomycin.

George Orwell died on 21 January 1950. The actress Vivien Leigh (d. 1967), star of the 1939 Hollywood blockbuster film *Gone with the Wind*, survived over 20 years of tuberculosis, thanks to streptomycin and other drugs. But survival rates increased. Beatle Ringo Starr spent two years in hospital under treatment for TB between the ages of 13 and 15. At the time of writing, he was a very healthy 86-year-old.

Over time, as with other drugs, bacteria acquired resistance to streptomycin. Fortunately, other molecules have been found to be active against TB. Streptomycin is there as a possible back-up, but usually multi-drug treatments are used, which one can think of as working on the principle that if one drug does not 'get' the bacterium, another one will. The most common of these treatments came into use in the 1970s, employing rifampicin, isoniazid, pyrazinamide, and ethambutol, which led to it being known as RIPE.

However, multi-drug-resistant TB has increasingly become a threat. This is in part due to more mobile populations – not least due to air travel. Another cause is that some patients do not complete their course of antibiotics, so although they are partly cured, they still have the bacterium in their bodies. TB has been a particular threat to AIDS victims, due to their weakened immune systems.

Rifampicin (37.2), another soil-derived antibiotic (first isolated from soil collected from a pine forest in the French Riviera), came into use in 1968. It works by blocking RNA synthesis.

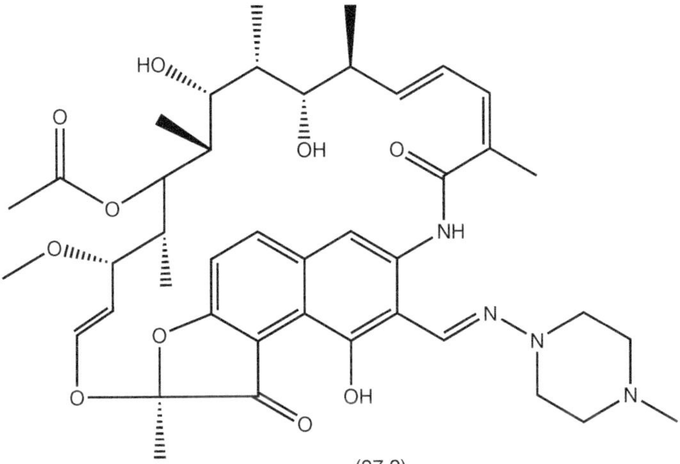

(37.2)

Another simple molecule that is used is *para*-aminosalicylic acid (4-aminosalicylic acid; (37.3)). Its structure is close to that of 4-aminobenzoic acid (aka *para*-aminobenzoic acid, PABA), a molecule that bacteria use in the synthesis of folic acid. For this reason, it was long thought that *para*-aminosalicylic acid had a similar mode of action to the sulfonamide family, but it is currently believed that *para*-aminosalicylic acid interferes with the folate pathway through the formation of molecules toxic to the bacterium.

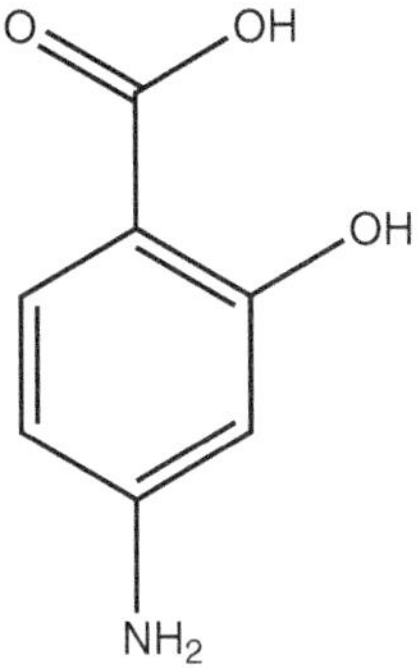

(37.3)

Another of these medications is isonicotinic acid hydrazide ((37.4): isoniazid). This inhibits cell wall synthesis. It is a prodrug, and one active molecule it produces is NO, as does the more recently discovered Pretomanid (formerly known as PA-824), which received FDA approval in 2019. Pretomanid (37.5) is used in combination with other drugs, notably linezolid (37.6) and bedaquiline (37.7). Linzeolid (Zyvox) is a member of the oxazolidinone family of antibacterial agents, which was originally developed to counter infections such as MRSA (the 'flesh-eating bug'), but which is sometimes used in the treatment of tuberculosis.

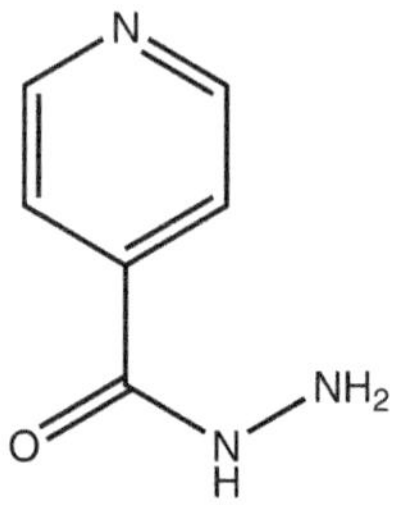

(37.4)

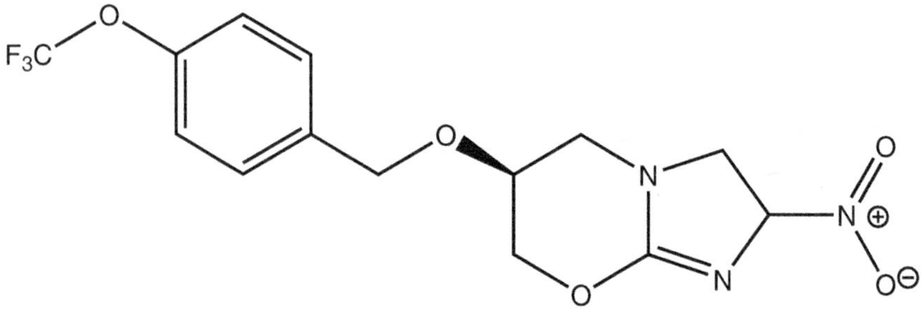

(37.5)

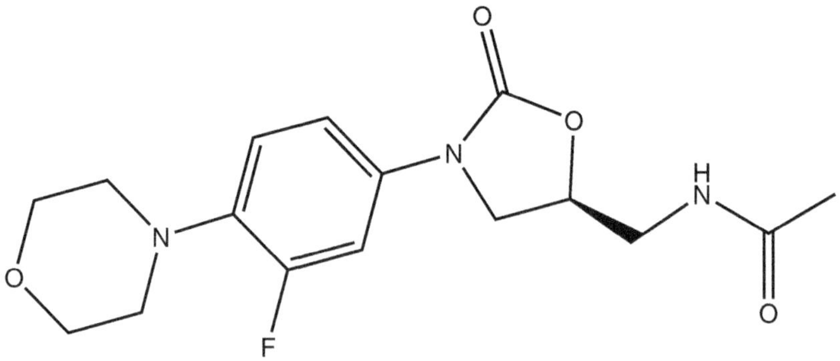

(37.6)

Two unusual molecules have come on to the scene to treat TB in the past decade, bedaquiline (37.7) and delamanid (37.8). Bedaquiline works by targeting the gamma subunit (ATP5F1C) of the proton pump for ATP synthase in the bacterium, a different target from other anti-TB drugs in the armoury, as it inhibits mitochondrial ATP production.

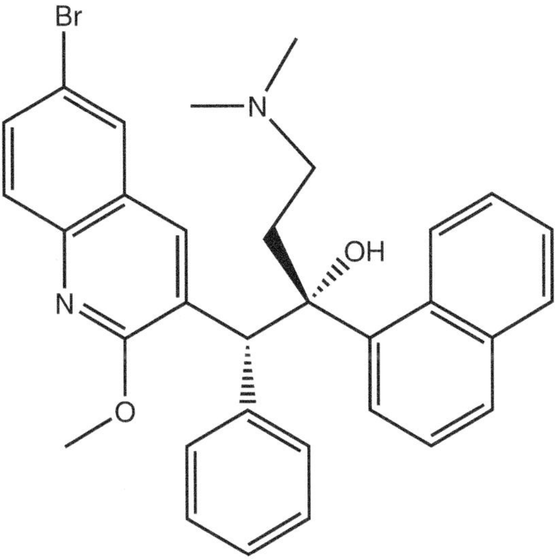

(37.7)

Delamanid works by inhibiting mycolic acid synthesis; its trials indicate improved outcomes in patients with multidrug-resistant TB, which is most encouraging.

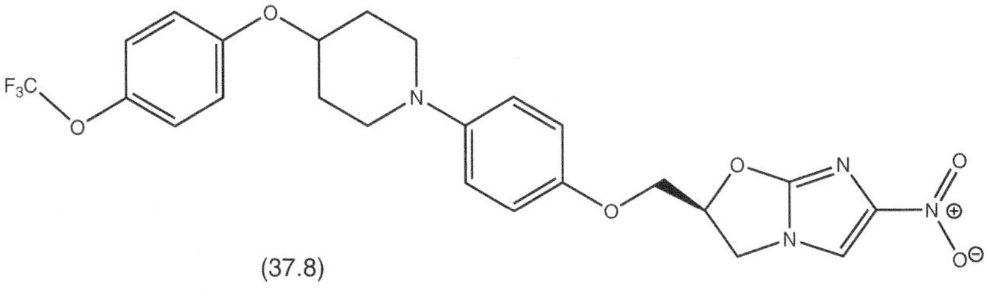

(37.8)

Indeed, many drugs are now known that target the bacterium in different ways, so that the use of a cocktail of many of these offers real hope.

38 Sulfanilamide and Other Sulfa Drugs

It is hard to realise how, less than a century ago, people could contract fatal infections in such trivial ways. After childbirth, women were especially vulnerable to puerperal sepsis ('childbed fever'). In June 1924, 16-year-old Calvin Coolidge Jr., son of the President of the United States, was playing tennis one day when he contracted a blister on the toe of his right foot. Septicaemia set in and nothing could be done to save him; he was dead in a week. In August 1930, W.W. ('Dodger') Whysall, a Nottinghamshire and England Test cricketer, was playing in the deciding Ashes Test match at the Oval. Three months later, he slipped on a dance floor, grazed his elbow, and died of septicaemia a fortnight later. However, six years later, when Franklin Roosevelt Jr., the son of another President of the United States (Franklin Roosevelt), fell ill with a septic sore throat, he made a full recovery after being injected with a new drug called Prontosil. And that was credited to a German scientist named Gerhard Domagk.

In 1914, Gerhard Domagk (1895–1964) was a medical student. He enlisted as soon as war broke out and that autumn he fought at the First Battle of Ypres. Being transferred to the Eastern Front, he wounded his leg. After convalescence, he went back to the front as a medical assistant in a mobile field hospital on the Russian Front. He did not forget how easily wounded soldiers contracted the bacterial infection 'gas gangrene' and swore that he would do all he could to cure such infections. After the war, he returned to his studies at university in Kiel, graduating in 1921. Moving into research, he ran a laboratory at the University of Munster, then moved to a laboratory of experimental pathology at I.G. Farbenindustrie, Elberfeld. He was investigating molecules that were effective against streptococcal infections and studying the use of azo-dyes as chemotherapeutic agents, in a systematic study of hundreds of compounds. In late 1932, after five years' work, Domagk found that a red dye 'labelled' KL730, known later as Prontosil (38.1), worked well against streptococcal infections in laboratory mice (in vivo), even at very low doses, though it did not work in test-tube tests against the bacterium (in vitro). I. G. Farben applied for a patent on Christmas 1932. Tests on humans with streptococcal infections – for which there was no other treatment – went well, and the drug was patented by the end of 1934 and came into use.

On 2 January 1935, the German patent on KL730 was granted. After the secret clinical trials, the drug was released in the market, and finally Domagk published his first paper on Prontosil in February 1935. Tests were carried out widely across Europe between 1935 and 1936, but Prontosil did not reach the USA until well into 1936. The most important British research on Prontosil was carried out in early 1936 by Leonard Colebrook, together with Méave Kenny, a colleague at Queen Charlotte's Hospital, on behalf of the Therapeutic Trials Committee of the Medical Research

DOI: 10.1201/9781003783565-38

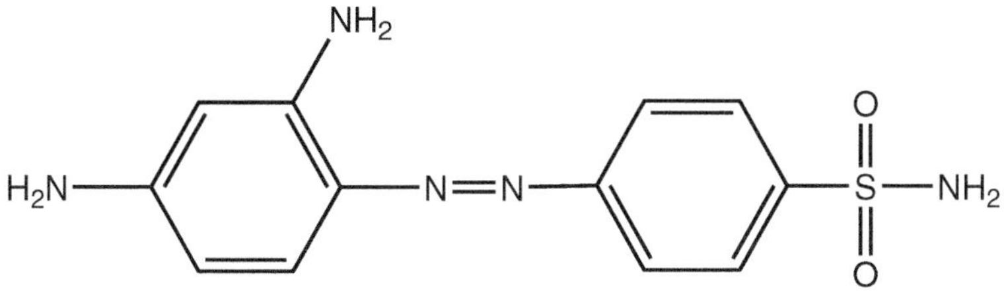

(38.1)

Council. Tests were first carried out on mice, then on patients suffering from puerperal sepsis. In comparison with earlier treatments, the mortality rate fell from over 20% to 4%. Meanwhile, back in Germany, 6-year-old Hildegarde Domagk fell from the stairs at home on 4 December 1935. She was carrying a large needle, which was driven into her hand up to the wrist, and then it broke off. The needle was removed by a surgeon, but a streptococcal infection set in, which spread through her arm. It seemed that amputation of the arm might be her only chance of survival, but Gerhard Domagk administered his daughter two successive treatments of Prontosil. Her temperature fell back to normal and the infection cleared up. A year later, the successful use of Prontosil on Franklin Roosevelt Jr. publicised the potential of this wonder drug to the USA.

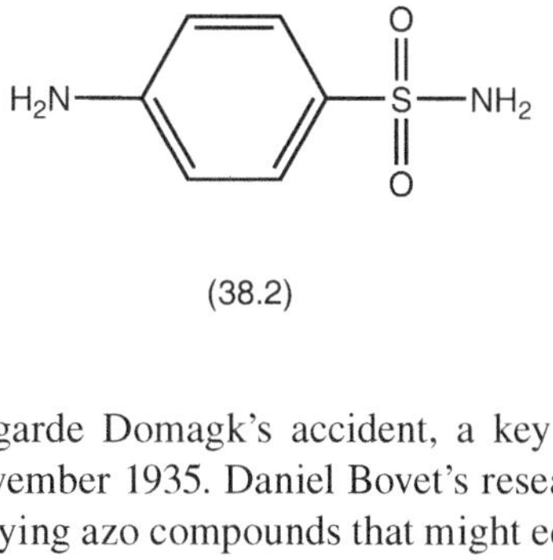

(38.2)

By the time of Hildegarde Domagk's accident, a key discovery was made in France, during 6 to 8 November 1935. Daniel Bovet's research group at the Pasteur Institute in Paris was studying azo compounds that might equal or surpass Prontosil. Bovet had the idea of studying a compound common to all the dyes they were examining, including sulfanilamide (4-aminobenzenesulfonamide). They found that sulfanilamide (38.2) was active both in vitro and in vivo. Prontosil was no longer the last word. Bovet's team had discovered that, in the human body, intestinal enzymes were converting Prontosil into sulfanilamide, the active molecule responsible for its activity when Prontosil was used as a medicine. In other words, Prontosil was a prodrug. There were no enzymes in the bacterial cultures upon which Prontosil had been tested by Domagk, which is why Prontosil was inactive in vitro.

Sulfanilamide had advantages over Prontosil. In one hand, the skin of patients was not turned red by the use of sulfanilamide. On the other hand, sulfanilamide was much cheaper when compared with Prontosil, and also a well-established compound. It had first been synthesised in Vienna in 1908 by the chemist Paul Gelmo, during his PhD research. It was patented in 1909. Therefore, the patent had expired by the time that Bovet's team made their discovery. Bovet later received the 1957 Nobel Prize for Physiology or Medicine.

Pharmaceutical manufacturers switched to using sulfanilamide, for which there is a straightforward synthesis of sulfanilamide from aniline (Figure 38.1). First aniline is acetylated to protect the easily oxidised amine group; next, sulfonated using chlorosulfonic acid; then ammonia substitutes the chloride, generating the sulfonamide group; finally boiling with dil. HCl (aq.) removes the protecting acetyl group, which is more easily hydrolysed than the sulfonamide group.

Sulfanilamide was very successful, but soon after its introduction there was a serious crisis. There was demand for an easily assimilated form of sulfanilamide for treating children. Hence, an American scientist working for a pharmaceutical firm

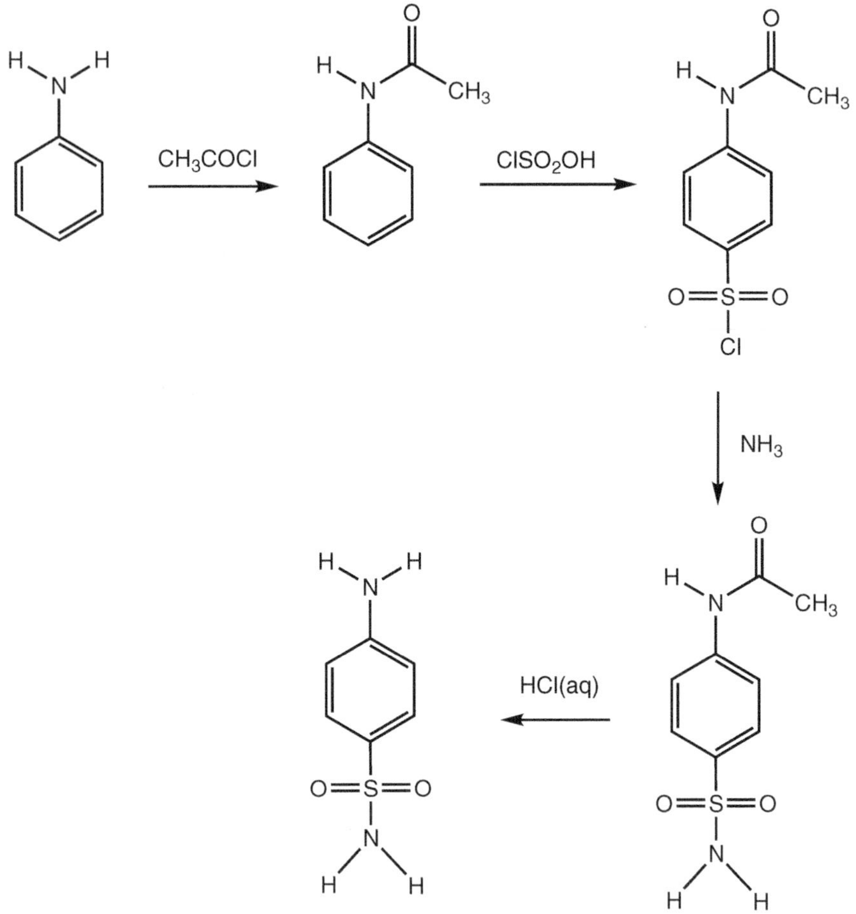

FIGURE 38.1 Synthesis of sulfanilamide from aniline.

created what was called an 'elixir of sulfanilamide' that would appeal to children. He dissolved sulfanilamide in the solvent diethylene glycol, then added a dye and raspberry fragrance. This 'elixir' was put on the market as a treatment for streptococcal infections of the throat. Neither he nor anyone else did any test on animals or examined the literature to check the safety of the chemicals they used. Diethylene glycol is toxic in the quantities used and there were 107 fatalities in people who used this 'elixir', mainly children.

Two important lessons were learned because of this. A firestorm of public outrage led to the passing of the 1938 Food, Drug & Cosmetic Act, which importantly mandated testing of new drugs before their introduction, with the power to demand for extra tests to be carried out. Ultimately, they had the power to prevent a drug from being introduced. The second lesson came to fruition in 1960. In 1938, Frances Oldham Kelsey, a 23-year-old PhD student, carried out animal tests on the 'elixir of sulfanilamide', showing that it was the diethylene glycol solvent (sweet but very toxic), that was responsible for the poisonings. She remembered the lessons of this case when, in 1960, she worked for the Food and Drug Administration in Washington and was asked to evaluate thalidomide (page 195).

The Nobel Prize Committee awarded Gerhard Domagk the prize in Medicine in 1939 for his discovery of the antibacterial properties of Prontosil. Now, when the 1936 Nobel Peace Prize went to the pacifist (and anti-Nazi) Carl von Ossietzky, Hitler forbade Germans to accept Nobel Prizes. After Domagk heard the news of his prize on 26 October 1939, he sent a guarded acceptance of the award to the Nobel Committee on 3 November, in case Hitler's veto only applied to the Nobel Peace Prize. Two weeks later, on 17 November, armed Gestapo surrounded Domagk's house and arrested him; he spent a week in gaol and was made to sign a letter turning down the award, which he eventually received in 1947.

How sulfanilamide works as an antibacterial was discovered in 1940 by Donald Woods, an Oxford University researcher. He discovered that para-aminobenzoic acid (PABA) blocked the antibacterial effects of sulfanilamide, noting that these two molecules had very similar structures. Bacteria use PABA in the synthesis of dihydrofolic acid, which is then used to make other molecules, including the purine bases necessary to construct DNA; PABA is thus absolutely vital to the bacteria's survival. More specifically, the enzyme dihydropteroate synthetase catalyses one step in the synthesis of dihydrofolic acid. Now sulfanilamide fits the active site of this enzyme, blocking the site to PABA, so that the bacterium cannot multiply and grow. This gives the immune system the time to clear up the infection.

Humans do not make dihydrofolic acid, unlike bacteria; instead, they take folic acid from their diet as a vitamin, obtained from green vegetables. Therefore, we do not contain any dihydropteroate synthetase, so that is why sulfanilamide is safe for humans.

Following on the success of sulfanilamide, chemists naturally made other sulfonamides in order to discover better medicines. The most famous of these is M&B 693 (sulfapyridine), synthesised at the Dagenham (Essex) laboratories of May and Baker (aka M&B). Chemists, including George Newbery and Montague Phillips, as part of a research team directed by Arthur Ewins, were making large numbers of sulfonamides for testing. Fortuitously, one day in October 1937, someone noticed a bottle of 2-aminopyridine on a shelf in the lab and decided to make its

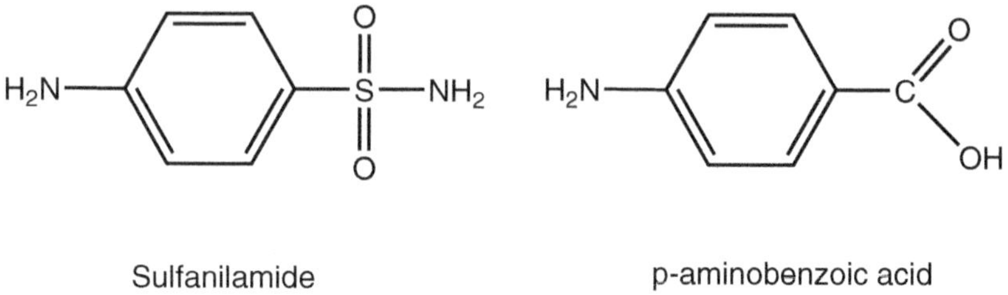

FIGURE 38.2 A comparison of the structures of sulfanilamide and para-aminobenzoic acid.

sulfonamide derivative, so that what became known as M&B 693 was synthesised on 2 November 1937. M&B 693 was tested upon pneumococcal and streptococcal mice by Lionel Whitby (later Sir Lionel Whitby) at the Middlesex Hospital. The trial produced very encouraging results, so the drug was moved on to human trials. These were carried out by G. M. Evans and Wilfrid Gaisford, at Dudley Road Hospital, Birmingham, in which 100 patients with lobar pneumonia were treated with M&B 693; the mortality rate was reduced from 78% to 8%. So sulfapyridine (38.3) went on the market on October 1938, with May and Baker licensing Merck to make it in the USA. It became the standard treatment for pneumonia and was credited with saving the lives of 33,000 victims of pneumonia annually in the United States alone.

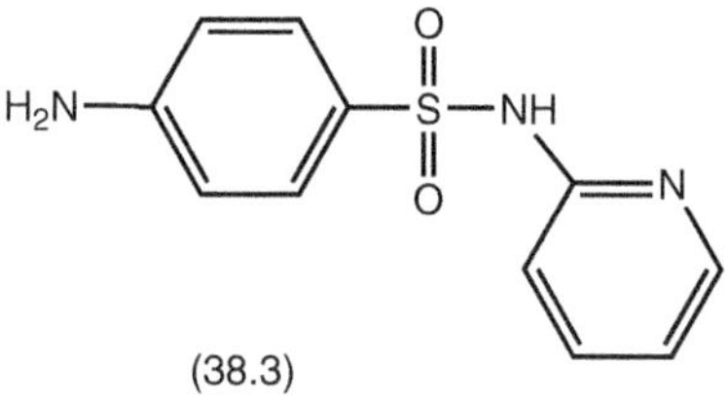

(38.3)

The most famous of the lives saved from pneumonia is probably that of Winston Churchill, the British Prime Minister. Between 28 November and 1 December 1943, he met the President of the United States, Franklin Delano Roosevelt, and the General Secretary of the Central Committee of the Communist Party of the Soviet Union, Joseph Stalin, at the Teheran conference, concerned with finalising the Allies' strategy for the war against Nazi Germany. After this conference, he spent an hour sitting on an airfield in a cold wind on his way to General Eisenhower's villa near Carthage. After this, he complained of a sore throat. Next day, his temperature rose to 101 °C. A portable X-ray machine revealed a shadow on the lung; it was pneumonia. Churchill was given tablets simply referred to as M&B which cleared up the pneumonia. It is not now clear whether the drug used was either M&B 693 (sulphapyridine) or M&B 760 (sulfathiazole), though it is more likely to have been the former. Churchill issued a communiqué saying:

> Excellent nurses and the highest medical authorities in the Mediterranean arrived from all quarters as if by magic. This admirable M & B, from which I did not suffer any inconvenience, was used at the earliest moment and after a week's fever the intruders were repulsed.

Other drugs such as sulfathiazole ((38.4): M&B 760) and sulfadiazine (38.5) followed in the wake of M&B 693. The silver salt of sulfadiazine has been used in burn creams.

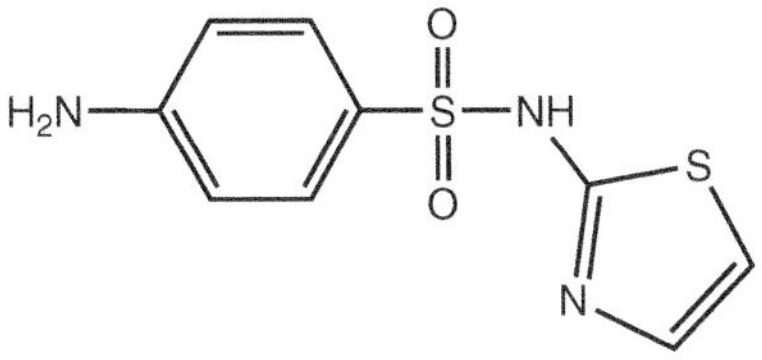

(38.4)

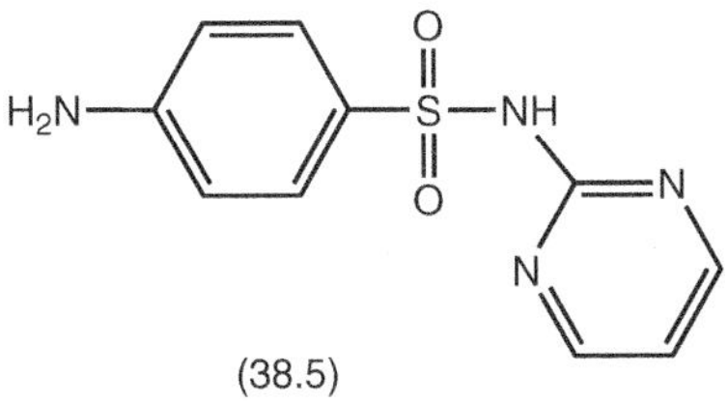

(38.5)

During World War II, each G. I. carried a first aid pouch containing a powdered sulfa drug, often sulfanilamide, and a dressing bandage. Whenever anyone was wounded and no medic was at hand, the sulfa powder was sprinkled into the wound. In fact, even the medics carried sulfa drugs.

Today, sulfa drugs are not used as much as they were in during the mid-twentieth century. Some bacteria are now immune to them, whilst there is the possibility of them causing liver damage and certain allergic reactions, including Stevens Johnson syndrome. On the other hand, some newer molecules have been synthesised, and these tend to be used to treat infections of the urinary tract and also certain sexually transmitted infections like Chlamydia.

Although more restricted than once was the case, a mixture of sulfamethoxazole and trimethoprim, known as Co-trimoxazole, has been used to treat a variety of infections. It has been recommended for HIV-exposed infants and children in certain situations. It can be an effective broad-spectrum antibiotic against a variety of illnesses since sulfamethoxazole (38.6) and trimethoprim (38.7) target successive steps in the biosynthesis of tetrahydrofolic acid (Figure 38.3). Hence, lower doses of each molecule can be used in the drug. It's a kind of double-whammy treatment, but concerns about side-effects mean that it sees limited use.

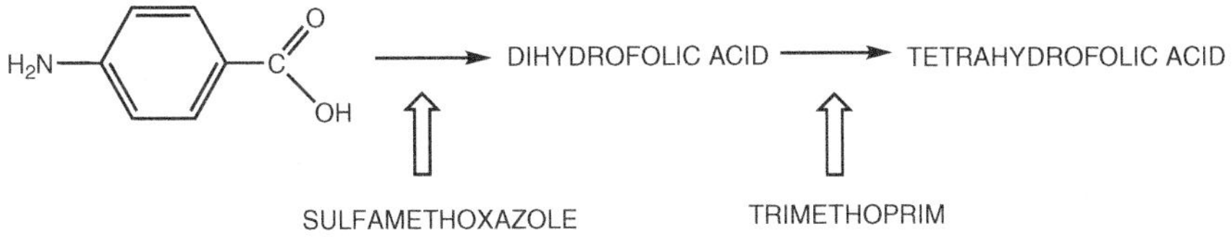

FIGURE 38.3 Sulfamethoxazole and trimethoprim target successive steps in the biosynthesis of tetrahydrofolic acid.

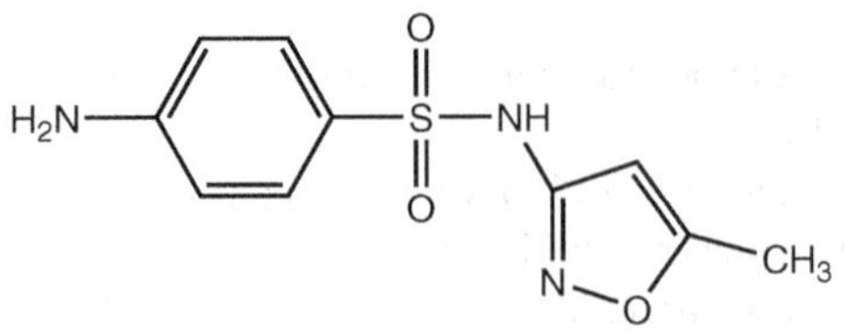

(38.6)

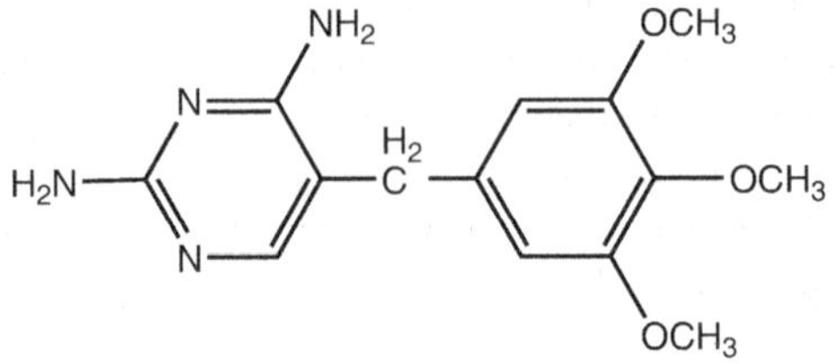

(38.7)

Sulfa drugs were first used as antimalarials in the 1930s but were superseded by other agents like chloroquine (see pp. 94–95 – now in its turn less effective). A combination of sulfadoxime (38.8) and pyrimethamine (Fansidar ®) has been used to treat chloroquine-resistant *Plasmodium falciparium* malaria, similarly, targeting the two enzymes in the synthesis of tetrahydrofolic acid. Side-effects mean that it is only employed for severe cases, and is not currently used in the UK.

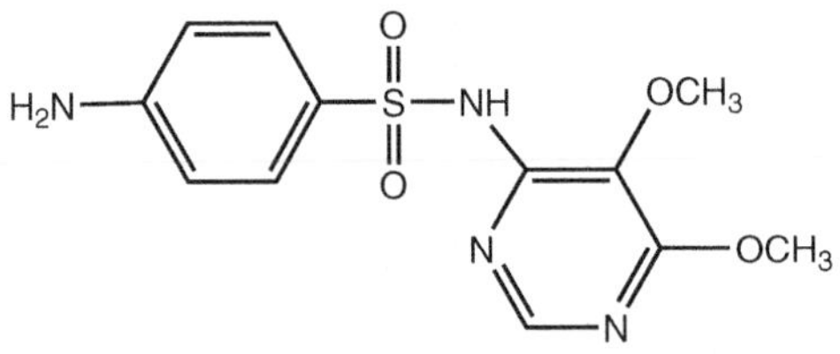

(38.8)

Historically, it is important to remember that sulfonamides were the first widely used and successful examples of chemotherapy – synthetic chemicals used to fight diseases. They were also the first example of the use of a competitive enzyme inhibitor as a drug.

39 Taxol

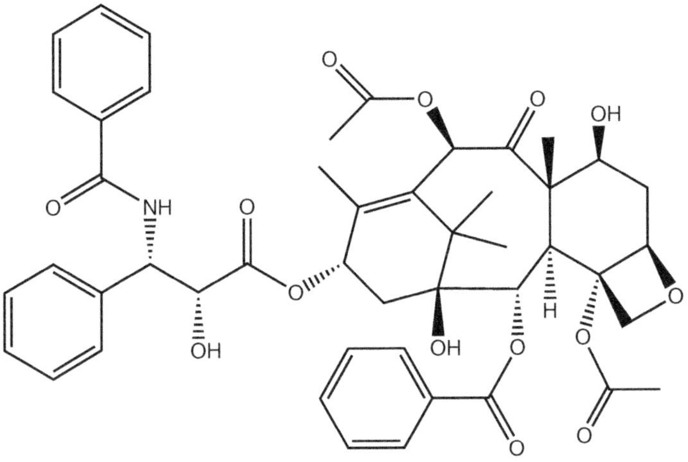

(39.1)

Over time, many diseases have become treatable, or even died out, thanks to medicines and vaccination. But cancer remains, it still carries the mark of what is often perceived to be an utterly deadly disease.

In the first part of the twentieth century, a number of breakthroughs in medicinal chemistry took place which involved medicines discovered from natural sources, prominent examples being penicillin, vancomycin and the tetracyclines. Thus, the National Cancer Institute (NCI) in Washington, USA, together with the US Department of Agriculture (USDA) devised a programme under Dr Jonathan L. Hartwell (who had a background in natural products and wrote a book called *Plants Used Against Cancer*) to screen plants for chemicals as possible anti-cancer agents. This programme ran for nearly 25 years and screened some 114,000 plant extracts. Botanist Dr Arthur Barclay, assisted by three college students, collected some 650 plant samples in August 1962; these samples included some barks, twigs, leaves and fruit from the Western (or Pacific) Yew, *Taxus brevifolia*. Two years later, in 1964, it was discovered by Mansukh C. Wani and Monroe E. Wall, at the Research Triangle Institute in North Carolina, that extracts from the bark killed leukaemia cells, so the search continued for the chemical responsible for this phenomenon. They isolated it a couple of years later, and named it taxol (39.1), on account of its origin in the yew tree, Latin name *taxus*. They assigned its structure in 1971.

Animal testing came next. Taxol was tested (1978) on human tumour cells that were transplanted into mice, followed by encouraging results. Subsequently, Phase 1

DOI: 10.1201/9781003783565-39

clinical tests on humans began in 1981, followed by Phase 2 tests in the mid-1980s, again leading to encouraging results against breast cancer, small-cell lung cancer and ovarian cancer. Taxol received FDA approval for treatment of drug-resistant ovarian cancer in 1992, for breast cancer in 1994 and for non-small cell lung carcinoma in 1999. It should be noted that although taxol was given its name by the scientists who discovered it, in 1992 the pharmaceutical company Bristol-Myers Squibb, who had acquired the exclusive right to develop the drug commercially, managed to (controversially) copyright the name taxol. Hence, taxol should be referred to by the generic name *paclitaxel*. But most chemists continue to refer to it as taxol.

During taxol's development, it was discovered how the compound worked, in what turned out to be an unusual way. Most cancer drugs restrict cell division and growth by preventing the tubulin proteins from assembling into microtubules, thereby stopping cell growth (and division). Dr Susan Horwitz led the investigation where it was discovered that taxol works in a different way. By binding to β-tubulin, which causes cell division, taxol promotes growth of stable microtubules and thus inhibits mitosis and the growth of the cancer.

Testing took some time. Taxol's low solubility in most solvents was one of the reasons; in the end, a mixture of ethanol and a surfactant (Cremophor EL) was found suitable for this purpose. Then, one of the first patients who received taxol in a trial died of an anaphylactic shock, so it took time to find a better way of administering the drug.

Another major environmental issue involved the Pacific yew tree, source of taxol. Removing bark to extract it for taxol kills the tree. Moreover, the bark from a complete tree contains about 350 mg of taxol, enough for one dose. However, around 3 g of taxol was needed for a complete course of treatment. Thus, half a dozen mature trees would have to be sacrificed per patient. And the Pacific yew tree is very slow growing, taking around 200 years to reach maturity. Adding to this, the Pacific yew tree was home to the endangered Northern spotted owl (*Strix occidentalis*). On the one hand were environmentalists; on the other, cancer patients.

In the end, the solution to this problem was discovered on the other side of the Atlantic, in France. Pierre Potier (1934–2006) was the director of the French National Centre of Scientific Research (CNRS) at Gif-sur-Yvette, 15 miles southwest of Paris. Potier was celebrated for looking for medicines in nature, which he termed *le magasin du bon dieu* – the Good Lord's shop. And Potier had a particular interest in anti-cancer drugs, as his first wife, Marie-France, had died from breast cancer in 1968. He had already developed vinorelbine (Navelbine), a semi-synthetic relative of the anti-cancer vinca alkaloid medicines – vincristine and vinblastine – obtained from the rosy periwinkle (*Catharantus roseus*).

At the end of 1979, a new road was being constructed across the campus at Gif, where some European yew trees *Taxus baccata* L. were in the way and so they had to be cut down. Potier decided to screen parts of the trees – leaves, roots, bark and heartwood – for taxol-like content. In the needles, the team discovered a compound called 10-deacetylbaccatin III (39.2), which contained the complicated taxane core. Not only was the European yew fast-growing, but its needles were a renewable resource, which could be harvested without killing the tree. The availability of the ready-made

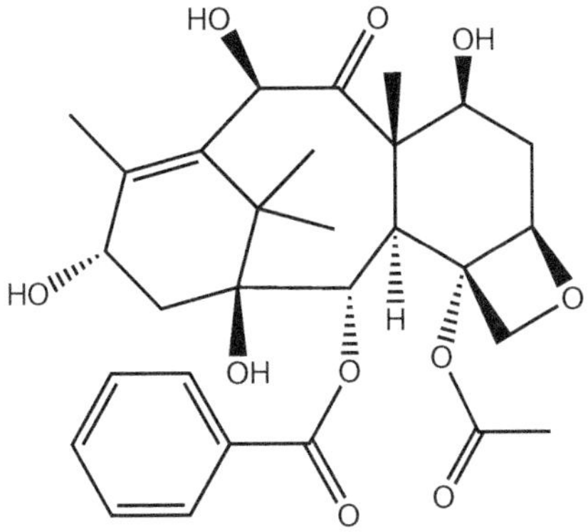

(39.2)

taxane core (which is difficult to make in the laboratory) enabled the synthesis of taxol in just four steps, by attaching the requisite side-chain to it. This process is known as semisynthesis, where nature is used to 'do the hard work', while the finishing touches are applied in the laboratory. Thus, taxol could be produced in increased quantity, to meet the expanding demand for the drug. A number of chemists have come up with ingenious multi-stage syntheses of taxol, the shortest of these consists of 37 stages with 0.4% overall yield. This makes the total synthesis very expensive, and therefore it does not compete economically with the four step semisynthetic route starting from 10-deacetylbaccatin III. Taxol is now mainly manufactured by *Taxus* cell culture methods and by semisynthesis from deacetylbaccatin III.

But this is not all. Potier's team came up with a semisynthetic analogue of taxol, differing in two peripheral groups. This molecule is known as taxotere (docetaxel) and is readily synthesised from deacetylbaccatin III. It has advantages over taxol, notably being more soluble. Taxotere (39.3) was approved for use in 1995.

The yew tree produces taxol differently from chemists in a laboratory, of course, by using a series of enzyme-controlled reactions. Of particular interest is the process responsible for creating the complicated ring system that forms the core of the taxol molecule by converting geranylgeranyl diphosphate into taxa-4,11-diene (Figure 39.1). This cyclization reaction is carried out by the enzyme taxadiene synthase, whose structure was reported in January 2011.

Once the 'taxol core' is formed, then other enzyme-catalysed steps take place, involving stages such as hydroxylation, acylation, adding a side-chain and the formation of the four-membered oxetane ring. The taxane oxetanase enzyme responsible for the last-named step was only recently identified and reported (2024–2025). Understanding the whole biochemical pathway has taken over 30 years.

The yew tree has long been associated with death, indeed the taxol alkaloids from it have been known to be a poison. In book VI of his Gallic Wars, Julius Caesar

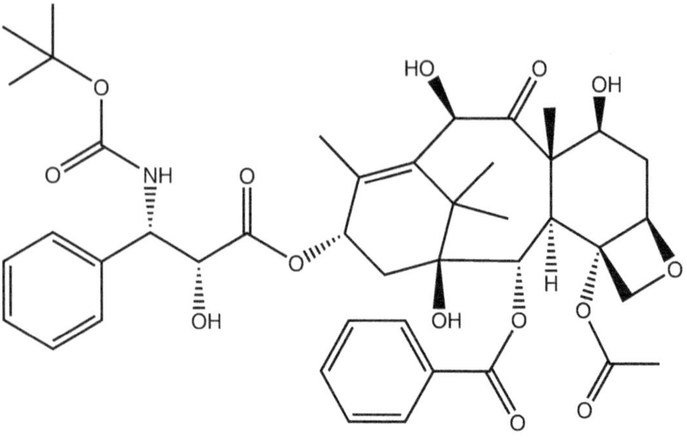

(39.3)

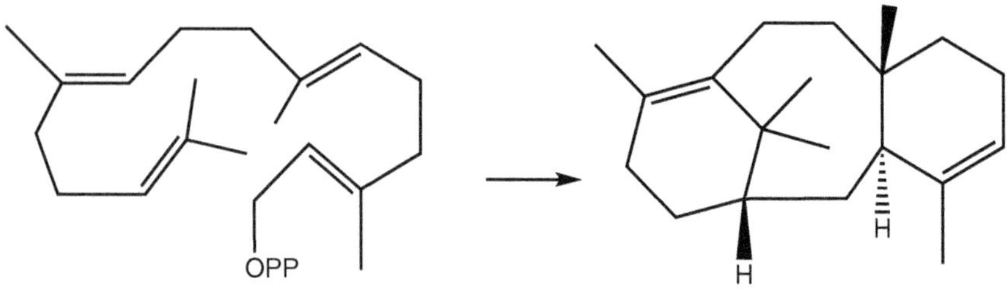

Geranylgeranyl diphosphate Taxa-4(5),11(12)-diene

FIGURE 39.1 Conversion of geranylgeranyl diphosphate into taxa-4,11-diene.

related that a rebel chieftain called Cativolcus committed suicide in 53 BC with the 'juice of the yew tree'. Six hundred years ago, the English archers used bows made of yew to rain down a storm of lethal arrows upon the French troops at the battle of Agincourt. Sadly, people still use it as a means of committing suicide. With taxol, the yew tree has been able to do something very different. Moreover, taxotere gave my friend Maria another two years of her life.

40 Terylene

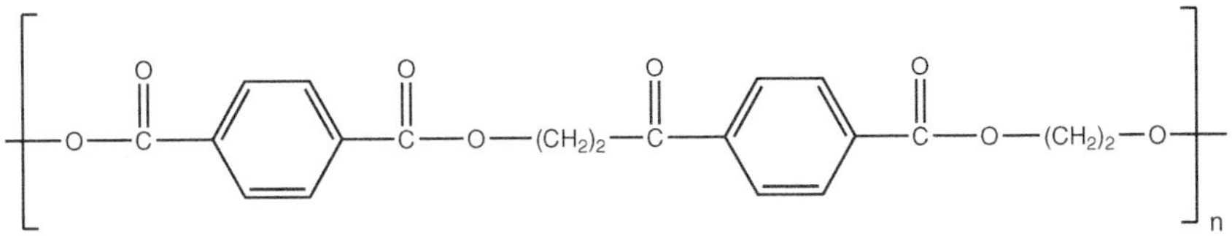

(40.1)

Just as nylon (pp. 121–125) is a polyamide that can be synthesised from a diamine and a diacid, Terylene (40.1) is also a polyester that can be made from a diol and a diacid – ethane-1,2-diol (40.2) and terephthalic acid, benzene-1,4-dicarboxylic acid (40.3) – respectively.

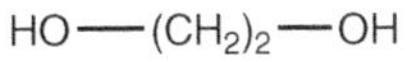

(40.2)

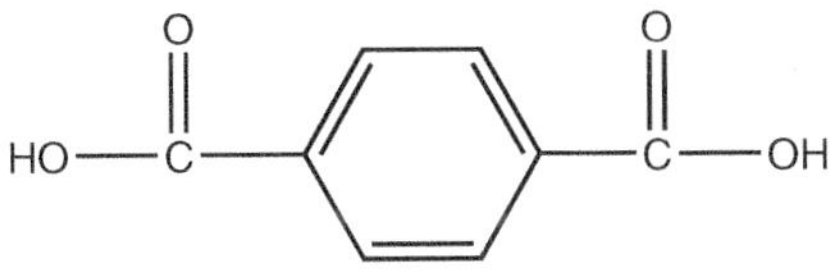

(40.3)

Analogous to what happens in the synthesis of nylon, an alcohol (–OH) group in the diol can react with the carboxylic acid (-COOH) group at one end of the diacid to form an ester linkage, eliminating a water molecule in the process. This ester still has an unreacted –OH group at one end, and a –COOH group at the other end, both of which can undergo further esterification – again with the elimination of a water molecule each time (Fig 40.1) – a process which can continue, eventually resulting in the polyester.

An alternative route involves a transesterification reaction between dimethyl terephthalate (DMT) and ethylene glycol.

PET was not the first polyester to be created – that distinction belongs to Carothers' team in Delaware during the 1930s, around the time they developed nylon (pp. 121–125) – it was nylon that had the properties that they were looking for, so polyesters were placed on the back burner. Poly(ethylene terephthalate), to use the chemical name of this giant molecule, was first developed in 1941 by two British chemists, Rex Whinfield and James Dickson, who worked in the research laboratory of the Calico Printers' Association in Accrington, near Manchester. While the name nylon is used on both sides of the Atlantic, Terylene – a name chosen by ICI (Imperial Chemical Industries) – is very much in use in the British Isles, and Dacron is the

DOI: 10.1201/9781003783565-40

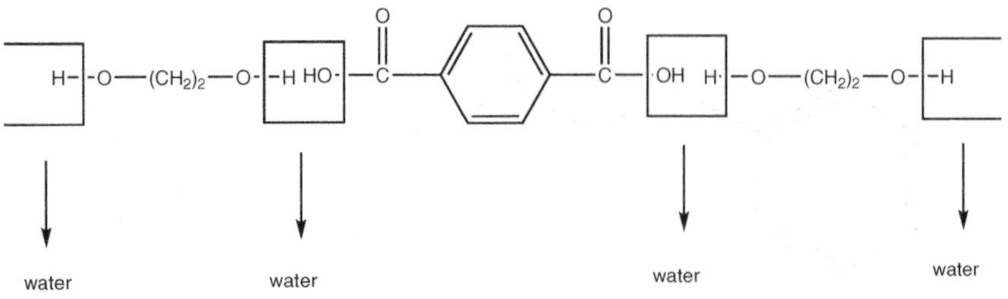

FIGURE 40.1 Synthesis of Poly(ethylene terephthalate).

name universally used for this polymer in the USA, thanks to Du Pont, not to mention Lavsan in the USSR. DuPont created polyester films and trademarked (1952) the name Mylar for these, a word still in use.

Polyesters such as poly(ethylene terephthalate), often abbreviated to PET (for obvious reasons), made an immediate impact, as it had several desirable properties. Polyesters were shrink- proof and could be 'heat-set'. Besides being chemically inert, it mixed well with other fabrics, like wool. This meant that Terylene could be used in low-maintenance drip-dry shirts that did not require ironing ('wash and go'). Crimplene was a crease-resistant 'wash and wear' fabric made of terylene that was popular in the UK during the 1960s. It derived its name from the river Crimple in Yorkshire. In fact, the HQ of ICI Fibres at Wetherby, near Harrogate, bore the name Crimple House. Crimplene found itself used for a variety of fashionable attire in that decade, ranging from shirts popular with Mods to the catsuits worn in *The Avengers* by Emma Peel, played by Diana Rigg.

Today around two-thirds of PET manufacture is used in making synthetic textiles, the remaining third for bottles. These days, we cannot talk about plastics without mentioning recycling. It is reported that 30.5 million metric tonnes (MMT) of poly(ethylene terephthalate) was manufactured in 2019, the downside of which involves the generation of 56 MMT of solid waste and that much of that PET soon gets put out as waste. The use of PET in bottles depends upon their ease of melting and reshaping, which makes their recycling relatively simple (but do not forget the need for sorting and washing, pre-recycling). However, poly(ethylene terephthalate) has many uses and recycling the Teflon found in materials like carpets or polyester-containing clothing is more problematic, so chemical recycling needs to be considered. Unlike poly(alkenes) such as polythene, polyester cannot be converted back to the monomer simply by heating it in the absence of air. However, polyesters such as Terylene, contain the more reactive ester groups within their polymer chains. These groups make depolymerisation easier than in poly(alkenes). As a result, polyesters can be broken down at relatively lower temperatures, for example by hydrolysis. In contrast, poly(alkenes) such as polythene contain no such reactive functional groups and therefore require more severe conditions for depolymerisation. Acid hydrolysis of PET generates terephthalic acid (TPA), which, of course, can be reused. Glycolysis is an older route for PET recycling. This process is a transesterification of PET with excess ethylene glycol (ethane-1,2-diol), usually between 180 °C and 240 °C, and a

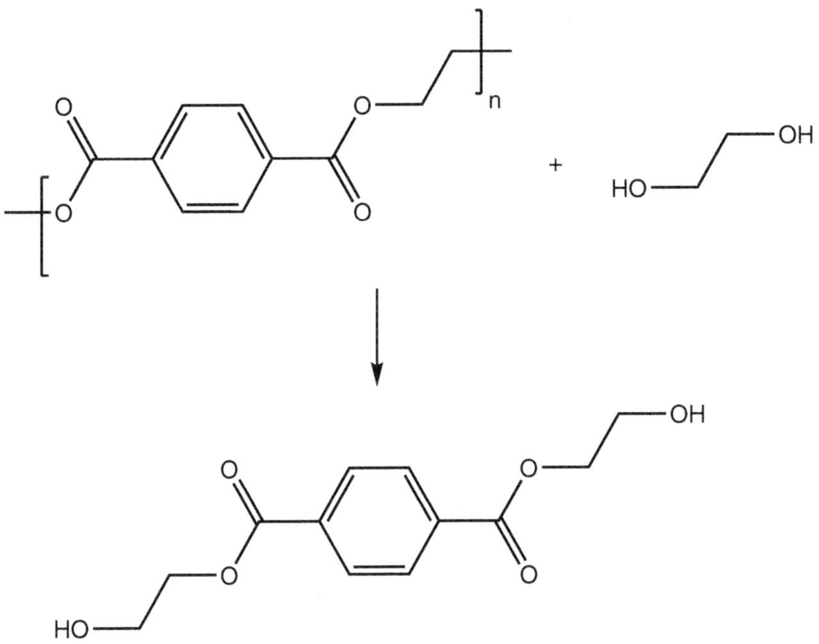

FIGURE 40.2 Depolymerisation of PET by ethane-1,2-diol affording bis(2-hydroxyethyl) terephthalate (BHET).

catalyst, often a metal salt. In this reaction, PET is ultimately depolymerised forming bis(2-hydroxyethyl) terephthalate (BHET), which has two terminal alcohol groups, as well as two ester linkages (Figure 40.2).

Of course, depolymerisation can also be achieved with simpler alcohols, methanol, for example. Such methanolysis of PET is typically carried out under pressure (20–40 atm.) at 180 °–280 °C, often using a catalyst (e.g. zinc acetate), leading to dimethyl terephthalate and ethylene glycol. And scientists have developed ways of turning PET waste into useful substances like epoxy and polyester resins, polyurethanes and corrosion inhibitors. Perhaps the most striking discovery relating to potential PET recycling was reported in 2025 by a group of researchers in the School of Biological Sciences at the University of Edinburgh, with input from AstraZeneca at Macclesfield. They discovered that the Lossen rearrangement, converting hydroxamate esters into isocyanates, and then into amines, which normally requires metal-based catalysts and high temperatures, could be carried out in living cells, with naturally present phosphate acting as a catalyst. They discovered that a strain of the *Escherichia coli* bacterium that was deficient in para-aminobenzoic acid (PABA, a nutrient required for *E. coli* growth) continued to grow when cultured with certain *O*-acyl substituted hydroxamate esters, proving that the Lossen rearrangement was generating PABA. After inserting two genes into *E. coli* –an aminobenzoate hydroxylase and an arylamine N-acyltransferase – it was found that the transformation worked just as well when using PET-derived esters. They converted PABA into

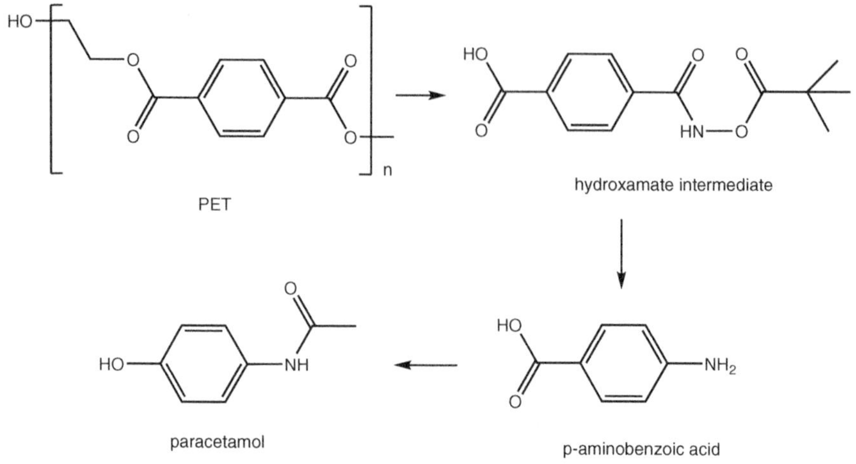

FIGURE 40.3 Sequence of conversion of PET into paracetamol.

paracetamol (known in North America as Tylenol) with a maximum yield of 92%. Overall, they had developed a one-pot room-temperature process using genetically engineered *E. coli* to turn PET into paracetamol (Figure 40.3). No one knows if this could ever become a commercially successful process but it definitely indicates a future direction of research.

41 Tetracycline

Some scientists are so interested in their subject that they never really retire from their jobs and continue thinking about science. Such a man was behind the discovery that led to the family of tetracycline antibiotics.

When Benjamin Duggar (1872–1956) turned 70, despite being a leading plant pathologist, he had to retire from his post at the University of Wisconsin and become an emeritus professor. However, he immediately became a consultant with a New York chemical company and could stay at the lab bench. Soil fungi were Duggar's speciality. In 1943, penicillin was hot property. It was just starting to be produced in large quantities and was being used as an antibiotic to treat casualties in the American armed forces, saving thousands of lives. Duggar knew that penicillin was extracted from a fungus and also that there were thousands of fungi in soil that produce antimicrobial agents to help them compete with other microbes and fungi, giving them an edge, if you like. Therefore, he was determined to obtain soil samples on his travels, which members of his team could screen for antimicrobial agents. Duggar's team struck gold (almost literally) in 1945, when they found a fungus in Missouri soil that made a golden yellow antibiotic, which was introduced commercially in 1948 by the Lederle Laboratories of American Cyanamid. Its colour led it to being given the name Aureomycin, and to the fungus that made it *Streptomyces aureofaciens*. Subsequently, the active molecule in Aureomycin was given the name chlorotetracycline (41.1).

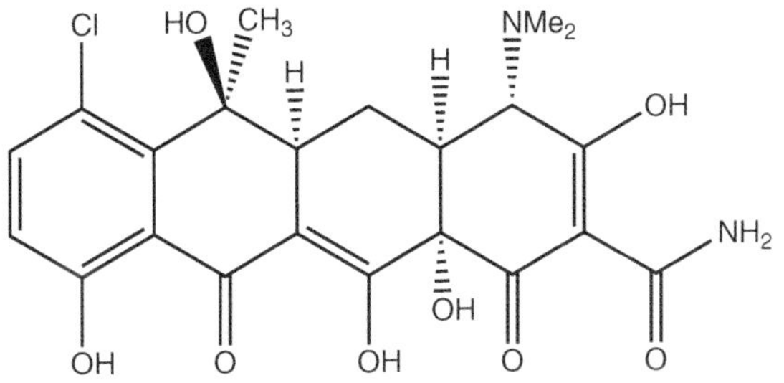

(41.1)

Soon afterwards, research in Pfizer Laboratories led to the isolation of another golden antibiotic, oxytetracycline (41.2), from *Streptomyces rimosus* in the soil near their plant in Indiana. It received approval for use as an antibiotic in 1951 and was given the commercial name Terramycin.

DOI: 10.1201/9781003783565-41

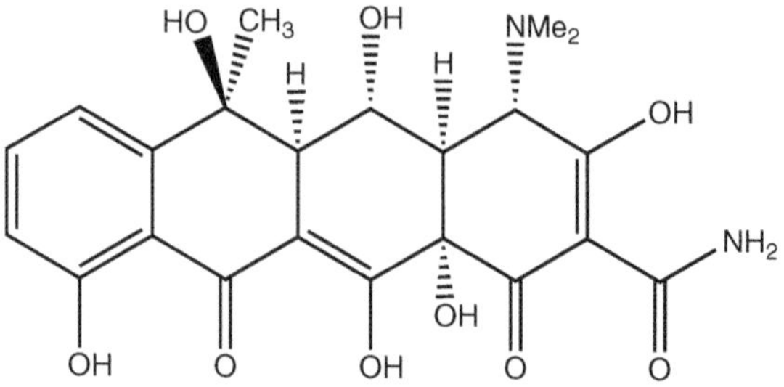

(41.2)

It took a while to determine the molecular structure present in these compounds, which was achieved through a collaboration between Pfizer and the great American organic chemist R. B. Woodward (1953). Famously, D. H. R. Barton, like Woodward, a Nobel Laureate, remarked, 'Woodward took a large piece of cardboard, wrote on it all the facts and, by thought alone, deduced the correct structure for Terramycin. Nobody else could have done that at the time'. The fact that they contained four rings fused together led the family to be known as tetracyclines.

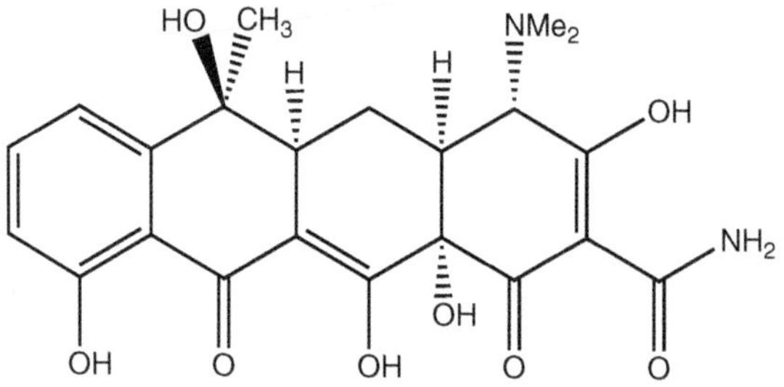

(41.3)

Tetracycline (41.3) itself arrived in 1953. In a selective reaction of chlorotetracycline with hydrogen, the C-Cl bond is replaced by a C-H bond, without either the benzene ring or the other C = C bonds being hydrogenated (Figure 41.1).

These tetracyclines became very widely used drugs, which did some things that the penicillins do not do. To start with, tetracyclines were active against both Gram-positive and Gram-negative bacteria compared to penicillin that only worked on Gram-positive bacteria. Furthermore, unlike penicillins (at that time), tetracyclines could be taken orally. Tetracyclines are effective antibiotics because they selectively inhibit protein synthesis in the pathogens they target. They bind to the 16S part of the 30S subunit of the ribosome of the bacteria and stop the amino-acyl tRNA from binding to the m-RNA-ribososme complex. This means that the polypeptide chain cannot lengthen, stopping bacterial growth.

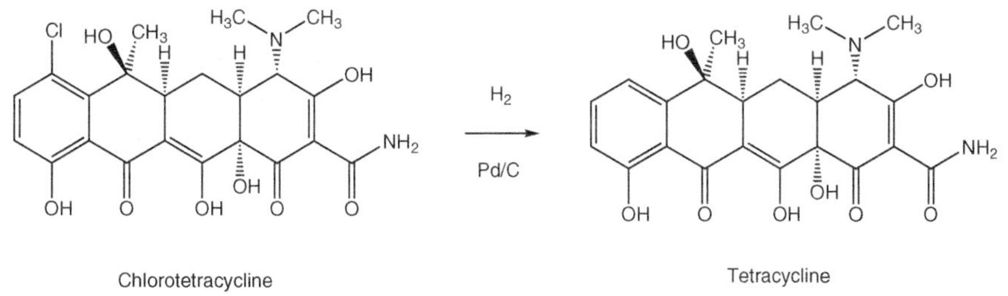

FIGURE 41.1 Catalytic Hydrogenation of chlorotetracycline affords tetracycline.

Unfortunately, tetracyclines became too widely used. Farmers put them into the food of animals like pigs, chicken and cattle. This did not just prevent diseases but also promoted animal growth. This led to the rise of strains of bacteria resistant to tetracyclines.

Tetracycline and the others already mentioned are known as 'first-generation tetracyclines'. Of these, tetracycline is the only one in widespread use today. A striking example of the power of tetracycline was observed in 1994, in the city of Surat on the west coast of India, which was hit by an epidemic of pneumonic plague, caused by the pathogen *Yersinia pestis*. This, of course, was caused by the pathogen responsible for the Justinian plague (AD 541–550) that devastated the Mediterranean region, and later by the Chinese pandemic in 1347, due to *Yersinia pestis* Medievalis, which decimated Europe as the Black Death. The plague in Surat killed over 50 people. But thanks to the United Nations International Children's Emergency Fund (UNICEF) promptitude in supplying more than 11 million tetracycline capsules that stopped the epidemic in its tracks.

A second generation of tetracyclines (notably doxycycline (41.4) and minocycline) came on the scene next, resulting from modifications to the C ring (third ring from the right in the diagrams) in the tetracycline structure, with doxycycline being the best known; developed by Pfizer, it was patented in 1957 and came into general use in 1967. These second-generation drugs were intended to circumvent the resistance problem.

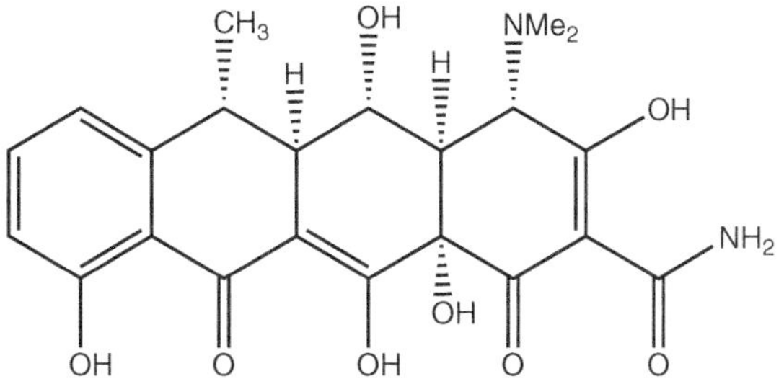

(41.4)

Doxycycline came into public consciousness after the attacks on the Twin Towers in New York on 11 September 2001. In the wake of the attacks, a number of letters containing anthrax spores were sent to two US Senators and to various media outlets. Twenty-two people were infected and five people died of inhalational anthrax. Bruce Edwards Ivins, a scientist at the US government's bio-defence laboratories at Fort Detrick, Maryland, came under suspicion, and committed suicide on 27 July 2008. In the wake of the anthrax attacks, people who were believed to have been exposed were given the drug Cipro (Ciprofloxacin), but doxycycline was recommended as an equally effective (and cheaper) alternative.

Doxycycline is effective against a wide range of diseases, including bacterial infections, infections of the respiratory tract, anthrax, traveller's diarrhoea, sexually transmitted diseases, *Yersinia pestis* (bubonic plague), certain rickettsial infections transmitted by arthropods, Lyme disease and severe acne. It is also being used as a prophylactic against malaria.

The complex ring system of the tetracyclines is not simple to make from scratch, so individual members are often made by semisynthesis, as with this synthesis of the second-generation tetracycline known as minocycline. Starting from sancycline (which itself is a product of semisynthesis), electrophilic nitration of the benzene ring occurs at the 7- and 9-positions. After separating the isomers, the nitro group in the 7-nitro compound can subsequently be converted to the dimethylamino derivative, minocycline.

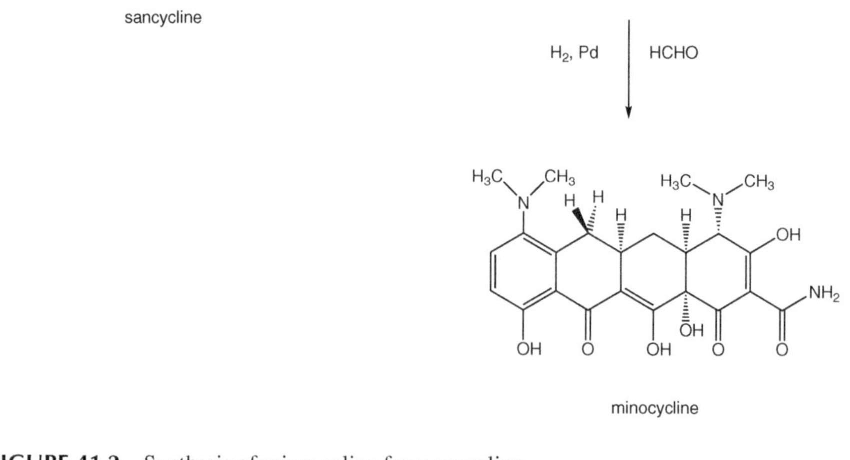

FIGURE 41.2 Synthesis of minocycline from sancyline.

In its turn, minocycline can be used in the synthesis of the third-generation tetracycline known as tigecycline. These types of drugs were first synthesised in the 1990s. If a second nitro- group is introduced at carbon 9 of minocycline, this can be reduced to a primary amine, which in turn undergoes acylation with an acyl chloride, to generate tigecycline. The presence of the glycylcycline side-chain is intended to lead to a reduction in tetracycline-specific efflux and ribosomal protection resistance, resulting in an expanded broad spectrum of antibiotic activity, including activity against resistant microorganisms, notably MRSA (methicillin-resistant staphylococcus aureus). Tigecycline (Tygacil) was approved for use across the EU and the USA in 2005–2006.

Perhaps, the most remarkable thing about the tetracycline family is that one of them, tetracycline itself, was being used as an antibiotic for around a millennium and a half before it was actually isolated. The late George Armelagos was an American anthropologist, based at Emory University. Nearly 50 years ago, he was examining human bones from Nubia, the present-day Sudan. When these bones were exposed to ultraviolet light, they gave out a strong yellow-green fluorescence, typical of tetracycline. Subsequent analysis of the bones by mass spectroscopy confirmed the presence of tetracycline. It appears that tetracycline was imbibed by these ancient Nubians of AD 350–550 – even small children – in their beer, a beer very unlike today's product, more like a fermented gruel, or possibly in their bread. The tetracycline was formed during the storage of the grain in mud bins, where *Streptomyces* soil bacteria could flourish and synthesise tetracycline. This dietary tetracycline made the Nubians very healthy – examination of their bones demonstrated freedom from infections.

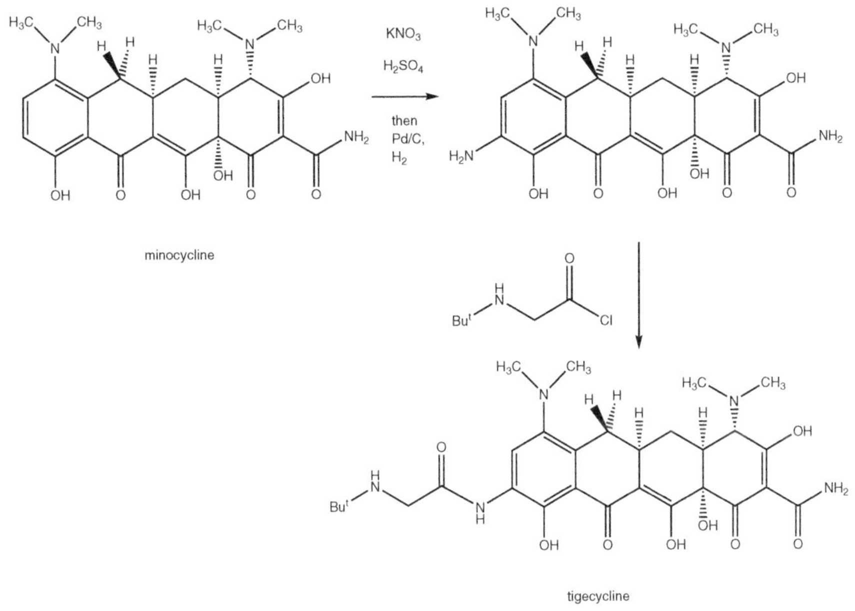

FIGURE 41.3 Synthesis of tigecycline from minocycline.

42 Tetraethyl Lead

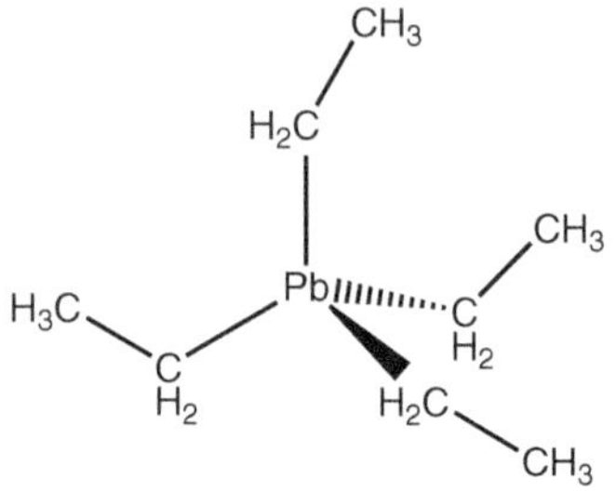

(42.1)

For half a century, tetraethyllead (42.1), also known as lead tetraethyl, $Pb(CH_2CH_3)_4$, was a substance of worldwide importance. However, it sank as quickly as it had risen. Lead was not new. It was a soft and unreactive metal that could be extracted easily from its ores. Since it had been mined from antiquity, the ancient Romans were familiar with it and had many uses for it, whether in pipes, paint, vessel and *sapa*, their sweetener. Later generations found more uses for lead, whether in roofing, in easily cast printing type, or in lead bullets.

The nineteenth century saw the development of chemistry as a scientific discipline, with advances in ways of making molecules and understanding of their formulae. The first organometallic compounds were prepared during this time. Tetraethyllead was one of them. It was first prepared in 1853 by the German chemist Carl Jacob Löwig (1803–1890), after he succeeded Robert Bunsen as the Head of Chemistry at the University of Breslau. Löwig used the reaction between a lead-sodium amalgam and iodoethane, essentially the one to be used by the Ethyl Corporation for its large-scale manufacture in the 1920s.

$$4Pb/Na + 4\ CH_3CH_2I \rightarrow Pb(CH_2CH_3)_4 + 4\ NaI + 3\ Pb$$

Tetraethyllead was not fully characterised until the pioneering English organometallic chemist George Buckton (1818–1905), who also synthesised dimethylmercury (see pp. 61–64), reacted diethylzinc with an excess of lead(II) chloride (1859).

Like many new chemicals, it was prepared for its scientific interest, 'because it could be made'. So, for 60-odd years, tetraethyllead was a molecule without a purpose, until motor cars came on the scene around the beginning of the twentieth century. As cars became more sophisticated, with more powerful engines being developed, problems of pre-ignition and knocking caused by some of the fuel-air mixture not burning at the same time as the rest got worse. Thomas Midgley headed a small team at the Research Division of General Motors Company to investigate

DOI: 10.1201/9781003783565-42

the problem. They examined many compounds until on 9 December 1921, a solution of $Pb(CH_2CH_3)_4$ in kerosene was found to provide knock-free combustion in an engine, even at a level of 0.025%. This was not an immediate solution, as it was soon found that lead-containing residues in the engine created problems. It took more research to establish that if 1,2-dibromoethane was added to the petrol, the lead deposits were removed from the engine as volatile $PbBr_2$, lead bromide.

Public sales of leaded petrol began in 1923, with the tetraethyllead eventually being made from a 1:1 Na: Pb alloy and chloroethane, essentially the method used by Löwig in his original synthesis in 1853. Immediately, problems arose, given the toxicity of $Pb(CH_2CH_3)_4$. Its volatility indicated that the vapour was readily inhaled, whilst the lipid-soluble liquid could be absorbed through the skin. 'Health and Safety' was yet to be invented. Safety precautions were negligible and ventilation standards were poor. There were stories of men carrying around buckets containing tetraethyl lead. Workers at the plants manufacturing tetraethyl lead suffered from delusions and exhibited violent behaviour. Stories reached the newspapers that headlined 'loony gas'. There were five fatalities reported in the autumn of 1924 at a Standard Oil plant in New Jersey, and some others elsewhere. Over 300 men became psychotic, as Herbert Needleman later pointed out – and Midgley suffered from lead poisoning and had to have a month's holiday in Florida in February 1923 to recover – not surprising, as he had stuck his hand in neat tetraethyllead for several minutes. About a year later, he repeated this at a press conference, designed to convince the reporters that tetraethyllead was quite safe – pouring tetraethyllead over his hands.

On 20 May 1925, a day conference called by the US Surgeon General took place in Washington D.C. The morning session was taken up by advocates of leaded petrol, employees of the Ethyl Corporation or people they had sponsored, assuring the audience of its safety, and this continued into the afternoon session. Robert Kehoe, Professor of Pathology at the University of Cincinnati and Director of the Kettering Laboratory at the University of Cincinnati, as well as Medical Director of the Ethyl Corp. and a corporate officer at GM, spoke up on behalf of tetraethyllead. It might reasonably be assumed that he had a vested interest.

Alice Hamilton was not called to speak until around 16.30 that day, well down the afternoon. Alice Hamilton (1869–1970) was a professor at Harvard University and regarded as the leading expert on industrial toxicology and environmental health in the United States, someone who argued strongly against the adoption of tetraethyllead. She pointed out that lead was a slow-acting, cumulative poison. She said that she was unwilling to believe that tetraethyllead was the only substance that could be used as an antiknock agent. She was quite correct in this. In 1920–1921, General Motors researchers had discovered that benzene and ethanol could be used as antiknock agents, but they could not be monetarised as could tetraethyllead. Hugh S. Cumming, the US Surgeon General, who had suspended sales of leaded petrol throughout the USA in May 1925, chose a committee of seven health experts to decide whether leaded petrol was safe to use. They had until 1 January 1926 to make their report, which is just seven months. But it took a month for the members of the committee to be appointed; another month for the committee to design their study; three more months for them to get access to the garages and industrial plants; and more time to arrange for insurance (They are looking into rather toxic chemicals, remember?). Thus, it was October until the investigation started. They examined 252 men, some who used lead-containing petrol and some who did not as control groups. All had lead in their bodies – after all, lead was routinely present in water pipes and paint at that time. In fact, some people working with tetraethyllead-containing petrol had more lead in their bodies. The committee concluded that Ethyl gasoline showed signs of danger, but not enough to recommend a ban, thanks to the investigation being carried out with an impossible timescale. They had insufficient time to recommend that. Ethyl Corporation's expensive PR men had won and therefore, in May 1926, leaded fuels were back on the market. The very similar tetramethyllead (42.2) was also used as an 'anti-knock' additive, but to nowhere near the same extent as the ethyl analogue.

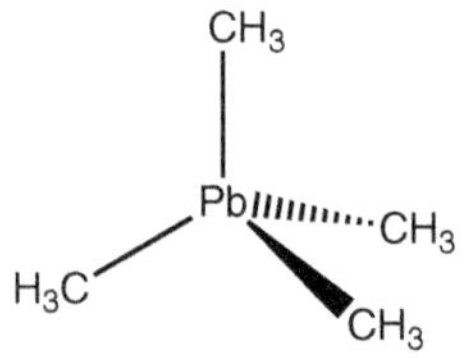

(42.2)

As we have seen before, many health experts were unconvinced that tetraethyllead containing fuels were safe or that the lead emitted into the atmosphere would not affect people. Whilst consulting for Shell in the late 1960s, chemist and eco-guru James Lovelock used his electron capture detector to discover that tetraethyllead, probably leaking from carburettors, was present in the passenger compartment of motor cars at levels up to 300 times the maximum permitted industrial levels. He included this in a report submitted to Shell, although this information was never released. Not only did lead compounds build up in the gutters, but in times to come it would be found that lead could be transported hundreds, even thousands, of miles. It would be years before the hazards of lead-containing fuels would be proved, but proved these hazards would be.

Principally responsible for this breakthrough was the American scientist Clair Cameron Patterson (1922–1995), who proved that a lot more lead was entering the environment in the preceding century than in preindustrial times. A key point of Patterson's argument was that many scientists had greatly underestimated the issue of lead pollution, because they used 'blank' background samples that were contaminated. Armed with this information, he went on to show that lead concentrations in Greenland ice had increased by a factor of at least 200 between 800 BC and AD 1969, with a sharp rise after 1940, attributable to the lead alkyls in petrol. Subsequently (in 1994), he went back almost 8,000 years, and pointed out that lead levels in ice, from the ancient Greek and Roman civilisations, were much greater than 'natural' values, which demonstrated how much lead was mined by the Greek and Roman Empires. Patterson drilled down into the ice for his samples – the deeper he went, the longer it had been since the ice (and the lead from that time) had been 'laid down'.

Subsequent experiments akin to Patterson's on a Swiss peat bog found that lead levels there peaked in 1979, before dropping off again. The Swiss had introduced leaded fuel in 1947. The researchers measured the isotopic composition of the lead in the sample and found that the values of $^{206}Pb/^{207}Pb = 1.1492 \pm 0.0002$ and $^{206}Pb/^{204}Pb = 17.876 \pm 0.022$ closely compared with the values for the tetraethyl-lead used in the petrol used in Berne, Switzerland, in the 1970s ($^{206}Pb/^{207}Pb = 1.14$ and $^{206}Pb/^{204}Pb = 17.83$). This showed that the increase in lead levels from 1936 to 1979 was therefore attributable to 'gasoline lead'.

Another American scientist, Herbert Needleman (1927–2019), linked high bone lead levels with lower IQ and increased inattention, aggression and delinquency.

Impetus for the removal of lead from fuels was provided by Rachel Carson's book *Silent Spring* (1962). The introduction of catalytic converters to car exhaust systems, in order to improve air quality, was an added reason for this, as the lead could poison the catalyst. Alternative materials such as ethanol and benzene were used to improve octane ratings, compensating for the removal of lead. At more or less the same time, other campaigns were driving the removal of lead from solder used in canning, as well as from paints – in the USA, lead-containing paint was banned from new houses in 1978.

Following the passage of amendments to the Clean Air Act in 1970 concerning air pollutants, the US Environmental Protection Agency began to enforce the removal of tetraethyllead from petrol from 1973, with a total ban coming into effect in 1996. At that time, the Environmental Protection Agency commented: 'The elimination of lead from gas is one of the great environmental achievements of all time'. Japan was the first country to ban leaded petrol in 1986; the European Union (including the UK) banned the sale of leaded gasoline in 2000, the same year as China. Algeria was the last nation to phase out leaded petrol in 2021.

An unexpected dividend of the removal of lead from petrol and homes was to follow. After World War II, blood lead levels in American children rose to a peak in the early 1970s, then steadily declined. Some 20 years later, when babies born in the 1970s reached the average age for criminality, violent crime rates followed a similar trajectory.

Tetraethyllead is an example of a chemical that became of major importance for over 50 years, but never should have been, and caused enormous damage. As someone once remarked, 'It took just two years to introduce, but sixty years to remove'.

43 Tetrodotoxin

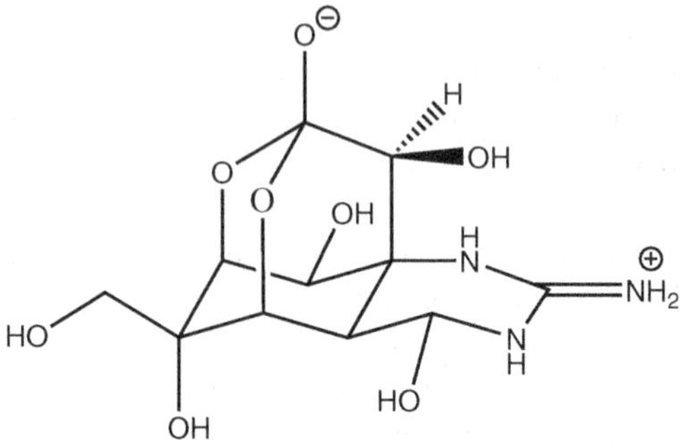

(43.1)

The name is Bond, Chemical Bond. Most people would associate Ian Fleming's James Bond stories with technology rather than chemistry, but *From Russia with Love* ends poised in mid-air, as Rosa Klebb kicks James Bond's right calf with a boot bearing a steel tongue, coated with an unknown poison. This causes Bond to crash unconscious to the floor. It is only in *Dr No* that we find out that he survived a poison revealed to be tetrodotoxin (43.1).

Tetrodotoxin is most familiarly found in the puffer fish, notably *Tora fugu*. The puffer fish gets its name from its ability to inflate itself into a spiky ball when threatened, expanding to several times its normal size by swallowing water. The spikes are rich in tetrodotoxin. *Fugu* flesh is a great delicacy in Japan, but since several parts of its anatomy are tetrodotoxin-rich, notably the liver, ovaries and intestine, preparing fugu for eating is a skilled business to make sure that all the viscera are removed. 'All' is the operative word; the lethal dose of tetrodotoxin is less than a milligram; it is 10,000 times more toxic than cyanide. It is not decomposed by cooking and there is no antidote. The best course of treatment is to rapidly administer an emetic, in the hope of vomiting up any of the toxin that has not been digested.

Would-be *fugu* chefs have to pass a two-part exam, with both theory and practical components, the latter involving preparing fugu with the deadly parts removed. Only about 1/3 of the people sitting the exam pass, the others die (only joking). Even so, 50 or more Japanese die every year from *fugu* poisoning. And it is said that some of the most popular *fugu* chefs leave just a soupçon of tetrodotoxin in their preparation, so that diners experience that tingling sensation without the risk.

Tetrodotoxin is not confined to the fugu fish, it is also found in some worms, including a flatworm, certain crabs and toads, as well as at least two species of newt, the eastern newt and the western or rough-skinned newt. Then, there are also cuttlefish and octopuses, notably the blue ringed octopus (familiar from another Bond film,

DOI: 10.1201/9781003783565-43

Octopussy). Captive-bred puffer fish do not contain tetrodotoxin, though they can subsequently acquire it from wild puffer fish. Animal toxins are usually very species-specific, but not tetrodotoxin. This and other evidences suggest that puffer fish (and the other bearers of the poison) acquire tetrodotoxin from an exogenous source, believed to be bacteria. It serves one or two purposes, depending on the species – as a defence against predators, and as a way of killing prey.

Tetrodotoxin is poisonous because it blocks the entrance to voltage-gated Na^+ channels in nerve membranes. It prevents the sodium ions from entering cells, so that nerve signals do not get transmitted. The fish does not get poisoned by its own poison because its sodium ion channel has a single amino acid mutation – phenylalanine or tyrosine is replaced by asparagine – so that tetrodotoxin cannot bind to it.

There is what has been called an 'arms race' between the rough-skinned newt, *Taricha granulosa*, who has a lot of tetrodotoxin in their skin, protecting it, and its predator, the common garter snake *Thamnophis sirtalis*. The snakes have developed resistance to tetrodotoxin, very likely caused by genetic mutations coding for the sodium ion channels in the snake, so that tetrodotoxin cannot bind to the channels, thus leading the snakes to become tetrodotoxin resistant and able to eat the poisonous newts.

Meanwhile, what happens when a human gets poisoned by tetrodotoxin? Its effects were vividly explained by Ian Fleming, describing the cold numbness creeping up Bond's body after Rosa kicked him. He lost feeling in his hands, then his breathing began to fail. Initial numbness in one part of the body spreads as nerves and muscles are paralysed in other parts and the victim becomes immobile. If you swallow tetrodotoxin, the tingling first affects the lips and tongue, gradually spreading to the face and then the rest of the body. According to Fleming, Bond survived because he was administered artificial respiration and was then treated for curare poisoning. In real life, this would not have worked on Bond, as the arrow poison curare works by blocking the action of the neurotransmitter acetylcholine. But acetylcholine is not involved in tetrodotoxin poisoning, for which there is no antidote. Anyway, James Bond survived, and was sent to Jamaica as a rest cure, where he was fortunate to meet Honeychile Rider (memorably played by Ursula Andress in the film version of *Dr No*).

Puffer fish were generally unknown to the Western world until the eighteenth century. The celebrated English explorer Captain James Cook was off the island of New Caledonia (roughly between Fiji and the Australian city of Brisbane) on his second round-the-world voyage with *HMS Resolution* on 7 September 1774 when he, along with the expedition's naturalist Johann Forster and his son, dined off the liver and roe from a puffer fish. During the night, they were visited by the feeling of numbness in their limbs and loss of feeling, subsequently being violently sick. They began vomiting that probably helped get rid of the remaining poison from their systems; even so, they were probably fortunate to have eaten only a tiny dose. Their fellow-passengers, the ship's pigs, were less fortunate, as they were fed the remains of the pufferfish, which killed them. There's a record that two crew members of the Dutch warship *Postillion* died in less than 20 minutes after eating puffer fish off the Cape of Good Hope one day in September 1845. Pufferfish have lately been identified in the Mediterranean, and there is concern about the possibility of tetrodotoxin-bearing shellfish in the region, as well as the eastern Atlantic Ocean, possibly connected with the warming of seawaters due to climate change.

Despite its toxicity, tetrodotoxin has been investigated as an effective painkiller for some years to treat severe pain, notably in cancer patients, as a means of long-lasting relief, because of its role in blocking pain-signalling channels.

The latest news from the world of tetrodotoxin concerns the blue-lined octopus *Hapalochlaena fasciata* and its sex life. It is a tiny species, where the adult female is about the size of a golf ball while the adult male can be as little as 20% of that. That makes them vulnerable to the female, who are prone to eating the males during copulation. The males have one weapon on their side, as they inject the females with tetrodotoxin during sex, which has the effect of immobilising the female for around an hour during mating.

44 Thalidomide

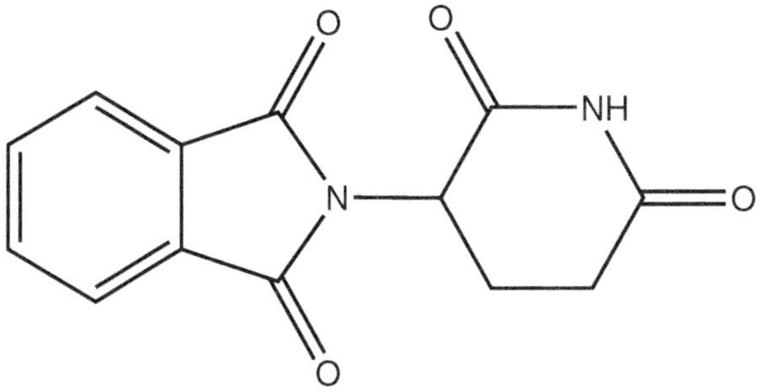

(44.1)

Possibly if a molecule's story is about trust, or the lack of it, that molecule is thalidomide (44.1). We trust the doctors and nurses who treat us, not to mention the medications that they administer. In the 1960s, thalidomide came to embody trust misplaced. Between 1957 and 1961, several thousand women taking thalidomide, in many parts of the world, gave birth to babies with serious defects, of which the most noticeable were with undeveloped limbs, often when hands and feet were attached directly to the body, a condition known as phocomelia. The lives of these survivors have often been heroic, as they have overcome the problems they have faced. And many babies whose mothers were given thalidomide were not born at all. Thus, thalidomide is known as a teratogen, something that induces birth defects.

Thalidomide was first made in the early 1950s. A West German firm, Chemie Grünenthal, found that thalidomide had some useful properties. It was tested on rats. Thalidomide had no effect on pregnant rats but it was given to pregnant women for treatment of the nausea associated with 'morning sickness'. Chemie Grünenthal marketed thalidomide as an over-the-counter sedative named 'Contergan', which unlike barbiturates, was not toxic. Hence, no prescription from a doctor was needed. In 1957, it was licensed to the Distillers Company, who put it on the market as 'Distaval' in the UK, and its use spread to some 45 other countries, notably in Europe and Australia. In January 1959, Chemie Grünenthal licensed thalidomide to the Richardson-Merrell company in the USA and the latter filed a new drug application with the FDA on 12 September 1960. The company had expected that there would be no complications and that thalidomide would be approved for use in the USA and Canada early in 1961.

At this point, a remarkable woman enters the story. Vancouver-born Frances Kathleen Oldham (1914–2015) studied for a M.Sc. in pharmacology at McGill

DOI: 10.1201/9781003783565-44

University in Montreal, and then moved to the University of Chicago to carry out research for a PhD. During her second year at Chicago, her supervisor (Professor Geiling) asked her to examine a medicine called the elixir of sulfanilamide, which had caused over 100 deaths in the USA, mainly young children (see pp. 170–171). She showed that the toxicity was due to the solvent diethylene glycol, and that sulfanilamide was otherwise quite safe. She joined the university faculty after obtaining her PhD, and then went on to continue her research. A few years later, while carrying out tests on antimalarial chemicals, she observed that pregnant rabbits did not metabolise quinine as well as normal rabbits could. Moreover, rabbit embryos could not break quinine down at all, and quinine could cross the placental barrier. Her memory of this research served her well when it came to considering approval of thalidomide. In 1943, she married a fellow faculty member, Fremont Ellis Kelsey, and became known as Dr Frances Oldham Kelsey.

Dr Frances Oldham Kelsey moved to a post with the Food and Drug Administration (FDA) in Washington, D.C. in August 1960. Just one month later, her first drug to review landed on her desk – thalidomide. Under FDA law, she had 60 days to review the application; she and her colleagues had serious concerns about the drug's safety. The animal tests were just on rats – none on mice or rabbits. Clinical evidence was limited – there were no serious studies on risks to pregnant women. So, they returned the application to Richardson-Merrell, who had to enclose the required information when resubmitting it. The 60-day period recommenced. At that time, a letter appeared in the 31 December 1960 issue of the *British Medical Journal,* from Dr A. Leslie Florence of Aberdeen. Florence reported four cases of peripheral neuritis (irreversible numbness in the limbs) in patients taking thalidomide; Kelsey saw this letter and asked Richardson-Merrell for more evidence from animal tests, demanding evidence that thalidomide was safe for use during pregnancy. Evidence was mounting that taking thalidomide was unsafe during pregnancy. Doctors in West Germany, where thalidomide had first come into use, had already noted birth deformities (phocomelia) in babies born to women taking thalidomide, and in spring 1961, children with malformed limbs were born to both the sister and the wife of a young Hamburg lawyer called Karl Schulte-Hillen. After the lawyer contacted Professor Widukind Lenz, head of the children's clinic at Hamburg University, Lenz made enquiries and by that November he knew of 14 such cases. Meanwhile, William McBride, an Australian obstetrician, realised that three women who had been taking thalidomide had given birth to deformed babies who died after birth, so he wrote a letter reporting this to the *Lancet*, which was published on 16 December 1961. Professor Lenz contacted the Grünenthal company. When the story emerged in the leading West German newspaper, *Welt am Sonntag,* Grünenthal withdrew thalidomide from the German market and informed licensees in other countries. Richardson-Merrell withdrew its FDA application on 8 March 1962 and consequently withdrew the drug from Canada. Thus, thalidomide was never marketed in the USA, and the US Congress passed legislation which required tests that showed drugs were safe for pregnant women (as did the governments of many other countries). Frances Oldham Kelsey, a heroine, had saved countless lives. A few months later, on 8 August 1962, President John F. Kennedy presented Frances Oldham Kelsey with the President's Award for Distinguished Federal Civilian Service, the highest honour that could be given to an American civilian. She later became head of the investigational drug branch of the FDA, where she continued working until the age of 90.

Some 5,000 thalidomide babies survived childhood; it is believed that at least twice as many died from the drug. Many survivors have gone on to have successful and fulfilling lives, despite their handicap. One reason traditionally suggested for thalidomide acting as a teratogen was that thalidomide might intercalate (slide) into the double helix structure between the paired DNA bases, as its structure resembles those of guanine and adenine. This could interfere with the synthesis of integrin proteins, thereby inhibiting angiogenesis and preventing blood vessels from forming to supply growing embryonic tissue, so that certain limbs would not develop properly. It is now known that thalidomide and its analogues promote teratogenesis by binding to a protein called cereblon (CRBN), thereby causing it to remove specific proteins that promote embryonic development, notably the transcription factors SALL4 and p63. These transcription factors are involved in the development of limbs and internal organs, and hence, normal growth of the embryo does not happen anymore. Thalidomide also inhibits the formation of normal blood vessels.

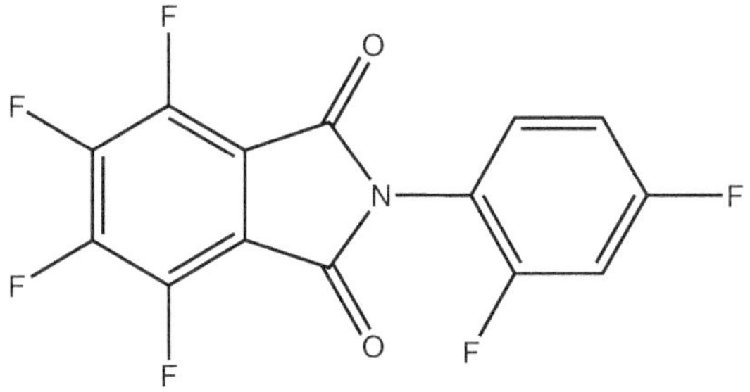

(44.2)

Professor Neil Vargesson of the University of Aberdeen has been studying thalidomide, along with certain derivatives and metabolites, for two decades. Thalidomide doesn't just affect human embryos, it also causes limb defects in rabbits and chicks and fin defects in zebrafish.

Using developing chick embryos, Vargesson's group found that these metabolites have either anti-inflammatory or antiangiogenic properties. None of the anti-inflammatory compounds had any effect on limb development. Only one molecule that they studied, CPS-49 (44.2), a fluorinated analogue of some thalidomide metabolites with antiangiogenic properties, produces the same limb defects as thalidomide does. It inhibits the growth of new blood vessels in the foetus. Limb formation in the human foetus occurs within five to eight weeks, when the limbs have a rapidly changing vessel pattern, although by this time, the vascular system in the rest of the body is complete. Thalidomide kills cells and stops limbs from forming, which is why only women taking thalidomide during the 34- to 50-day window following the last menstruation were at risk of giving birth to deformed babies.

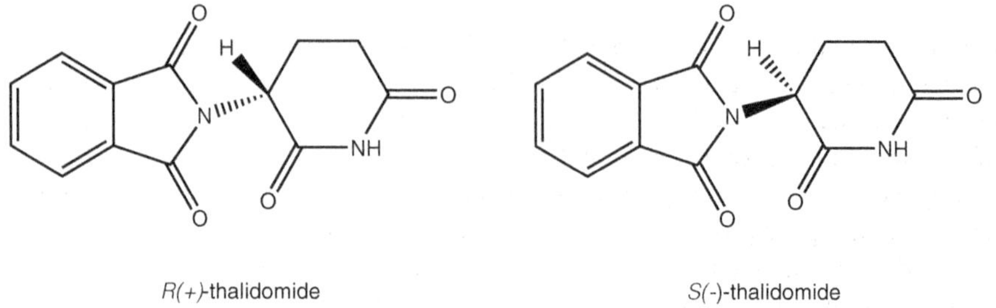

FIGURE 44.1 Enantiomers of the thalidomide molecule.

The thalidomide molecule contains one chiral carbon atom, and therefore it exists as two enantiomers (Figure 44.1). Both these isomers have sedative properties, but only the (*S*)-isomer is a teratogen (a producer of birth defects). Medicinally, it was used as a racemic mixture of the isomers. However, the isomers quickly interconvert in vivo (the half-life at 37°C and pH 7.4 is about 9 hours), so that even if only the pure (*R*)-enantiomer were administered, both isomers would soon be found in the body, and birth defects could still occur.

Even after thalidomide was withdrawn from the market, research continued, and it was found to have unexpected clinical benefits, and therefore it can still be used as a drug. In 1965, an Israeli doctor, Jacob Sheskin, examined a leprosy patient who had a very painful complication of erythema nodosum laprosum (ENL; also known as Hansen's disease). This patient was bedridden and on the verge of death – an utter insomniac, upon whom no sedative worked. Dr Sheskin remembered that thalidomide was a sedative and hence gave the man two thalidomide tablets. The patient slept for 20 hours and, remarkably, was able to get out of bed. Two more tablets eradicated the pain completely and his leprosy sores began to heal. Further study confirmed the treatment. Thalidomide is now the drug of choice worldwide for this form of leprosy. The FDA has given its approval to this use of thalidomide, marketed as 'Thalomid', under very restricted conditions. Thalidomide has also been used to treat multiple myeloma and is effective in some forms of cancer therapy because it shuts down the blood supply to tumours. It can still be used as a sedative. It can be used under carefully controlled conditions because it is a good treatment for leprosy and multiple myeloma. But it must never again be administered to pregnant women.

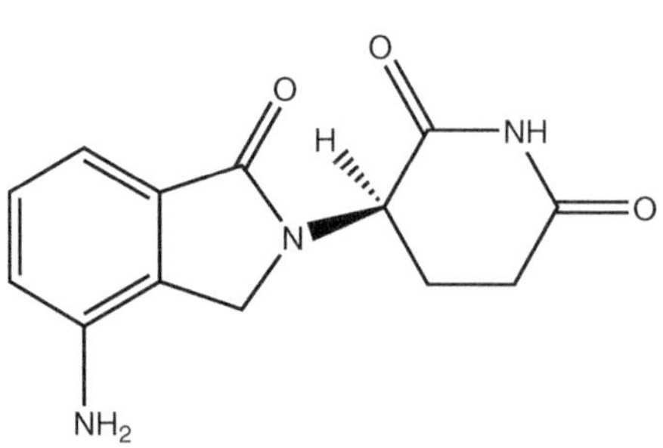

(44.3)

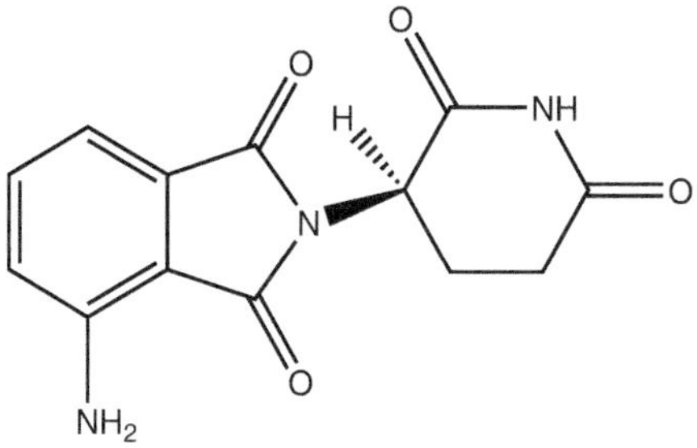

(44.4)

Some thalidomide derivatives are also important medications. In particular, it has been found that some amine-substituted thalidomides have anti-cancer activity against haematological malignancies, including multiple myeloma. Notable among these are Lenalidomide (44.3) and pomalidomide (44.4), produced by Celgene Corporation, today a part of Bristol-Myers Squibb. They have strong immunomodulatory effects and are referred to as IMiDs. Lenalidomide is a first-line treatment for multiple myeloma, *inter alia*, and Pomalidomide is used against multiple myeloma.

45 Vanillin

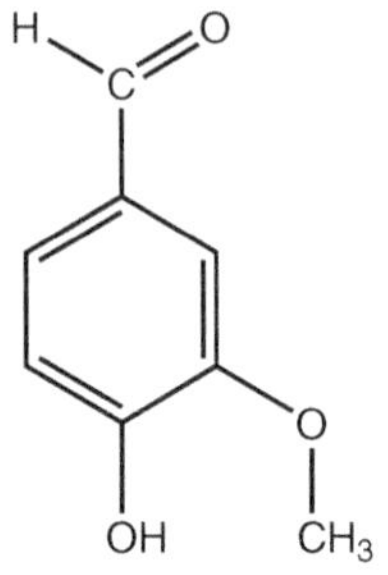

(45.1)

Vanilla was brought to Europe in the wake of Columbus's expedition, so it has only been known outside its native range for about 500 years, in which time it has become the world's most popular flavour. The Aztecs used it, in combination with chilli peppers, to season their chocolate drink, *xocolatl* (which may be where the word chocolate comes from). Vanilla is obtained from a climbing orchid which produces green seed pods, which look rather like runner beans. *Vanilla planifolia* Andrew is the main species grown in tropical and subtropical regions. It is native to Mexico, though most of the natural vanilla grown nowadays comes from Madagascar. A second species, *Vanilla tahitensis,* is grown exclusively in Tahiti and Papua New Guinea as a source of vanilla flavour. Other species are known, for example, *Vanilla cribbiana* Soto Arenas, known as Maya vanilla, which is native to Guatemala and Mexico.

The main ingredient of vanilla is the molecule vanillin (45.1). A vanilla pod contains about 2% of vanillin, and thus 50 kg of pods, a vast quantity, are needed to produce 1 kg of pure vanillin. The growing beans do not smell of vanilla. After the growing pods have been dehydrated in the sun, they turn black. The green beans contain molecules of vanillin bound to a glucose molecule as a glycoside. During the ripening period, enzymes hydrolyse the linkage, freeing the separate molecules, including vanillin, short for 4-hydroxy-3-methoxybenzaldehyde, which is responsible for the vanilla flavour (Figure 45.1).

The vanilla bean makes vanillin through a sequence of enzyme-catalysed reactions, starting with the amino acid phenylalanine. An example of the specificity of enzymes is that just one single enzyme vanillin synthase (VpVAN) is responsible for the conversion of ferulic acid to vanillin.

Vanillin makes up 98% of natural vanilla extract; the remaining 2% comprised of around 200 other chemicals, in tiny individual amounts. However, some of these other ingredients themselves have strong aromas, and make a significant contribution

DOI: 10.1201/9781003783565-45

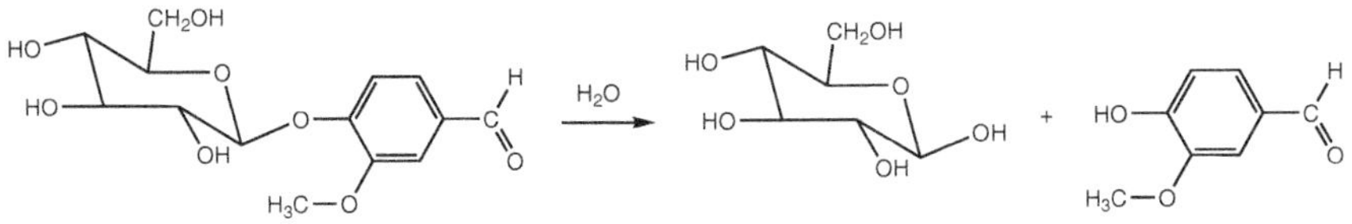

FIGURE 45.1 Formation of vanillin by hydrolysis of its glycoside.

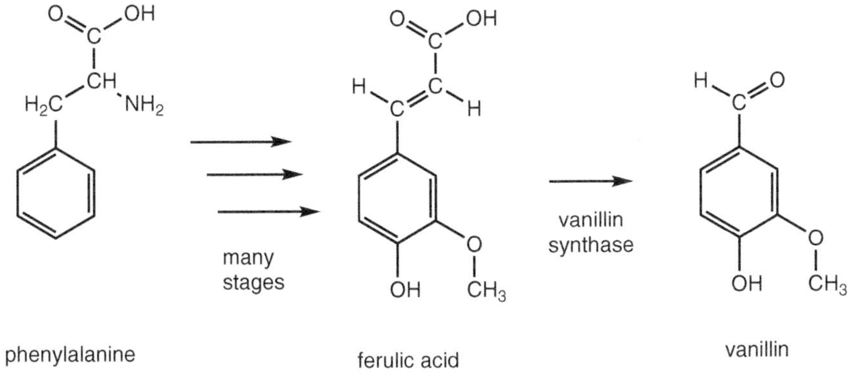

FIGURE 45.2 Enzymatic synthesis of vanillin from phenylalanine.

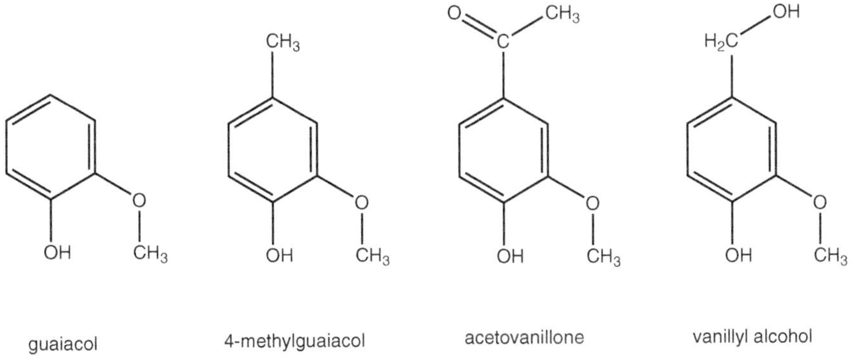

FIGURE 45.3 Some key odorants in Mexican vanilla beans.

to the overall smell of vanilla extract, so that the extract smells somewhat different to pure vanillin. So an examination of the aroma of Mexican vanilla beans (*Vanilla planifolia*) from the Tuxtepec region showed that guaiacol, 4-methylguaiacol, acetovanillone and vanillyl alcohol – all with vanillin-related structures (Figure 45.3) – were important to the overall smell, although present at much lower concentrations in these vanilla beans than vanillin.

In contrast, a study of Tahitian vanilla beans (*Vanilla tahitensis*) revealed that the key odour-active compounds were vanillin, anisaldehyde, anisyl alcohol and anisyl acetate (Figure 45.4).

Vanillin was first isolated (from vanilla pods of course) by Nicholas-Theodore Gobley in 1858. Its structure was worked out in 1874, and then chemists got down to finding ways of synthesising it, as the demand for vanillin cannot simply be met by nature (as the demand continues to grow). An important nineteenth-century industrial route starts with eugenol, the molecule responsible for the smell of clove oil. Eugenol isomerises under alkaline conditions, and the resulting isoeugenol can be oxidised to vanillin (Figure 45.5).

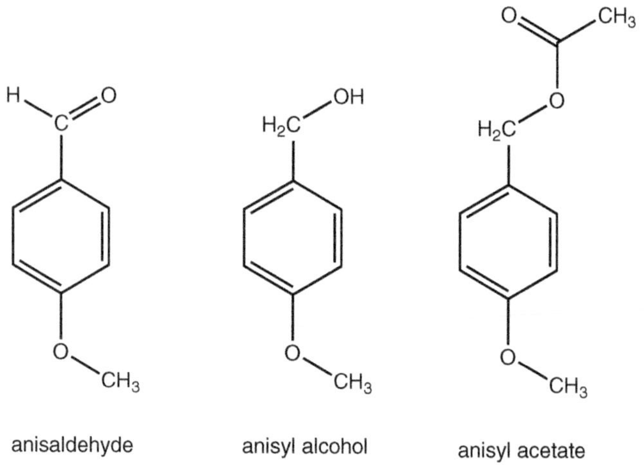

FIGURE 45.4 Some key odorants in Tahitian vanilla beans.

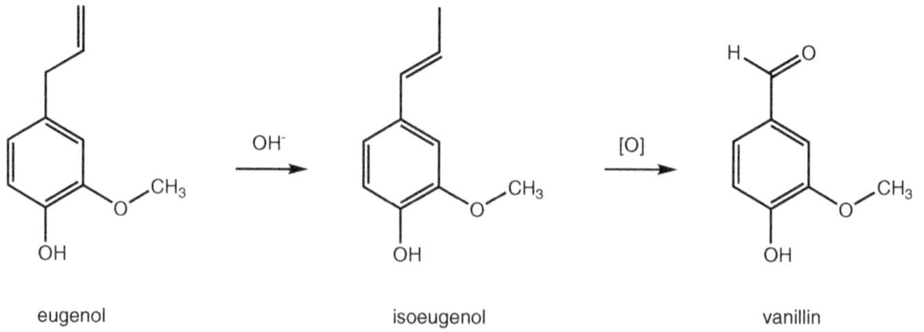

FIGURE 45.5 Conversion of eugenol to vanillin.

The principal industrial route for the manufacture of vanillin for many years involved the lignin wastes formed in the manufacture of paper. The polymer lignin strengthens wood and plant cell walls. Burning wood generates small organic molecules like vanillin and guaiacol – take a look at the structure of lignin (Figure 45.6). They are responsible for the smell of 'smoked' food and also add to the flavour of whisky (and other wines and spirits) stored in wooden casks. Heating lignin waste with alkali under oxidising conditions forms vanillin. However, the caustic waste formed is environmentally undesirable, and therefore this process is now used much less.

Most industrial vanillin today is made (Figure 45.7) from the reaction of petrochemically derived guaiacol, which is reacted with an alkaline solution of glyoxylic acid (oxoacetic acid).

The product of the first step, hydroxyl(4-hydroxy-3-methoxyphenyl)acetic acid, can then be reacted with O_2, in the presence of a catalyst like MnO_2, forming the corresponding keto-acid, 4-Hydroxy-3-methoxyphenylglyoxylic acid. When this is acidified, the acid loses CO_2 in a decarboxylation reaction, generating the desired vanillin. Around 85% of synthetic vanillin production comes from this route, with

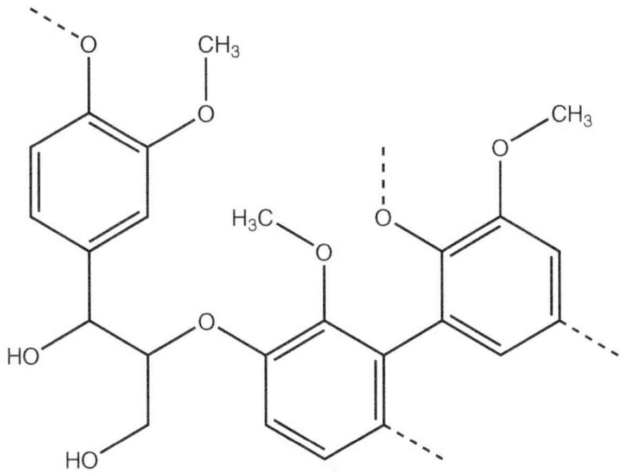

FIGURE 45.6 A representative fragment of a lignin molecule.

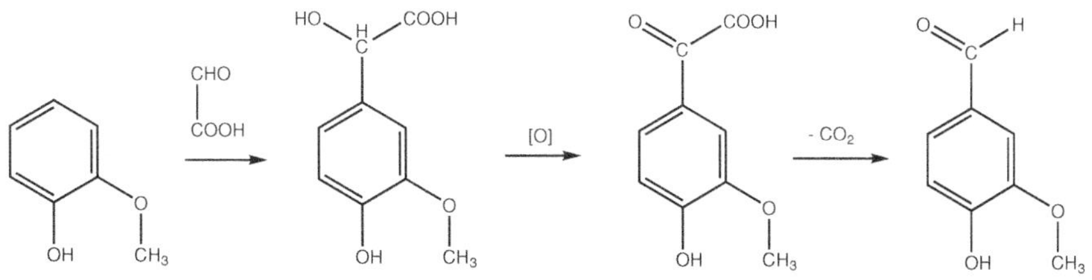

FIGURE 45.7 Industrial synthesis of vanillin from guaiacol.

most of the remaining vanillin originating from lignin. Another route being developed produces vanillin by fermenting glucose.

Vanillin has many uses, most obviously in food, from confectionary – cakes, pastries, chocolate and desserts – to ice cream and yogurts. It has been used in perfumes since the late nineteenth century. It is also used as a reagent to make some important pharmaceuticals, such as trimethoprim (45.2), an antibiotic used particularly in the treatment of certain urinary tract infections, and L-DOPA ((45.3); L-3,4-dihydroxyphenylalanine), used to treat Parkinson's disease by increasing dopamine levels in the brain.

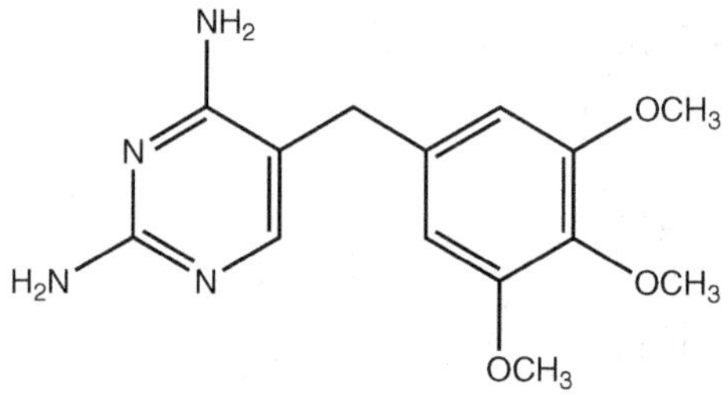

(45.2)

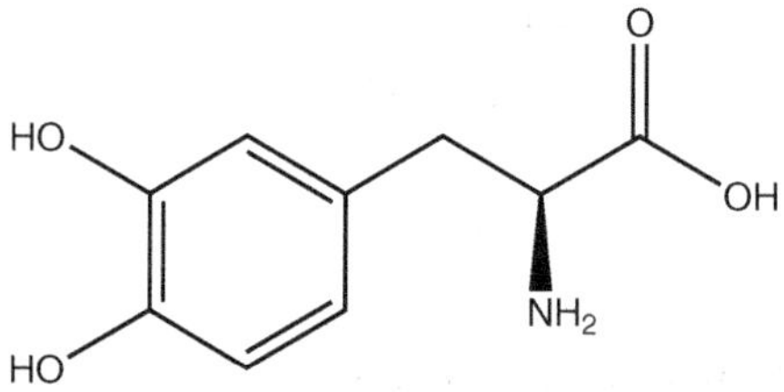

(45.3)

Vanillin is in huge demand. Less than 1% of the vanillin used comes from vanilla beans, and therefore naturally sourced vanillin costs around 200 times that of synthetic vanillin. Numerous food companies have said that they will only use natural vanillin, helping to push prices upwards.

Criminals have tried to pass off their synthetic vanillin as the 'real thing'. One way to do this is to add compounds which are impurities in the real vanilla essence, like vanillic acid (45.4) and 4-hydroxybenzaldehyde (45.5). But the fraudsters are more sophisticated than that.

One way of distinguishing between natural and synthetic vanillin is to measure the level of radioactive ^{14}C isotope present. ^{12}C and ^{13}C are stable and not radioactive, but ^{14}C has a half-life of 5,730 (± 40) years. A living vanilla plant is in equilibrium with the surrounding air and, hence, will have the same levels of ^{14}C as the air; so too will vanillin made from it. This will be much higher than vanillin made

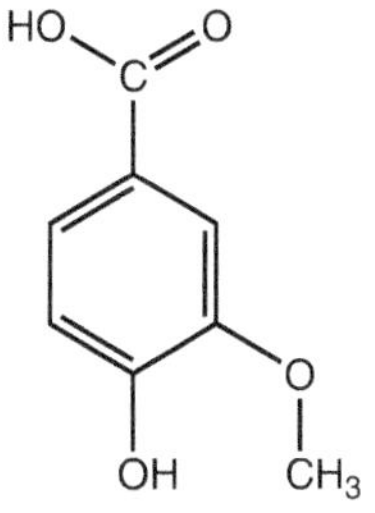

(45.4)

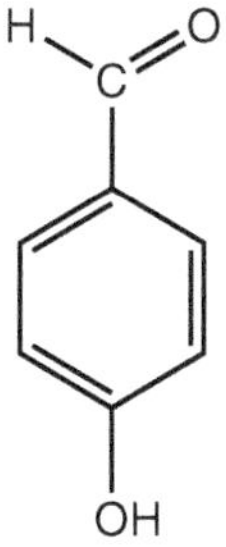

(45.5)

from petrochemicals, which have been in the ground for a long period, reducing the decaying radiocarbon content. Fraudsters responded by spiking their vanillin molecules with ^{14}C atoms.

Analytical chemists responded by studing stable carbon isotopes. The ratio of $^{13}C/^{12}C$ in a molecule varies, depending upon whether they use a C_3 or C_4 photosynthetic cycle. This ratio is expressed as a quantity $\delta^{13}C$. This compares the isotope ratio in the sample material to the isotope ratio in the standard (originally based on Pee Dee Belemnite limestone).

$$\delta^{13}C\% = \frac{(^{13}C/^{12}C)_{sample} - (^{13}C/^{12}C)_{standard}}{(^{13}C/^{12}C)_{standard}} \times 1000$$

The "C_3 " photosynthetic cycle, where the first step involves the 3-carbon molecule 3-phosphoglyceric acid, is associated with most food plants grown in temperate climates, notably Europe, such as potatoes, wheat, barley, oats, rice, sugar beet, several fruits and most trees. They generally have $\delta^{13}C$ ratios, which vary between -22 to -35 ‰ (average ~ −25.1 ‰). The "C_4 cycle", where the first step involves the 4-carbon molecule malate, is associated with food crops linked with hot climates such as pineapple, sugar cane, maize, sorghum, millet and tropical grasses that lead to markedly different ^{13}C:^{12}C ratios, with values in the range -8 to -20 ‰ (average ~ −10.8 ‰). These European diets are richer in C_3 plants, whilst C_4 plants tend to dominate North America. Since 'you are what you eat', a person's body ^{13}C:^{12}C ratio will reflect their diet. *V. planifolia* has an unusual metabolism, producing vanillin with $\delta^{13}C$ values intermediate to those for C_3 or C_4 plants, in the range of -18.2 ± 0.6 ‰.

The fakers' response was to add extra ^{13}C to their molecules to bring the $\delta^{13}C$ values into the right range. But they could not easily add ^{13}C atoms to the benzene ring – chemically it is so much easier to change the carbons in either the methyl or aldehyde groups rather than in the six-membered benzene ring. Because of this, the 'peripheral' carbon sites will have an overabundance of ^{13}C compared with the other six. Spectroscopists have used ^{13}C NMR technique involving 'Position-Specific Isotope Analysis' to show up unnatural ^{13}C levels.

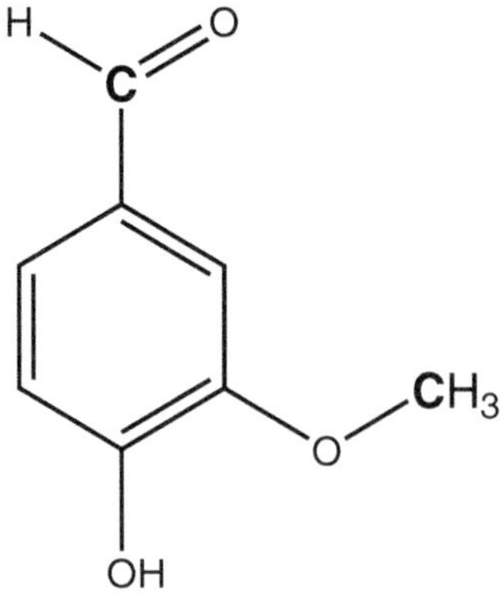

Figure 45.8

FIGURE 45.8 Vanillin, with emboldening showing the easiest site to introduce ^{13}C atoms.

The similar molecule ethylvanillin (45.6) does not occur in nature but has a similar yet stronger smell. Vanillin was the second 'synthetic' molecule to be used in a perfume, Aimé Guerlain's *Jicky* (1889) [the first was coumarin in Houbigant's *Fougère Royale* (1882)]. Jacques Guerlain, nephew of Aimé, used ethyl vanillin to create *Shalimar* in 1925. The odour threshold of ethyl vanillin is quoted as 0.000007 mg/m^3 of air, in comparison with a value for vanillin itself in the range of 0.005 – 0.000001 mg/m^3 of air.

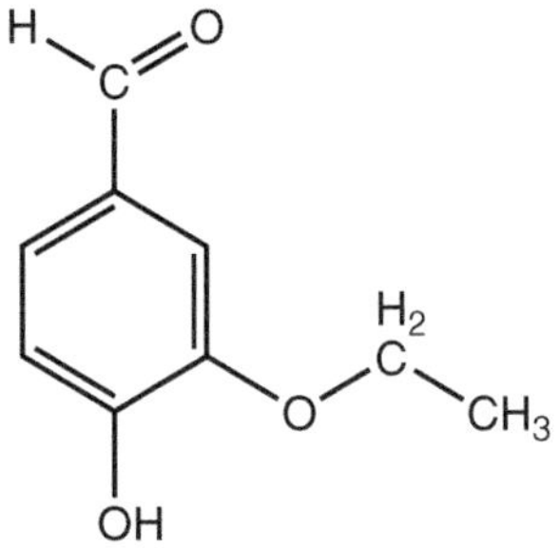

(45.6)

46 Vaska's Compound

Le hazard favorise l'esprit preparé.
(Pasteur)

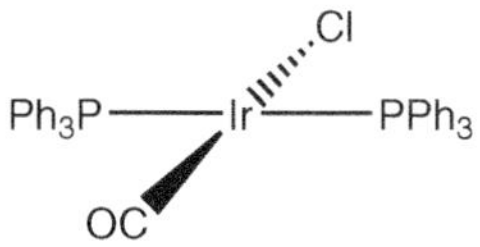

(46.1)

Lauri Vaska (1925–2015) received his primary and secondary education in Estonia, his native country. In the immediate aftermath of the German retreat from Estonia, he was in a refugee camp. He then went on to study chemistry to degree level mainly at a university in Gottingen (1946–1949). Vaska emigrated to the United States and carried out research leading to a PhD at the University of Texas, under the supervision of George W. Watt, working on osmium chemistry. From 1957 to 1964, Vaska was a Fellow at the Mellon Institute, in Pittsburgh, where he made the discoveries concerning $[IrCl(CO)(PPh_3)_2]$ (46.1) associated with his name. From 1964 to 1990, he was a professor at Clarkson University in New York, where he remained an emeritus professor until his death.

He studied the reactions between $(NH_4)_2[IrCl_6]$ and triphenylphosphine in hot alcoholic solvents at 100°C and above, obtaining $[IrHCl_2(PPh_3)_3]$ and $[IrH_2Cl(PPh_3)_3]$, complexes typical of the hydrido-complexes of the platinum metals being discovered at that time, particularly by chemists like Joseph Chatt. However, in what appears to have been a fortuitous discovery, extended reaction between $IrCl_3 \cdot xH_2O$ and triphenylphosphine in ethylene glycol at 190 °C was found by Vaska to lead to the subject of this chapter, $[IrCl(CO)(PPh_3)_2]$. This type of reaction (Figure 46.1), where the CO molecule is 'abstracted' from the hot alcohol molecule used as the solvent, is not uncommon in platinum-metal chemistry. Salts of $[IrCl_6]^{2-}$ are an alternative starting material, whilst the solvent dimethylformamide (DMF) can also act as a source of CO.

At the same time, similar iridium complexes were being studied by the Italian chemist Maria Angoletta, though her work, published in *Gazzetta Chimica Italiana,* was not known to Vaska. Although Vaska did not use it, an obvious way of making $[IrCl(CO)(PPh_3)_2]$ is by substituting two carbonyl groups in $[IrCl(CO)_3$ with PPh_3:

$$[IrCl(CO)_3] + 2\ PPh_3 \rightarrow [IrCl(CO)(PPh_3)_2] + 2\ CO$$

DOI: 10.1201/9781003783565-46

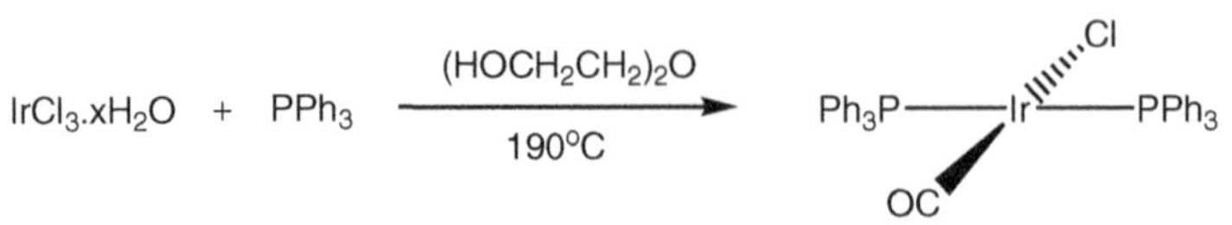

FIGURE 46.1 Synthesis of Vaska's Compound.

Vaska did some of the research by himself and some with his student John DiLuzio. He soon discovered the ability of $[IrCl(CO)(O_2)(PPh_3)_2]$ to react reversibly with both H_2 and O_2, as well as its important reactions with other small molecules, many being pioneering examples of oxidative addition reactions. The structure of the oxygen adduct, $[IrCl(CO)(O_2)(PPh_3)_2]$ (46.2), was of particular interest. The iridium compound mimics haemoglobin in binding oxygen reversibly, and at that time it was not known how haemoglobin did this. Subsequently, it emerged that haemoglobin binds O_2 end-on, but that takes nothing away from the significance of this work reported between 1963 and 1965.

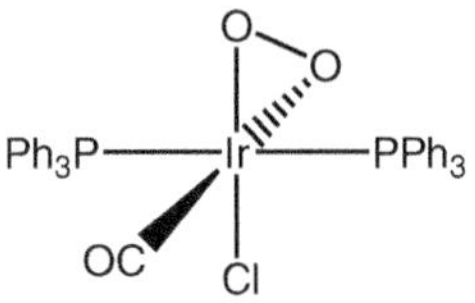

46.2

$[IrCl(CO)(PPh_3)_2]$ reacts with a wide variety of molecules, including H_2, Cl_2, HCl and CO, as well as CH_3I, SO_2 and $HgCl_2$ (Figure 46.2).

This type of reaction is now generally known as an oxidative addition reaction. These are often found in the chemistry of square-planar Rh(I) and Ir(I) complexes – such as $[IrCl(CO)(PPh_3)_2]$ – which have 16-electron configurations. In these reactions, both the coordination number of the metal and its oxidation state increase by two units; so does the electron count, so that the complexes now obey the 18-electron rule, typical of organometallic compounds. Complexes of transition metals in low oxidation states with strong π-acceptor ligands, notably CO, have nine orbitals in their 'valence shell', which can hold 18 electrons.

Insight into the bonding changes taking place can be obtained by monitoring the position of the CO stretching frequency in the IR spectrum (Table 46.1).

This is affected by the balance between CO-to-metal σ-donation and metal-to-ligand π-backbonding, in which electron density goes from occupied metal d orbitals to empty π antibonding orbitals on the CO ligand.

The higher the oxidation state of the iridium, the fewer d electrons there are available for donation into the antibonding orbitals of CO, the 'backbonding'. This means that the C-O bond is strengthened – the stronger the C-O bond, the higher is ν_{CO} in the infrared spectrum.

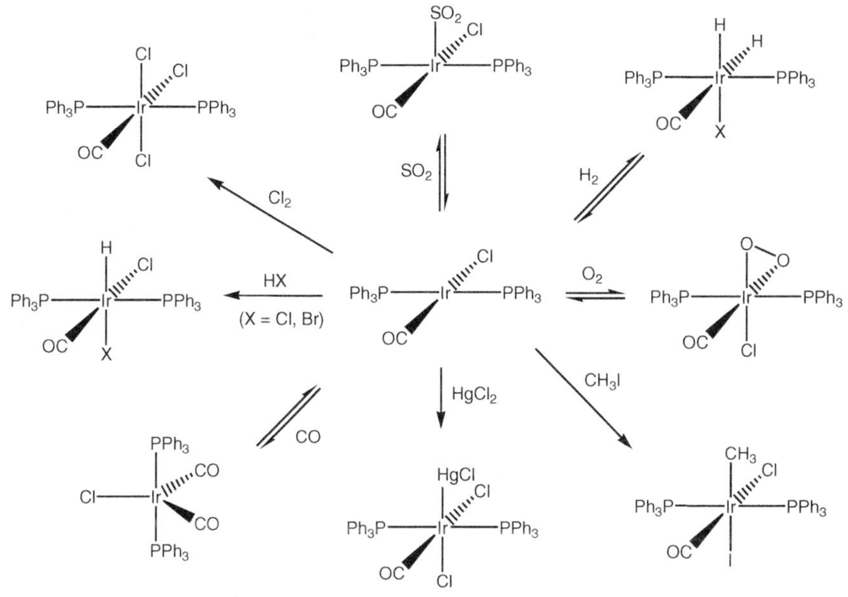

FIGURE 46.2 Reactions of Vaska's Compound.

TABLE 46.1
νCO (cm^{-1}) for Vaska's compound and its adducts

Molecule added	Formula of product	ν_{co} / cm^{-1}
	$IrCl(CO)(PPh_3)_2$	1967
CO	$IrCl(CO)_2(PPh_3)_2$	1934, 1988
O_2	$IrCl(O_2)(CO)_2(PPh_3)_2$	2015
SO_2	$IrCl(SO_2)(CO)(PPh_3)_2$	2021
D_2	$IrClD_2(CO)(PPh_3)_2$	2034
HCl	$IrHCl_2(CO)(PPh_3)_2$	2046
CH_3I	$IrCl(CH_3)I(CO)(PPh_3)_2$	2047
Cl_2	$IrCl_3(CO)(PPh_3)_2$	2075

Thus, when a molecule of CO adds to [$IrCl(CO)(PPh_3)_2$], the product [$IrCl(CO)(PPh_3)_2$] is still a compound of Ir(I); iridium's oxidation state remains unchanged. The CO stretching frequency is hardly changed, from 1967 cm^{-1} in the parent molecule [$IrCl(CO)(PPh_3)_2$] to an average value of 1961 cm^{-1}

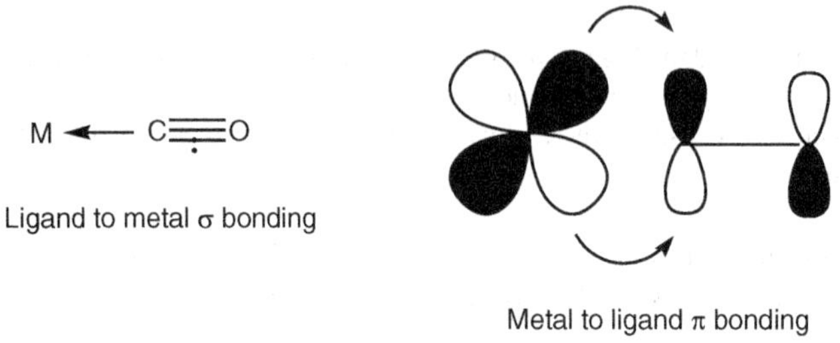

FIGURE 46.3 σ and π bonding of CO to a transition metal such as iridium.

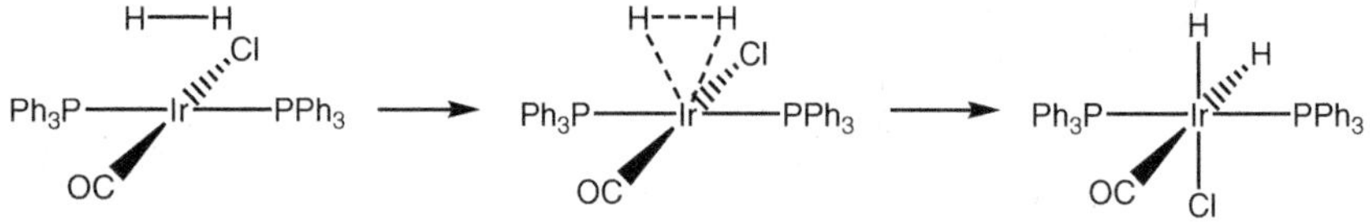

FIGURE 46.4 Formation of the dihydrogen adduct of Vaska's Compound.

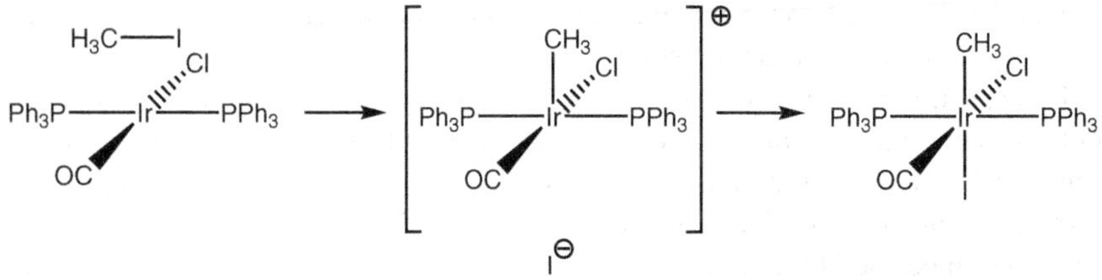

FIGURE 46.5 Formation of the iodomethane adduct of Vaska's Compound.

in $[IrCl(CO)_2(PPh_3)_2]$. On the other hand, when one molecule of Cl_2 adds to $[IrCl(CO)(PPh_3)_2]$, the product $[IrCl_3(CO)(PPh_3)_2]$ is an Ir(III) compound, so there is a shift in ν_{CO} of over 100 cm^{-1}, from 1967 cm^{-1} to 2075 cm^{-1}.

Crystallography also sheds light on the bonding. The structure of the dioxygen adduct $[IrCl(CO)(O_2)(PPh_3)_2]$ was redetermined in 2008 (the crystal in the original determination may have partly decomposed) and revealed an O-O bond length of 1.47Å – this should be compared with the values of 1.21 Å in the O_2 molecule and 1.49 Å in the peroxide ion, O_2^{2-}. This suggests that the iridium's oxidation state has increased substantially above (+1), in agreement with analysis of the IR data. As its oxidation state increases, the iridium is losing electron density, donated into the antibonding orbitals of the O_2 ligand too, weakening – and lengthening –the O-O bond. $[IrCl(CO)(O_2)(PPh_3)_2]$ (46.2) is a case where the added molecule has maintained its original connectivity on binding to iridium, similar to the case with CO and SO_2. There are other cases where the incoming molecule makes a concerted addition, before the molecule breaks up, as is the case with H_2 and Cl_2 (Figure 46.4).

In contrast, CH_3I adds via a S_N2-type reaction; the CH_3I dissociates, the slow step involving a five-coordinate cationic species (Figure 46.5).

By 1966, the name Vaska's compound was being used to refer to $[IrCl(CO)(PPh_3)_2]$; as evidenced by three papers written in 1966 and published in 1967 (see the bibliography of this chapter under 'Collman').

In his research, Lauri Vaska pioneered the concept of oxidative addition reactions, which came to underpin an understanding of the catalytic processes that were discovered in the coming years, reactions involving complexes – notably of Rh(I) – that came into being (see Wilkinson's complex, pp. 223–227).

47 VX and the Earlier Nerve Agents

Nerve agents were first synthesised in Nazi Germany, shortly before the outbreak of World War II, but the discovery of the first nerve agent was an accident, as was the discovery of VX much later. In 1932, a German PhD student named Gerde von Krueger, working at the Friedrich-Wilhelms University of Berlin under the supervision of Willy Lange, was researching cheap synthetic alternatives to nicotine, used as an insecticide at that time. She carried out the reactions between alkyl halides and Ag_2FO_3P (silver monofluorophosphate), obtaining very low yields of the organophosphorus compounds $(RO)_2P(=O)F(R = Me, Et)$, dimethyl- and diethyl monofluorophosphate. The researchers reported that the compounds led to difficulty in breathing, blurred vision and a painful oversensitivity of the eyes towards light.

In Frankfurt, a few years later, Gerhard Schrader of I. G. Farben was studying sulphur- and phosphorus-containing compounds as possible insecticides. Two days before Christmas 1936, he synthesised a clear, colourless liquid, $[(EtO)P(=O)(CN)(NMe_2)]$, and found that very dilute solutions of it killed 100% of leaf lice. This compound, now known as tabun (47.1), was accompanied by some unpleasant properties, due to which spillage of just a tiny drop caused shortage of breath and meiosis – contraction of the pupils of the eye. For a while, Schrader could hardly read, except in daylight, nor was he able to drive home. Schrader and his assistant had to stop work for three weeks. Animal tests showed that tabun was far too toxic to mammals to be used commercially as an insecticide. At that time, any scientific discoveries with possible military applications had to be reported to the Wehrmacht, the German armed forces, who wanted better chemical warfare agents than those available in World War I. Field trials immediately began, followed by construction of a pilot plant, making tabun at Dyrhenfurth, near Breslau (now Wroclaw, in Poland), in December 1939. This proved to be a slow project, extending two and a half years on account of the toxicity of both tabun and the chemicals used to make it. During the project, at least 10 workers were killed during the plant's operation.

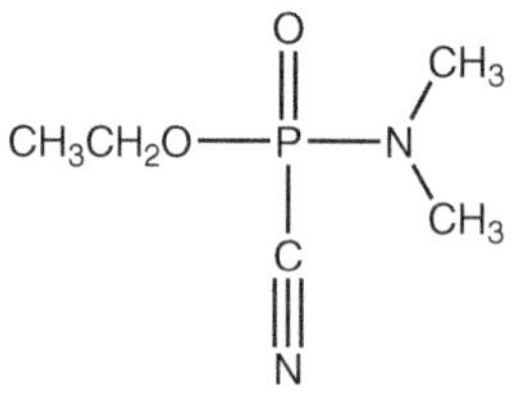

(47.1)

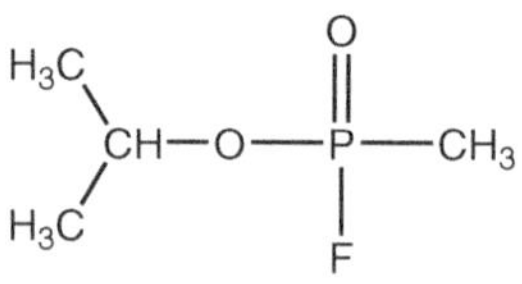

(47.2)

 DOI: 10.1201/9781003783565-47

Two years after tabun, Schrader synthesised the more toxic sarin, $[MeP(=O)F(OCHMe_2)]$ (47.2). Once again, the Germans started building a pilot plant for sarin synthesis at Dyrhenfurth, following up with a production plant near Frankfurt. Still incomplete at the end of World War II, both these plants were captured by Soviet forces, giving Russia a nerve agent capability. The related cyclosarin (47.3) was also synthesised and studied as a possible nerve agent. Another nerve agent, which was given the name soman (47.4), was synthesised towards the end of the war in 1944.

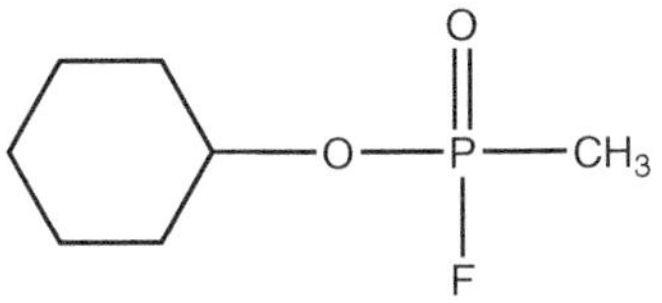

(47.3)

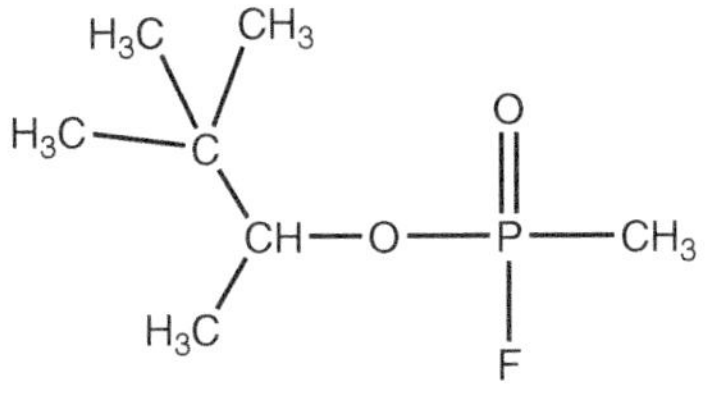

(47.4)

Meanwhile, in 1941, the British chemist Bernard Saunders, working at the University of Cambridge, had synthesised some esters of monofluorophosphoric acid (which were similar to Lange's published work), notably diisopropyl fluorophosphate (DFP), $[(Me_2CHO)_2P(=0)F]$ (47.5). He found that exposure to a small amount of DFP vapour could similarly lead to meiosis. It was found to be less toxic than sarin and tabun; as a weapon it was considered to have no advantages over the World War I compounds phosgene or mustard gas.

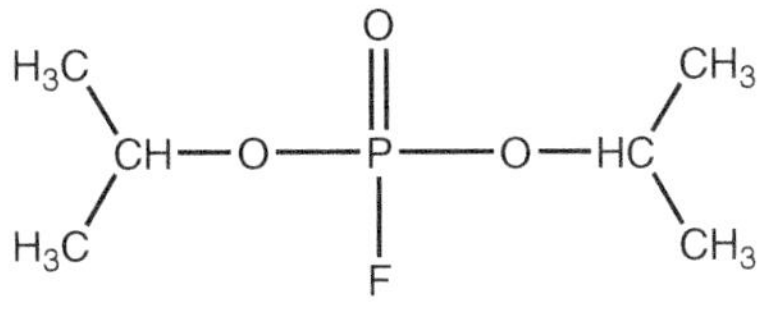

(47.5)

The Nazis did not use nerve agents during World War II, partly for logistical reasons, and also possibly because they expected the Allies to have something similar, as Lange's studies had been published in the open literature. With the passage of time, nerve agents became more accessible to other countries, including Iraq. Sarin has been the weapon of choice on a large scale as it is much more volatile than VX, (the boiling points of sarin and VX are around 158 °C and 298 °C, respectively) and quicker acting. Denser than air, it permeates areas like underground bunkers. It is also non-persistent, dissipating quickly.

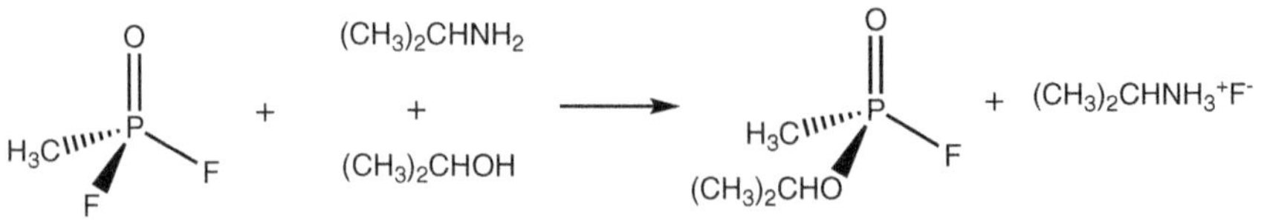

FIGURE 47.1 Formation of sarin in a binary weapon.

It is believed that Iraqi forces used sarin against Kurds in Halabja on 16-17 March 1988, resulting in some 5,000 fatalities. The sarin was part of a 'cocktail' also involving a mustard agent and cyanide. An exploded Iraqi 155 mm shell (used as an improvised roadside bomb) showed that it had been a binary weapon that contained the precursors for sarin. Because sarin tends to decompose on keeping, it is ideal for use in binary weapons, which contain the nerve agent's precursors that are mixed immediately before the weapon is used. In the case of sarin, a binary weapon contains separate containers of methylphosphoryldifluoride and a mixture of propan-2-ol and isopropylamine. Once the shell or rocket is fired, the containers break, the reactants mix and sarin is formed (Figure 47.1).

A few years later, on 20 March 1995, ten members of the Japanese *Aum Shinrikyo* 'doomsday cult' attacked commuters on the Tokyo subway system. They placed bags containing sarin in the carriages, using umbrellas with sharpened tips to puncture the bags. Hundreds of people showed symptoms of sarin poisoning, with 12 fatalities and about 500 hospitalisations. Many more would have died had the sarin not been impure and had prompt medical attention not been available.

When a civil war broke out in Syria in 2011, a land which had not adopted the Chemical Weapons Convention, reports emerged of the government forces using chemical weapons, particularly chlorine gas and sarin. When a rocket landed at Khan al-Assal near a rebel-held area on 19 March 2013, a spray of sarin droplets killed over 20 people, injuring over a hundred others. Just a few days later, on 29 March, a helicopter dropped three grenade-type weapons on the city of Saraqueb, killing a woman. Tissue samples from her body revealed degradation products of sarin, particularly the hydrolysis product O-isopropyl methylphosphonic acid. One grenade failed to explode and was found to contain sarin. Then on 21August 2013, an attack on rebel-held areas of the Ghouta suburb in Damascus using multiple sarin-filled rockets caused massive fatalities (estimates vary from around 300 to well over a thousand). On 4 April 2017, a chemical attack killed more than 80 people in the Syrian town of Khan Shaykhun, again sarin-containing bombs were used.

The next nerve agent to be discovered was VX (47.6), another accident, though the reason is familiar. Working for Imperial Chemical Industries (ICI) in 1952, Dr Ranajit Ghosh was trying to make a new pesticide. Again, it was far too toxic for its intended purpose, so details were shared with the Chemical Weapons Research Centre at Porton Down (coincidentally, near Salisbury). At around that time, it was independently synthesised by the Swedish scientist Lars-Erik Tammelin.

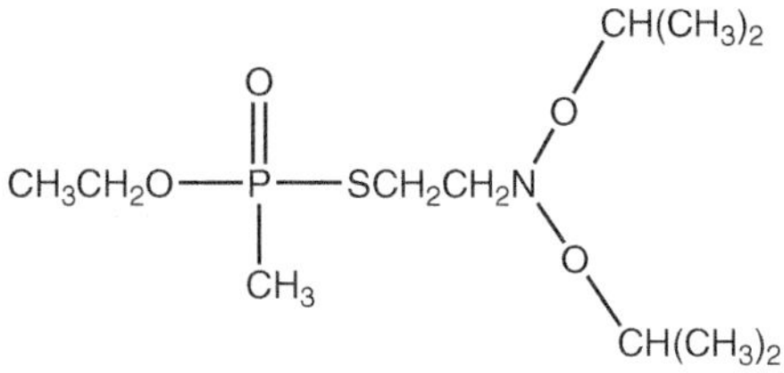

(47.6)

Britain shared information about its nerve agent work with the USA. It is said that VX was part of the deal between Britain and the USA in exchange for information on thermonuclear weapons (though this is disputed). The United States began manufacturing their own VX in 1961. The Russians had some intelligence of what the West was up to and produced their own version of VX. Not an exact copy, but an isomer (47.7) – maybe they just had the molecular formula of VX to go on?

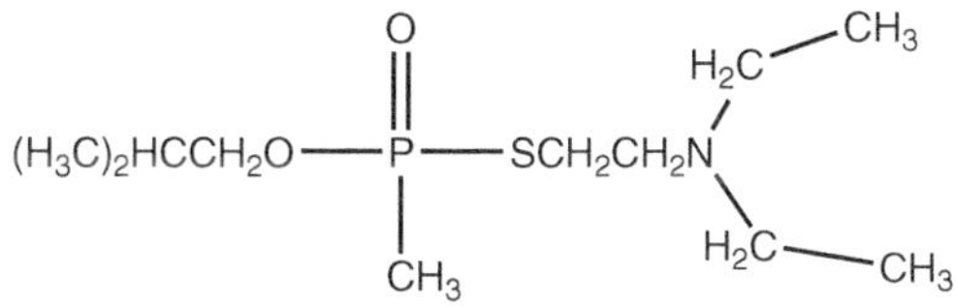

(47.7)

Although less volatile than them, VX is more toxic than the German nerve agents – not just as aerosols, but also because it is absorbed through the skin. During testing, an accident indicated the danger posed by VX – an escape of VX from the chemical weapon proving ground at Dugway (Utah) on 13 March 1968 killed over 4000 sheep grazing 27 miles away in the Skull Valley area.

After the introduction and testing of VX, the Americans prepared a considerable number of weapons containing VX, particularly artillery shells and bombs and rockets. Soon after the Dugway episode, the United States, under President Richard Nixon, renounced the first use of chemical weapons in November 1969. Subsequently, nations worldwide supported the Chemical Weapons Convention, which banned the creation, manufacture and storage of chemical weapons. The United States signed it in 1993 and ratified it in 1997, as did Russia. But this agreement has not stopped some small nations – or individuals – from using nerve agents.

Unlike sarin, VX does not seem to have been used in mass attacks, rather for the assassination of individuals. The first case occurred on 12 December 1994, when two members of the Japanese *Aum Shinrikyo* cult (one of them named Akira Yamagata) injected VX into the neck of Takahito Hamaguchi, a former member of their cult. Within ten minutes, Hamaguchi went into a coma, dying ten days later. Subsequent analysis of Hamaguchi's blood showed the presence of three VX metabolites – ethyl

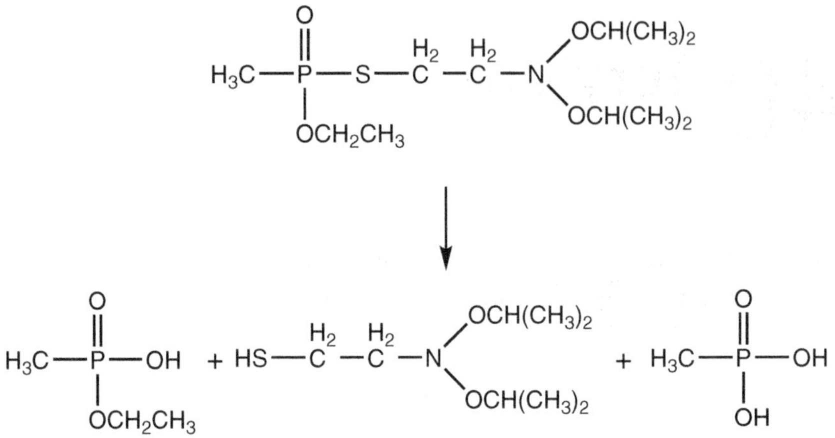

FIGURE 47.2 Formation of blood metabolites of VX.

methylphosphonate, methylphosphonic acid and diisopropyl-2-(methylthio)ethylamine (Figure 47.2).

Another assassination using VX took place at Kuala Lumpur International Airport on 13 February 2017, when a man called Kim Choi was assaulted by two women – one approached him from behind and splashed a liquid onto his face, whilst the other woman placed a damp cloth on his face. Choi told the airport staff about what had happened, but then started to feel unwell, sweating and salivating, having difficulty in breathing. The pupils of his eyes narrowed to pinpoints, whereupon the classic antidote atropine was administered. It was a case of 'too late' – Choi died while still on the way to hospital. There was much more to the story than appeared at first sight. 'Kim Choi' was really Kim Jong Nam, half-brother of Kim Jong Un, the leader of North Korea. As the eldest son of Kim Jong Il, the previous North Korean leader, Kim Jong Nam was the 'heir apparent' until 2001, when Kim Jong Un was elevated to replace him. Kim Jong Nam was later exiled. Autopsy detected a very low level of acetylcholinesterase in Kim's body, and VX was detected on his face. His rucksack contained a dozen vials of atropine, suggesting that he was afraid of being attacked. The two attackers – a 28-year-old Vietnamese woman named Doan Thi Huong and a 25-year-old Indonesian woman named Siti Aisyah – said that they had been recruited for a 'stunt' for a TV programme. They may have had tiny amounts of VX on their hands, which they washed off very soon afterwards (it is only slowly absorbed), though others have suggested that a binary form of VX was used, with each woman having a different precursor on their hands (this seems unlikely). The North Korean government is suspected of being responsible for this assassination.

Nerve agents interfere with nerve transmission in the body's key organs, such as the heart and lung. Nerve fibres have gaps in them called synapses; messages are carried across synapses using a neurotransmitter molecule called acetylcholine (ACh), which is released when the message arrives. When acetylcholine gets to the

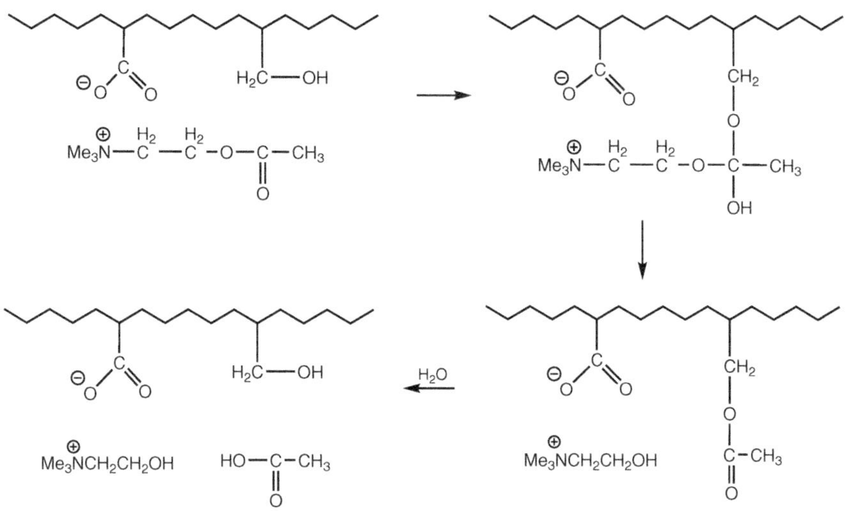

FIGURE 47.3 Acetylcholinesterase (AChE) breaking down and acetylcholine (ACh) molecule.

other side of the synapse, it docks in a 'receptor', which triggers a response, a nerve impulse, sending the message on. It is important that the ACh molecule that has just arrived be released by the receptor and then destroyed, else there will be a build-up in ACh concentration, causing more and more nerve impulses, resulting in overstimulation of the nervous system, with high doses causing seizures and possibly death from respiratory failure. An enzyme called acetylcholinesterase (AChE) breaks down ACh molecules.

Figure 47.3 successively shows acetylcholine (ACh) approaching a serine residue in the enzyme acetylcholinesterase, with the acetyl group being transferred to the serine, choline being released, and finally the serine group being freed from the acetyl group by hydrolysis, thus regenerating the enzyme.

So, what does a 'nerve agent' do? To take a representative example, a molecule of sarin binds to the same serine group as ACh does (Figure 47.4), forming an adduct in which the hydrogen from the serine -OH is transferred to the P=O group of the sarin, analogous to what happens to the C=O group in acetylcholine. However, this initial adduct quickly decomposes, a process known as 'aging', in which a molecule of HF is eliminated, leading to a product which cannot be regenerated, thus blocking the active site of the enzyme, an enzyme that can no longer function. Hence, ACh can no longer be broken down, followed by life-ending consequences.

As an antidote to a nerve agent, there is more than one option. Soldiers can take the compound pyridostigmine bromide (PB) as a prophylactic, in advance of battle. PB is a reversible inhibitor of AChE. It binds to some of the AChE, protecting it from possible nerve agent exposure, so that after the attack, the PB slowly unbinds from AChE over a few hours, releasing the AChE so that it can get back to doing its

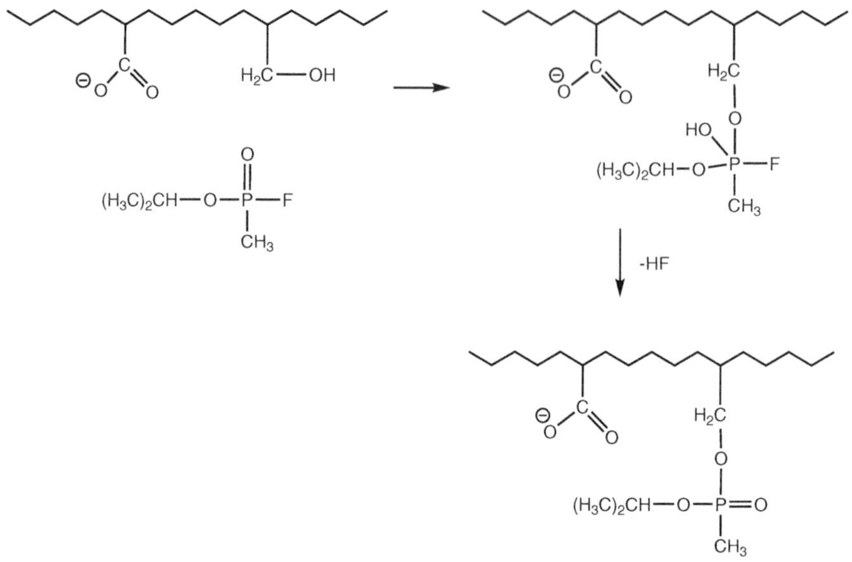

FIGURE 47.4 Sarin blocks the active site of AChE.

job. Injecting a mixture of the alkaloid atropine and 2-pyridinealdoxime methiodide (PAM) is an alternative. Atropine's role is to block the receptor for acetylcholine, whilst the PAM displaces nerve agent binding to the AChE, so that the AChE can get back to its job of breaking down ACh. However, PAM only works with sarin and VX, and must be administered quickly, before the adduct 'ages'. In the case of soman and tabun, another oxime, obidoxime, is the treatment.

48 Water

Water is chemically hydrogen oxide (or dihydrogen oxide), but no one ever calls it that. In ancient times, water was regarded as an elemental substance, of course. Its true nature was not appreciated until much later. The elements hydrogen and oxygen were not isolated until the late eighteenth century. It was observed by scientists like Henry Cavendish that burning hydrogen produced a colourless liquid, which after a while was seen to be water, though the implications of this were not immediately realised, until Lavoisier caught on, naming this inflammable gas *hydrogène*, the water former. Very soon it was discovered that electrolysis, a brand-new experimental technique, split water into hydrogen and oxygen.

The water molecule is V-shaped, not linear, because it contains two pairs of non-bonding electrons, in addition to the two electron pairs in the O-H bonds. These four electron pairs are arranged approximately tetrahedrally around the oxygen atom. The oxygen atom is much more electronegative than the hydrogen atom, causing the O-H bonds to be rather polar (48.1). This has a significant effect on the properties of water.

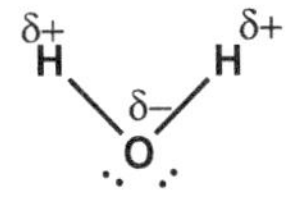

(48.1)

The O-H bond length is 0.958 Å and the H-O-H bond angle is 104.45° for an isolated H_2O molecule in the gas phase, the distortion from pure 'tetrahedral' due to the repulsions involving lone pairs being slightly stronger than those involving bond pairs.

Water manages to be commonplace and extraordinary at the same time. It is singular in that it is the only substance that can exist as a solid, a liquid and a gas at normal Earth temperatures. This is certainly not expected from a molecule the size of water. If you study Table 48.1, it shows the melting and boiling points of water, along with three other molecules of comparable size and mass – methane, ammonia and hydrogen fluoride – which contain the same number of electrons, as well as the noble gas argon, made of atoms of comparable mass.

The boiling point of a substance depends upon its intermolecular (or interatomic) forces. Noble gases like argon are made of individual atoms. Overall, these atoms are not polar, but at a given instant, the electron cloud may be asymmetric, polarising the atom, causing an attraction to another 'polarised' atom. That attractive force increases as the number of electrons in the atom increases. This is why the boiling points of the noble gases increase as the atoms get heavier. A similar effect can be seen with the alkanes, hydrocarbons with the formula C_nH_{2n+2}, such as methane.

DOI: 10.1201/9781003783565-48

TABLE 48.1
Properties of water, methane, ammonia, hydrogen fluoride and neon compared

	CH_4	NH_3	H_2O	HF	Ne
M_r	16	17	18	20	20 (A_{rr})
Covalent bonds per molecule	4	3	2	1	
Number of polar H-X bonds	0	3	2	1	
Lone pairs per molecule	0	1	2	3	
M.p. (°C)	−182.5	−77.7	0	−83.4	−248.6
B.p. (°C)	−161.5	−73.4	100	19.5	−246.1
Electronegativity of central atom	2.5	3.1	3.5	4.1	

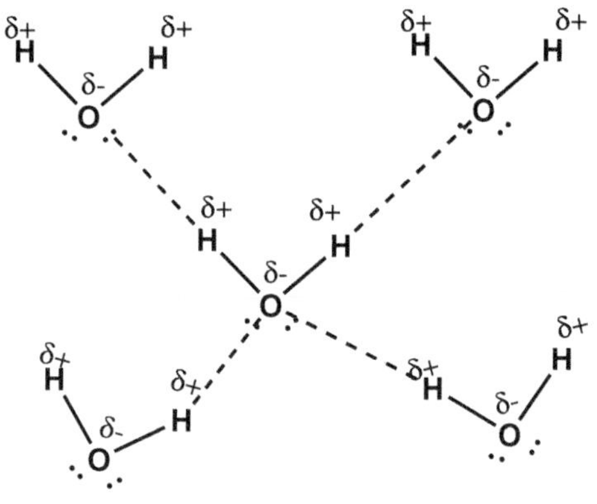

FIGURE 48.1 A water molecule can form hydrogen bonds (dotted lines) to four neighbouring water molecules.

As the molecules get bigger, the boiling points increase; this is why the mixture of alkanes that we know as crude oil can be separated in oil refining by fractional distillation. These intermolecular forces are mainly what are known as 'London' forces (after the scientist who enunciated the principle) but are more often called Van Der Waals' forces.

Table 48.1 shows that water has melting and boiling points that are at least a hundred degrees higher than the others. Boiling points, in particular, are a measure of intermolecular forces. Let's begin with methane. The electronegativities of carbon (2.5) and hydrogen (2.2) are similar, so the covalent C-H bonds are essentially non-polar and the intermolecular forces are simply Van der Waals' forces. In contrast, the three succeeding elements are significantly more electronegative than hydrogen and therefore the C-X bonds are polar enough to participate in 'hydrogen bonds', where there is an attraction between the hydrogen (bearing a slight positive charge) in one molecule and a non-bonding pair (known as a 'lone pair') of electrons in a neighbouring molecule. This causes NH_3, H_2O and HF to have significantly higher melting and boiling points than CH_4.

Then, why are the melting and boiling points of H_2O distinctly higher than those of NH_3 and HF? The answer lies in the numbers. A molecule of ammonia has got three polar N – H bonds but only one non-bonding pair ('lone pair'). So, on average, a NH_3 molecule can only participate in two hydrogen bonds, as there are not enough lone pairs for more. On the other hand, a HF molecule has got three lone pairs of electrons, but only one polar H – F bond. So again, an HF molecule can only participate in two hydrogen bonds, there are not enough polar H-F bonds for more. But a water molecule has two polar O – H bonds and two non-bonding pairs – 'just right' – so that a water molecule can participate in four hydrogen bonds.

This means that more energy is needed to separate the water molecules from each other when it is boiled. At the distances between gas-phase water molecules, there are no hydrogen bonds – which make boiling water a more endothermic process than boiling NH_3 and HF – let alone CH_4.

Ice, like the liquid form of water, also has a hydrogen-bonded structure, linking the individual molecules. This open structure has four hydrogen bonds per water molecule (in the liquid phase, it is believed to average around 3.3). Water molecules are slightly further apart in ice than in the liquid state, consequently ice is less dense than water and hence, which it floats, of course. Likewise, when an ice cube tray is filled with water, then put in the ice-making compartment of the fridge, as the ice freezes the ice cubes expand beyond the top of the tray.

The greatest density of liquid water occurs just above its freezing point, at 4°C. This means that generally there is liquid water at the bottom of a 'frozen' lake, and thus fish can survive at the lake bottom.

In the liquid and solid forms, of course, water could be said to have a giant structure, because of the way in which the molecules associate via hydrogen bonds. This should not be confused with 'polywater', a scientific aberration that arose in the 1960s. In 1962, a Russian chemist named Nikolai Fedyakin reported some anomalous properties of water that had been condensed in quartz capillary tubes, for example, that it boiled at 250°C. It had a low freezing point, an increased density and very high viscosity, compared to 'ordinary' water. This work was taken up by a respected Russian physical chemist, Boris Derjaguin, or Deryagin (1902–1994) and details of further experiments were published in Russian scientific journals. These results did not attract much attention in the West until Derjaguin came on a lecture tour of England, talking about this 'anomalous water' and his work was taken up by the influential (and pro-Soviet) chemist J. D. Bernal. Study of this phenomenon began in England and the United States. A number of scientists began to take this seriously. Concern was expressed that if this polywater escaped, it could polymerise all the normal water. The bubble did not burst until around 1970, when spectroscopy showed the presence of foreign material such as silica, as well as some very common ions (e.g. Na^+, K^+, Cl^- and SO_4^{2-}). Derjaguin graciously conceded in 1973 that he was mistaken, saying that the material he had isolated was contaminated by colloidal silica.

Water dissolves many substances, though not everything of course. It dissolves a lot of ionic compounds. Substances like sodium chloride, NaCl, are held together in a lattice by the strong attraction between the oppositely charged ions, Na^+ and Cl^-, which means that a good deal of energy is required to overcome the attraction. Water can dissolve this because it is a polar molecule. So there is an attraction between the oxygen atom, which bears a slight negative charge (δ-), and the positively charged

Na^+ ions; likewise an attraction between the hydrogen atoms, which bears a slight positive charge (δ+), and Cl^- ions. Since there are about six waters around each ion, these processes release quite a lot of energy, which is enough to compensate for the energy needed to separate the ions when the lattice is broken up.

In contrast, water does not dissolve a lot of organic molecules, including hydrocarbons such as the ones found in petrol ('oil and water don't mix'). These hydrocarbons are not polar and thus cannot hydrogen-bond to water molecules. Small alcohols like methanol and ethanol have O-H groups, which can hydrogen-bond to water molecules, and are therefore soluble in water. Larger alcohol molecules have long hydrocarbon-like chains, which is why they do not dissolve in water, because the hydrophobic effect of the long carbon-based chain outweighs the hydrophilic O-H group.

Water in the human body does many vital things. For example, it takes nutrients around the body and controls the temperature of the body. And at a much smaller scale, it is vital to the way cells work, playing an essential transport role. The blood that takes oxygen to cells for respiration (and transports the waste like carbon dioxide away) is around 81% water.

On average, a human body is about 60% water; in infants, it is much more, as high as 75%, but significantly less than 60% in old or obese people. Men on average have a greater percentage than women, due to greater muscle mass. Apart from bones (around 22%), organs tend to have nearly 70–80% water.

The origin of the earth's water is uncertain, but isotopic analysis suggests that it may have come from material like enstatite chondrite meteorites. Water is vital for human survival. And there is a lot of it on earth, but a good deal is in the wrong places, or the 'wrong sort of water'. Approximately 70.8% of the earth's surface is covered in water, primarily saltwater oceans (96.5%). Just some 3% is freshwater, but around 69% of that is in solid form in the polar ice caps – mainly in the Antarctic region, with some of it in glaciers – and about 30% is groundwater below the surface, potentially accessible, leaving only 1% in rivers and lakes. All that seawater is unusable for drinking, unless you have access to an (expensive) desalination plant. People in some parts of the world have to walk miles to obtain drinking water, whereas many others have drinkable water on tap, taking it for granted, as well as the fact that this water has been chlorinated to kill microorganisms, making it safe to drink. Wars may be fought over water in the coming years, just as they have been over oil.

49 Wilkinson's Catalyst

Having revolutionised organometallic chemistry (think, ferrocene) in the 1950s, Geoffrey Wilkinson (1921–1996), Professor of Inorganic Chemistry at Imperial College, London, moved into the area of rhodium chemistry. He found that hydrido-complexes of rhodium had catalytic potential and looked to extend this to complexes of tertiary phosphine ligands, which were being popularised by Joseph Chatt, another great British inorganic chemist. Chatt's team was using reactions between ethanolic solutions of $RhCl_3$ and trialkylphosphines. When rhodium trichloride combines with tertiary phosphines such as trimethylphosphine or triethylphosphine, the reaction yields octahedral $RhCl_3$ complexes like $RhCl_3(PMe_3)_3$. In many ways, these are classical Werner type complexes. These trialkylphosphines are very air-sensitive and not the easiest of chemicals to work with. Noting that triphenylphosphine was air-stable and much easier to handle, Wilkinson asked his research student, Fred Jardine, to react it with rhodium trichloride. The first time he tried, Fred got pale brown $[RhCl_3(PPh_3)_3]$ (49.1), analogous to Chatt's compounds. Whenever Fred tried to repeat this, he got burgundy-red crystals – Fred had made a Rh(I) complex, $[RhCl(PPh_3)_3]$ (49.2) - in fact, the rhodium(III) complex can be hard to isolate, as some triphenylphosphine usually acts as a reducing agent.

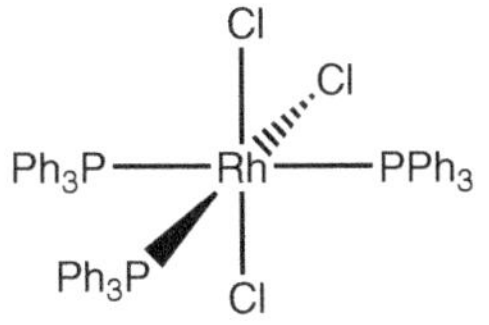

(49.1)

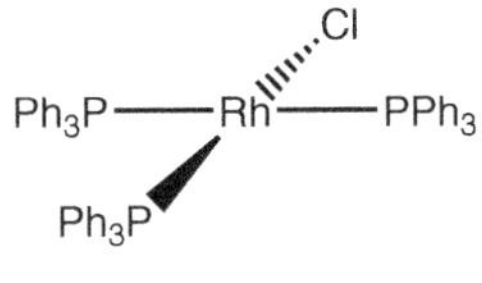

(49.2)

It was soon found that this coordinatively unsaturated Rh(I) compound – a truly serendipitous discovery – was an outstanding homogeneous hydrogenation catalyst, the first such to work at normal atmospheric pressure, its catalytic cycle being illustrated in Figure 49.1.

When dissolved in a suitable solvent (such as benzene-ethanol), one of the PPh_3 molecules dissociates, being replaced in the coordination sphere of the rhodium by a weakly bound solvent molecule, generating a molecule of $[RhCl(PPh_3)_2]$ (solvent)], effectively three-coordinate. When the atmosphere above the solution is hydrogen gas, a H_2 molecule coordinates, forming two Rh-H bonds and a Rh(III) compound, $[RhH_2Cl(PPh_3)_2]$ (solvent)] in an oxidative addition. The colour of the solution changes from red to yellow. $[RhH_2Cl(PPh_3)_2]$ is a 16-electron species and has a vacant coordination site, so it can add an alkene molecule, forming a six–coordinate

DOI: 10.1201/9781003783565-49

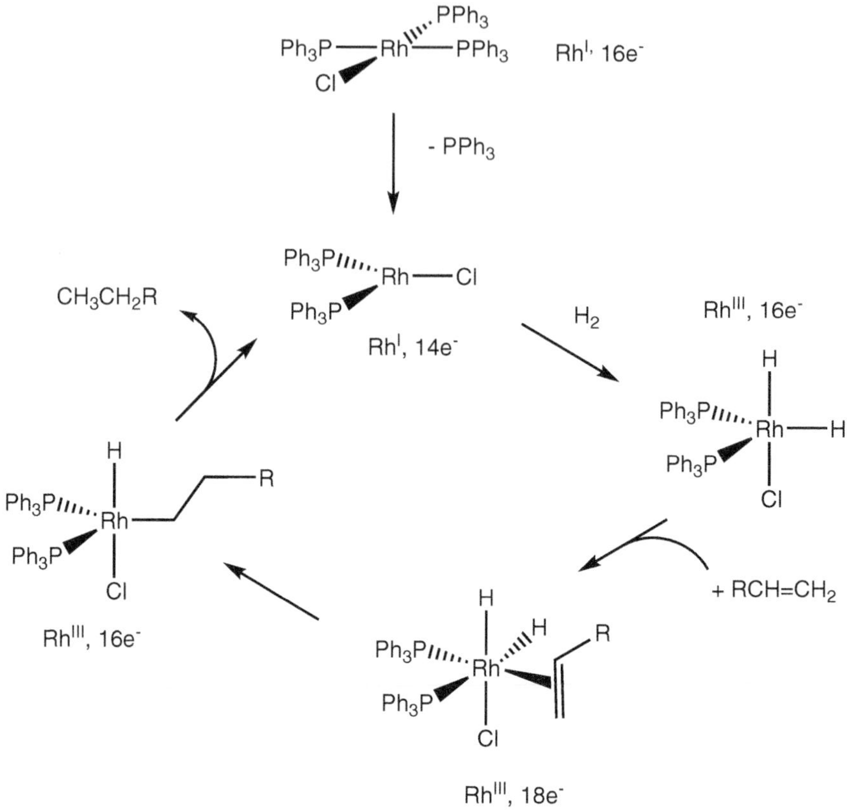

FIGURE 49.1 Catalytic hydrogenation cycle of a terminal alkene, with $[RhCl(PPh_3)_3]$ catalyst.

18-electron complex. Then, this complex rearranges – the bound alkene is 'inserted' into a Rh-H bond, becoming an alkyl group; another way of thinking of this step is that a hydrogen (a hydride group) is transferred to the alkene. Almost instantly, the other coordinated hydrogen atom is transferred to the alkyl group, turning it into an alkane molecule, which is released. This step can be termed a 'reductive elimination', with the active catalyst species '$RhCl(PPh_3)_2$' being released, and the catalytic cycle can begin again (Figure 49.1). In an alternative view of the cycle, the H_2 molecule is bound to $RhCl(PPh_3)_3$ before the phosphine ligand is lost.

$[RhCl(PPh_3)_3]$ does not reduce benzene rings, they are untouched. Similarly, it does not reduce polar carbonyl or carboxyl groupings, nor nitrile or nitro groups – these polar substances that are typically reduced by hydride ions (H^-).

Terminal alkenes are hydrogenated very rapidly, rather than branched or internal alkenes, because there is less congestion around the C=C bond, whilst reaction with terminal alkynes is much faster than with terminal alkenes. The stereochemistry of addition is *cis-* (Figure 49.2).

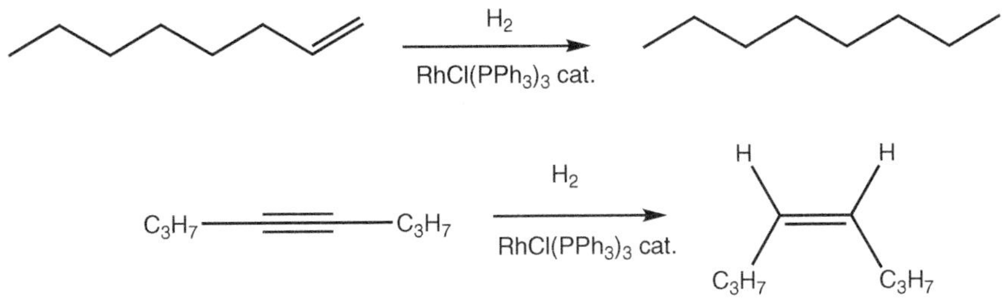

FIGURE 49.2 Hydrogenation of an alkene and an alkyne.

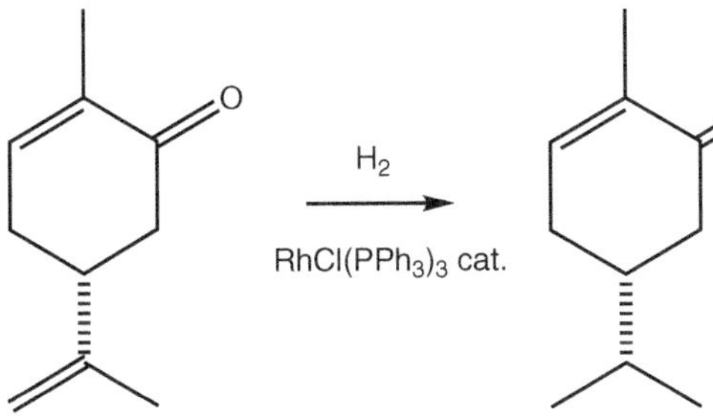

FIGURE 49.3 Reduction of carvone to dihydrocarvone.

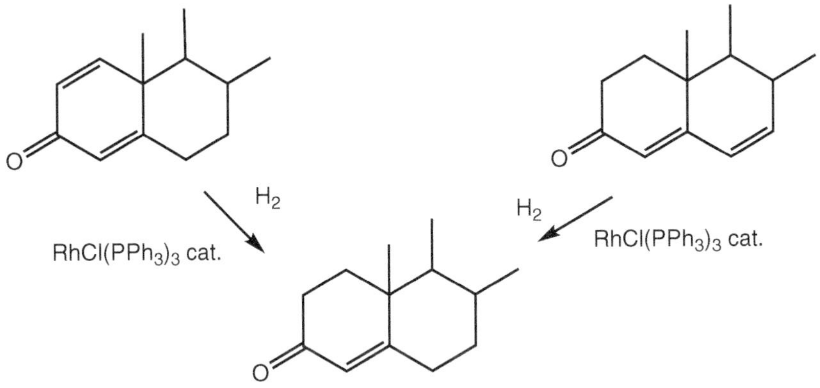

FIGURE 49.4 Selective hydrogenation of C=C double bonds using $RhCl(PPh_3)_3$ as catalyst.

In the reduction of carvone to dihydrocarvone, the least hindered C=C bond is reduced – the disubstituted one, not the trisubstituted C=C (Figure 49.3)

The following examples (Figures 49.4 and 49.5) again show the reduction of disubstituted C=C, rather than trisubstituted or tetrasubstituted.

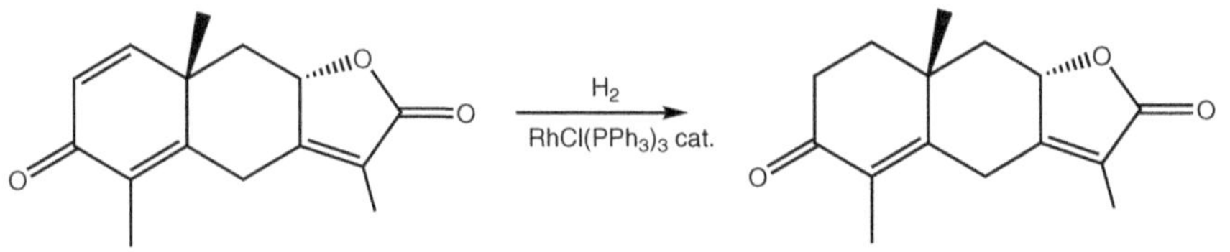

FIGURE 49.5 Further examples of selective hydrogenation of C=C double bonds using $RhCl(PPh_3)_3$ as catalyst.

This selectivity was most important in the synthesis of Ivermectin (Mectizan™; pages 81–83) from the naturally occurring insecticide avermectin. The avermectin molecule contains three C=C double bonds in the *E*- (*trans*-) configuration and two in the less hindered *Z*- (*cis*-) configuration. One of the latter is much more accessible than the other, and it is this one which is selectively hydrogenated by Wilkinson's catalyst.

Apart from its role in catalysis, many of the reactions of $RhCl(PPh_3)_3$ involve addition, often accompanied by a change in oxidation state. Such reactions are now known to be typical of Rh (I) and Ir (I) compounds. Thus, the additions of X-Y (X and Y = H or Cl) are classic oxidative additions forming Rh(III) compounds (Figure 49.6). An unexpected reaction is that with carbon disulphide, ending up in a thiocarbonyl complex, with the CS ligand.

Carbon monoxide easily replaces one tertiary phosphine ligand in $RhCl(PPh_3)_3$, forming *trans*-[RhCl(CO)(PPh_3)$_2$], which is the rhodium analogue of Vaska's compound (p. 207). This compound is also rapidly formed in decarbonylation reactions of aldehydes.

$$[RhCl(PPh_3)_3] + CO \rightarrow [RhCl(CO)(PPh_3)_2] + PPh_3$$

$$[RhCl(PPh_3)_3] + C_6H_5CHO \rightarrow [RhCl(CO)(PPh_3)_2] + C_6H_6 + PPh_3$$

Ethene is bound reversibly – but strongly enough for this coordinated ethene to be not reduced by H_2 - forming $[RhCl(C_2H_4)(PPh_3)_2]$:

$$[RhCl(PPh_3)_3] + C_2H_4 \rightarrow [RhCl(C_2H_4)(PPh_3)_2] + PPh_3$$

However, if the H_2 is bound before the ethene, the ethene can be hydrogenated.

Geoff Wilkinson would have been familiar with Lauri Vaska's work (see pp. 207–211) and could have predicted how $RhCl(PPh_3)_3$ would behave in many of these reactions. Geoff really knew the chemical literature. In the course of doing my PhD at Imperial, I would regularly visit his office. He wasn't my supervisor, but as the head of the inorganic sector, he had to sign my microanalysis request forms. I would go into his office, which was littered with chemical journals. You could tell that he read *Inorganic Chemistry* and the *Journal of the Chemical Society* carefully; there would be notes on the cover of each issue, reminding him who had published papers

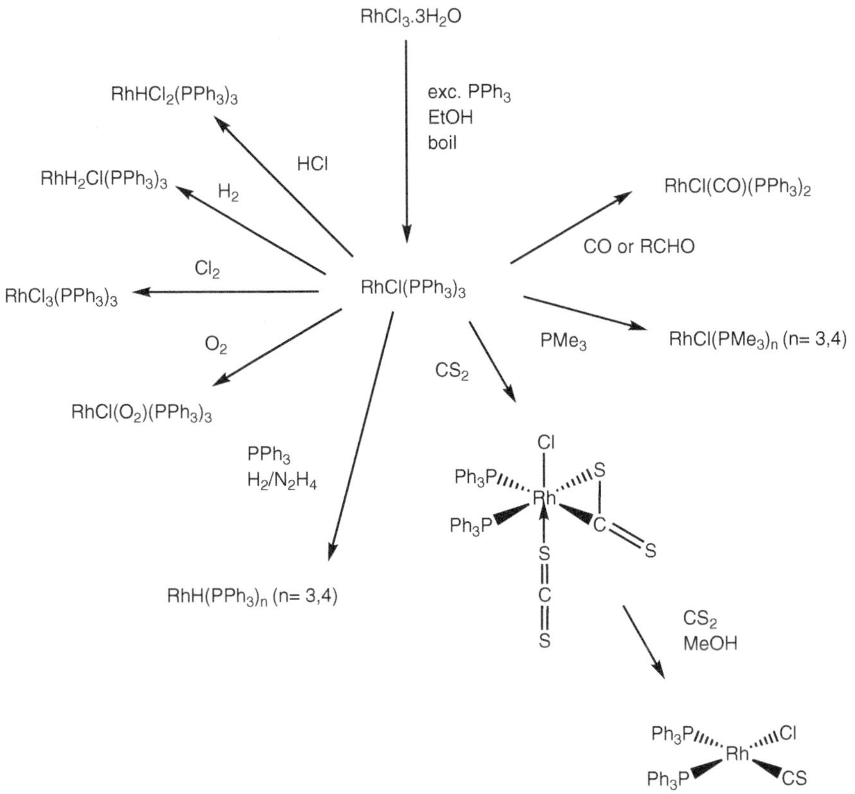

FIGURE 49.6 Reactions of $RhCl(PPh_3)_3$.

in it and what they were about. I'd ask for his signature (please) and Geoff would always ask if it was a new compound. This was not because he was some sort of stamp collector, he had genuine scientific curiosity; Geoff was a man who embodied 'blue-sky' thinking.

It is indicative of the intense research going on at the time that two other researchers – R.S. Coffey at ICI and Martin Bennett at University College, London, (Bennett had done his PhD research a few years earlier under Wilkinson's supervision) – also independently synthesised $RhCl(PPh_3)_3$.

Geoff carried out world-leading research in numerous areas – nuclear chemistry; organometallics, notably sandwich compounds (for which he was co-recipient of the Nobel Prize for Chemistry in 1973); the chemistry of rhodium and ruthenium; stable transition metal aryls and alkyls; imides – are examples which come to mind.

Nevertheless, it is the compound to which others have attached his name that will most readily come to the mind of chemists.

50 Zingerone

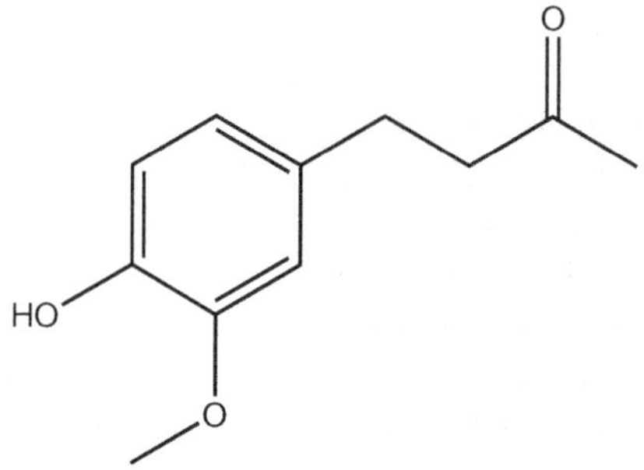

(50.1)

Ginger is indigenous to tropical southeast Asia, believed to originate in Austronesia, centred on the region of Taiwan. It is suggested that ginger was so vital to these communities that when they migrated, they took ginger with them (in their canoes), and thus ginger got planted in the gardens of their new homes. Ginger was essential to them, as a spice, a herb and a medicine. Thus, it reached the Philippines and what we now know as the Spice Islands; on to New Guinea and Java; to Sumatra and Malaysia. From there, ginger went further afield, arriving at places like India and Sri Lanka over 3,000 years ago, whence it reached the Middle East and the Mediterranean world, being described by Pliny the Elder in the first century AD (he said that a pound of ginger cost six denarii, three days' wages for a working man), and in the same century the writer Dioscorides reported that the supply came from Eritrea and Arabia. He also noted that Greek dieticians prescribed ginger for their patients, as it was warming, digestive, gently laxative and appetising, as well as helping in poison antidotes and in cases of cataracts (it was one of the ingredients in *Mithridaticum*, the universal antidote concocted by King Mithridates of Pontus).
Confucius is said to have been a habitual eater of ginger (though history is not clear whether this was among his sayings). Ginger reached England during the Middle Ages (it gets a mention in Shakespeare's *Henry V*) and reached as far as the Caribbean in the sixteenth century, where its cultivation began. Leading countries for its cultivation include India (the leader), Nigeria, China, Indonesia and Nepal, in descending order, though there are many other producers worldwide, including Taiwan, Japan, Sierra Leone, Haiti and Jamaica.

The main spicy components of fresh ginger – out of some four hundred compounds detected – are the gingerol family. Gingerol causes 'hotness' of ginger, but at a lower level than chillis, through activating the TRPV1 receptor, the same one targeted by the capsaicin in chilli peppers (p. 21) and by black peppers. Drying and

DOI: 10.1201/9781003783565-50

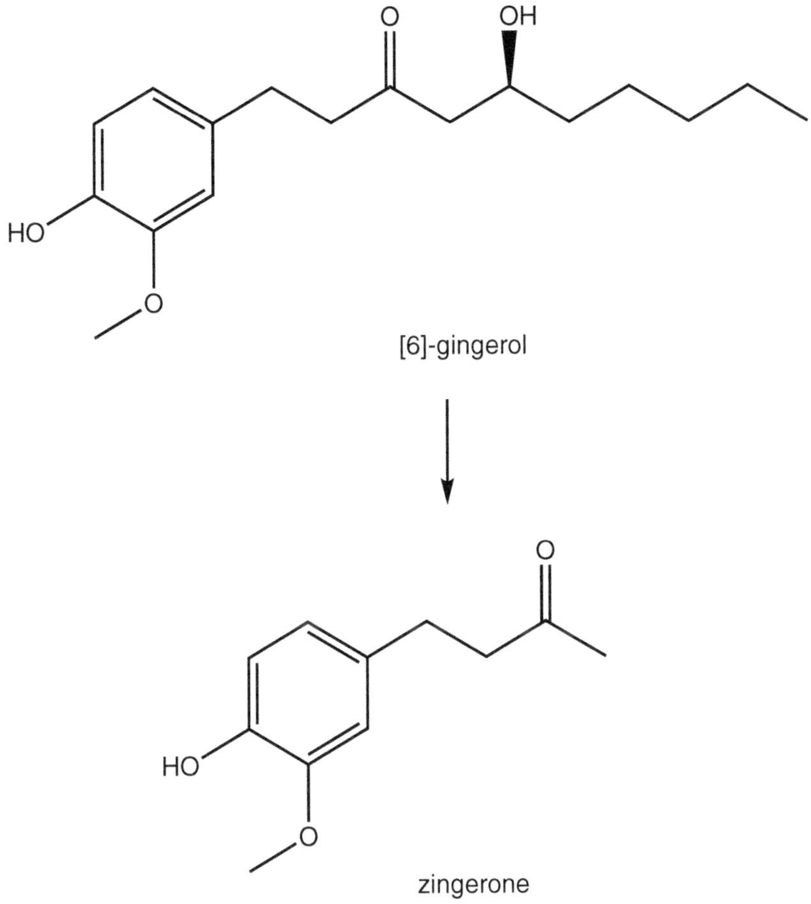

FIGURE 50.1 Formation of zingerone from [6]-gingerol.

cooking changes the balance of chemical present in ginger, with significant amounts of the less spicy zingerone (a shorthand for 4–(4-hydroxy-3-methoxyphenyl)-2-butanone) being formed (Figure 50.1).

Health benefits of ginger have long been perceived. Recent research has revisited the area, suggesting a number of areas where zingerone may be valuable, not least because it is an antioxidant. Thus, it is believed to promote bone health, countering osteoporosis. With this in mind, there are suggestions that zingerone could be a dietary supplement. Tests in mice are claimed to show that it can counter asthma and inflammatory pain, whilst tests on rats indicate a role for zingerone against oxidative stress. Additionally, whilst cyclophosphamide is an important anticancer drug used against a wide range of cancers – lung and breast cancer, ovarian and prostate cancer as well as leukaemia – it has serious side effects, as it attacks normal cells along with cancer cells. Rat testing suggests that zingerone could have a role to counter

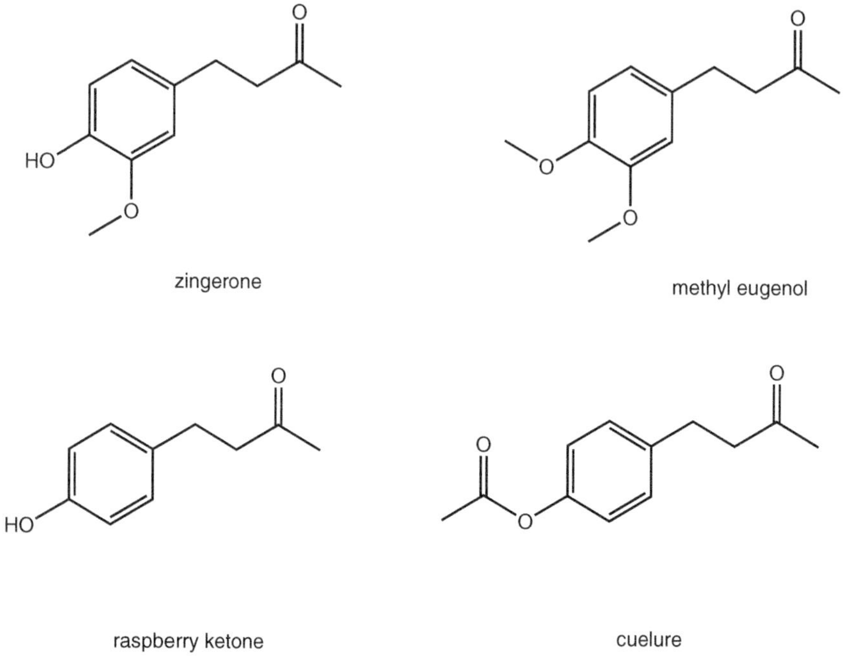

FIGURE 50.2 The structures of zingerone, methyl eugenol, raspberry ketone and cuelure compared.

these side effects. These results are promising, but it should be noted that so far they are primarily based on animal studies and have not been followed up with human studies.

One aspect of zingerone lies in its ability to attract fruit flies – significant horticultural pests – particularly some *Bactrocera* species in the rain forests of Malaysia. Solitary flowers of *Bulbophyllum patens* use it as a lure to bring these flies up close, trapping them between its hinged lip and its column so that it can be pollinated. Zingerone is the reward for the fruit fly, which transfers the pollinarium (the pollen grains) to another flower. Male melon flies store the zingerone and use it as a pheromone to attract females, whilst male *B. papayae* converts zingerone to the closely related zingerol, stores it and then uses it to attract females.

Zingerone has a structure resembling two major natural fruit fly attractants, methyl eugenol and raspberry ketone (4–(4-hydroxyphenyl)butan-2-one), as well as the synthetic 'cuelure'. It can attract some fruit flies unresponsive to those lures. It appears that all three functional groups in zingerone are important in attracting *Bactrocera jarvisi,* Jarvis's fruit fly. This knowledge may influence the design of future fruit fly lures.

Bibliography

1. 2-ACETYLPYRROLINE

J. P. Walradt, R. C. Lindsay and L. M. Libbey, *J. Agric. Food Chem.*, 1970, 18, 926–928 (popcorn flavour volatiles, including 2-acetylpyrazine)

R. G. Buttery, L. Ling and B. O. Juliano, *Chem. Ind.*, 1982, 958–959 (2-acetyl-1-pyrroline in cooked rice)

P. Schieberle and W. Grosch, *Z. Lebensm. Unters. Forsch.*, 1985, 180, 474–478 (volatile flavour compounds of wheat bread crust including 2-acetyl-1-pyrroline)

R. L. Brahmachary, M. Poddar-Sarkar and J. Dutta, *Nature*, 1990, 344, 26 (the aroma of tiger and rice)

P. Schieberle, *J. Agric. Food Chem.*, 1991, 39, 1141–1144 (primary odorants in popcorn, including 2-acetyl-1-pyrroline)

T. Hofmann and P. Schieberle, *J. Agric. Food Chem.*, 1998, 46, 616–619 (2-acetyl-1-pyrroline synthesis)

D. Park and J. A. Maga, *Food Chemistry*, 2006, 99, 538–545 (odorants in different popcorn hybrids)

A. Vanavichit and T. Yoshihashi, *Advances in Botanical Research*, 2010, 56, 49–73 (fragrance and aroma in rice)

A. B. Nadaf, K. V. Wakte and R. L. Zanan, *Journal of Plant Science & Research*, 2014, 1, 01–07 (2-acetyl-1-pyrroline biosynthesis)

R. L. Brahmachary and M. Poddar-Sarkar, *Curr. Sci.*, 2015, 108, 2178–2185 (fifty years of tiger pheromone research)

L. K. Greene, T. W. Wallen, A. Moresco, T. E. Goodwin and C. M. Drea, *Sci. Nat.*, 2016, 103, 37 (volatile urinary compounds of Arctictis binturong)

S. B. Soso and J. A. Koziel, *Molecules*, 2016, 21, 834 (2-acetyl-1-pyrroline in marking fluid of the Siberian Tiger)

K. Wakte, R. Zanan, V. Hinge, K. Khandagale, A. Nadafa and R. Henry, *J. Sci. Food Agric.*, 2017, 97, 384–395 (2-acetyl-1-pyrroline, basmati aroma scented rice (*Oryza sativa* L.))

I. Jinnakot and W. Jirapakkul, *J. Nutrition Sci. Vitaminol.*, 2019, 65, S231–234 (2-acetyl-1-pyrroline in jasmine rice)

N. Renuka, V. T. Barvkar, Z. Ansari, C. Zhao, C. Wang, Y. Zhang and A. B. Nadaf, *Scientific Reports*, 2022, 12, 3911 (2-acetylpyrroline biosynthesis)

2. A-234 AND THE NOVICHOK AGENTS

D. H. Ellison, *Handbook of Chemical and Biological Warfare Agents*, Boca Raton, FL, CRC Press, 2007 (includes alternative structures for Novichoks)

S. L. Hoenig, *Compendium of Chemical Warfare Agents*, New York, Springer, 2008 (includes alternative structures for Novichoks)

V. S. Mirzayanov, *State Secrets: An Insider's Chronicle of the Russian Chemical Weapons Program*, Parker, Outskirts Press, 2008

P. R. Chai, B. D. Hayes, T. B. Erickson and E. W. Boyer, *Toxicology Communications*, 2018, 2(1), 45–48 (Novichok agents: a historical, current, and toxicological perspective)

M. Peplow, *Chem. Eng. News.*, 2018, 96(12), 3–10 ('Nerve agent attack on spy used 'Novichok' poison')

M. Urban, *The Skripal Files*, London, Pan Macmillan, 2018

J. A. Vale, T. C. Marrs and R. L. Maynard, *Clinical Toxicology*, 2018, 56, 1093–1097 (Novichok: a murderous nerve agent attack in the UK)

The House of Commons, Statement by Prime Minister Theresa May, 'Salisbury incident', *Hansard*, Vol. 637, col. 620–21, Mar 12, 2018

H. Bhakhoa, L. Rhyman and P. Ramasami, *R. Soc. Open sci.*, 2019, 6, 181831 (theoretical study of the suspected Novichok agent A-234)

E. J. Hulse, J. D. Haslam, S. R. Emmett and T. Woolley, *British Journal of Anaesthesia*, 2019, 123, 457–463 (organophosphorus nerve agent poisoning: managing the poisoned patient)

K. Kuca, *Int. J. Mol. Sci.*, 2019, 20, 1222 (review: 'Novichoks')

D. Kaszeta, *Toxic*, London, Hurst & Company, 2020, pp. 157–166 on 'The Newcomers' and pp. 234–253 on the Salisbury attack

T. C. C. Franca, D. A. S. Kitagawa, S. F. de A. Cavalcante, J. A. V. da Silva, E. Nepovimova and S. P. Harvey, L. R. McMahon and F. J. Berg, *Heliyon*, 2020, 6, e03153 (hydrolysis and enzymatic degradation of Novichok nerve agents)

O. Yagmuroglu and B. Subasi, *Open Access Journal of Science*, 2020, 4(2), 47–50 (nerve agents: chemical structures, effect mechanisms and detection methods)

3. ALLYL ISOTHIOCYANATE

S.-i. Fujiwara, T. Shi-Ike, N. Sonoda, M. Aoki, K. Okada, N. Miyoshi and N. Kambe, *Tet. Lett.*, 1991, 32, 3503–3506 (synth.)

I. T. Johnson, *Phytochem. Rev.*, 2002, 1, 183–188 (glucosinolates in diet)

S.-E. Jordt, D. M. Bautista, H.-h. Chuang, D. D. McKemy, P. M. Zygmunt, E. D. Högestätt, I. D. Meng and D. Julius, *Nature*, 2004, 427, 260–265 (mustard oil excites sensory nerve fibres through the TRP channel ANKTM1)

P. Hartley, *The Colman's Mustard Cookbook*, Bath, Absolute Press, 2004 (history of Colman's mustard)

L. J. Macpherson, A. E. Dubin, M. J. Evans, F. Marr, P. G. Schultz, B. F. Cravatt and A. Patapoutian, *Nature*, 2007, 445, 541–545 (activation of TRPA1 ion channels by covalent modification of cysteines)

M. E. Cartea and P. Velasco, *Phytochem. Rev.*, 2008, 7, 213–229 (glucosinolates and health)

W. Everaerts et al., *Current Biology*, 2011, 21, 316–321 (TRPV1 mediates the noxious effects of mustard oil)

Y. A. Alpizar, B. Boonen, M. Gees, A. Sanchez, B. Nilius, T. Voets and K. Talavera, *Pflugers Arch – Eur J Physiol*, 2014, 466, 507–515 (allyl isothiocyanate sensitizes TRPV1 to heat stimulation)

A. Mazumder, A. Dwivedi and J. du Plessis, *Molecules*, 2016, 21, 416 (review of sinigrin)

C. E. Paulsen, J.-P. Armache, Y. Gao, Y. Cheng and D. Julius, *Nature*, 2015, 520, 511–517 (structure of the TRPA1 ion channel)

D. Güzey, *Mustard. A Global History*, London, Reaktion Books, 2019

Y. Suo, Z. Wang, L. Zubcevic, A. L. Hsu, Q. He, M. J. Borgnia, R.-R. Ji and S.-Y. Lee, *Neuron*, 2020, 105, 882–894 (how human TRPA1 channel senses benzyl isothiocyanate)

V. P. Thinh Nguyen, J. Stewart, M. Lopez, I. Ioannou and F. Allais, *Molecules*, 2020, 25, 4537 (glucosinolates reviewed)

J. Zhao, J. V. Lin King, C. E. Paulsen, Y. Cheng and D. Julius, *Nature*, 2020, 585, 141–145 (how electrophiles activate the TRPA1 receptor)

4. ARTEMISININ

D. L. Klayman, *Science*, 1985, 228, 1049–1055 (*Qinghaosu* (artemisinin): an antimalarial drug from China)

S. Krishna, L. Bustamante, R. K. Haynes and H. M. Staines, *Trends in Pharmacol. Sci.*, 2008, 29, 520–527 (review)
M. Dondorp et al., *New England Journal of Medicine*, 2009, 361, 5 (artemisinin resistance)
G. D. Brown, *Molecules*, 2010, 15, 7603–7698 (biosynthesis of artemisinin)
D. Chaturvedi, A. Goswami, P. P. Saikia, N. C. Barua and P. G. Rao, *Chem. Soc. Rev.*, 2010, 39, 435–454 (artemisinin and its derivatives)
P. M. O'Neill, V. E. Barton and S. A. Ward, *Molecules*, 2010, 15, 1705–1721 (mechanism of action of artemisinin)
Y. Tu, *Nature Medicine*, 2011, 17, 1217–1220 (discovery of artemisinin)
N. Klonis, M. P. Crespo-Ortiz, I. Bottova, N. Abu-Bakar, S. Kenny, P. J. Rosenthal and L. Tilley, *Proc. Natl. Acad. Sci. USA*, 2011, 108, 11405–11410 (artemisinin activity against Plasmodium falciparum requires hemoglobin uptake and digestion)
W. Sherman, *Magic Bullets to Conquer Malaria*, Washington, ASM Press, 2011
H. Cheeseman et al., *Science*, 2012, 336, 79–82 (genome region underlying Artemisinin resistance in malaria)
F. Lévesque and P. H. Seeberger, *Angew. Chem. Int. Ed.*, 2012, 51, 1706–1709 (continuous flow process for converting dihydroartemisinic acid to artemisinin)
Y. Tu, *Angew. Chem. Int. Ed.*, 2016, 55, 10210–10226 (Nobel lecture)
D.-Y. Xie, D.-M. Ma, R. Judd and A. L. Jones, *Phytochem. Rev.*, 2016, 15, 1093–1114 (artemisinin biosynthesis in Artemisia annua)
S. H. Kung, S. Lund, A. Murarka, D. McPhee and C. J. Paddon, *Front. Plant Sci.*, 2018, 9, 87 (approaches to semisynthesis)
F. Lu, X.-L. He, R. Culleton and J. Cao, *Chin. J. Nat. Med.*, 2019, 17, 331–336 (artemisinin, history, means of action and resistance)
C. M. Moore, J. Wang, Q. Lin, P. Ferreira, M. A. Avery, K. Elokely, H. M. Staines and S. Krishna, *Antimicrobial Agents and Chemotherapy*, 2022, 66, e0207921 (inhibition of Plasmodium falciparum ATPase 6, PfATP6)
T. V. Padma, *Nature*, 2023, 618, S26–S28 (combating drug resistance in malaria)
G. Das, H.-S. Shin and J. K. Patra, *Curr. Issues Mol. Biol.*, 2024, 46, 12099–12118 (artemisinin review)
Y. Xia, C. Shi, J. Lu, Z. Zhu, M. Li, Y. Pan, X. Huang, L. Zhang and A. Liu, *Int. J. Mol. Sci.*, 2025, 26, 8409 (artemisinin review, including anticancer activity)

5. ASPIRIN

E. Stone, *Philos. Trans.,* 1763, 53, 195–200 ('An Account of the Success of the Bark of the Willow the Cure of Agues')
K. M. Starko, C. G. Ray, L. B. Dominguez, W. L. Stromberg and D. F. Woodall, *Pediatrics,* 1980, 66, 859–864 (Reye's syndrome and salicylate use)
C. C. Mann and M. L. Plummer, *The Aspirin Wars*, New York, A. A. Knopf, 1991
D. Picot, P. J. Loll and R. M. Garavito, *Nature*, 1994, 367, 243–249 (COX-1)
P. J. Loll, D. Picot and R. M. Garavito, *Nature Structural Biology*, 1995, 2, 637–643 (aspirin activity and crystal structure of inactivated prostaglandin H2 synthase)
P. J. Loll, D. Picot, O. Ekabo and R. M. Garavito, *Biochemistry*, 1996, 35, 7330–7340 (COX-1)
R. G. Kurumbail, A. M. Stevens, J. K. Gierse, J. J. McDonald, R. A. Stegeman, J. Y. Pak, D. Gildehaus, J. M. Iyashiro, T. D. Penning, K. Seibert, P. C. Isakson and W. S. Stallings, *Nature*, 1996, 384, 644–648 (COX-2)
O. Lafont, *Revue d'histoire de la pharmacie*, 1996, 84, 269–273 (on Gerhardt's synthesis of acetylsalicylic acid)
C. Osborn, ed., *Aspirin*, London, Royal Society of Chemistry, 1998
W. Sneader, *British Medical Journal*, 2000, 321, 1591–1594 (discovery of aspirin: a reappraisal)

J. R. Vane and R. M. Botting, eds., *Therapeutic Roles of Selective COX-2 Inhibitors*, London, William Harvey Press, 2001

J. R. Vane and R. M. Botting, *Thrombosis Research*, 2003, 110, 255–258 (mechanism of action of aspirin)

K. D. Rainsford, ed., *Aspirin and Related Drugs*, London, Taylor & Francis, 2004

D. Jeffreys, *Aspirin, the Story of a Wonder Drug*, London, Bloomsbury, 2004

J. Karha and E. J. Topol, *Cleve. Clin. J. Med.*, 2004, 71, 933–939 ('the sad story of Vioxx')

S. Houlton, *Chemistry World*, December 2007, 56–59 (withdrawal of Vioxx)

P. Sheldon, *The Fall and Rise of Aspirin – The Wonder Drug*, Studley, England, Brewin Books, 2007

H. Krumholz et al., *BMJ*, 2007, 334, 120–123 ('What have we learnt from Vioxx?')

T. Nesi, *Poison Pills: The Untold Story of the Vioxx Drug Scandal*, New York, Thomas Dunne Books, 2008

K. C. Nicolaou and T. Montagnon, *Molecules that Changed the World*, Weinheim, Germany, Wiley VCH, 2008, pp. 22–27

K. M. Starko, *Clinical Infectious Diseases,* 2009, 49, 1405–1410 (salicylates and 1918–1919 flu pandemic)

G. Rimon, R. S. Sidhu, D. A. Lauver, J. Y. Lee, N. P. Sharma, C. Yuan, R. A. Frieler, R. C. Trievel, B. R. Lucchesi and W. L. Smith, *Proceedings of National Academy of Sciences,* 2010, 107, 28–33 (celecoxib binding to COX-1)

P. Patrignani and C. Patrono, *J. Am. Coll. Cardiol.*, 2016, 68, 967–976 (aspirin and cancer)

M. J. R. Desborough and D. M. Keeling, *British Journal of Haematology*, 2017, 177, 674–683 (from willow to wonder drug)

J. N. Wood, *Phil. Trans. R. Soc. B*, 2015, 370, 20140317 (commentary on Stone's 1763 paper)

M. R. Montinari, S. Minelli and R. De Caterina, *Vascul. Pharmacol.*, 2019, 113, 1–8 (comprehensive history of aspirin)

I. Raber, C. P. McCarthy, M. Vaduganathan, D. L. Bhatt, D. A. Wood, J. G. F. Cleland, R. S. Blumenthal and J. W. McEvoy, *Lancet,* 2019, 393, 2155–2167; J. G. F. Cleland, *Thromb. Haemost.,* 2022, 122, 311–314 (aspirin and the prevention of cardiovascular disease)

O. Werz, H. Stettler, C. Theurer and J. Seib, *Pharmaceuticals*, 2024, 17, 437 (on the 125th Anniversary of Aspirin)

6. ATROPINE

K. Mein, *Annalen der Pharmazie*, 1833, 6, 67–72 ('*Ueber die Darstellung des Atropins in weissen Kristallen*')

L. E. Shutt and J. B. Bowes, *Anaesthesia*, 1979, 34, 476–490 (atropine and aneasthesia)

R. S. Holzman, *Anaesthesiology*, 1998, 89, 241–249 ('*The Legacy of Atropos*')

J. Burdon, 'Nerve Gases', in: R. H. Waring, G. B. Steventon and S. C. Mitchell (eds.), *Molecules of Death*, London, Imperial College Press, 2nd edition, 2007, pp. 222–223 (atropine as nerve agent antidote)

M. R. Lee, *J. Roy. Coll. Physician Edinb.*, 2007, 37, 77–84 (*Atropa belladonna* and the Agutter case)

J. Emsley, 'Molecule of Murder', *Chemistry World*, August 2008, 64–66; *Molecules of Murder*, Cambridge, RSC Publishing, 2008, pp. 46–67 (the Agutter case)

S. J. Stellpflug, J. B. Cole, B. A. Isaacson, C. P. Lintner and E. F. Bilden, *West J. Emerg. Med.*, 2012, 13, 77–79 (attempted suicide using eye drops)

J. Patocka and R. Jelinkova, *Mil. Med. Sci. Lett. (Voj. Zdrav. Listy)*, 2017, 86, 58–69 (atropine in warfare)

S. Scholtz, L. MacMorris, F. Krogmann and G. U. Auffarth, *J. Eye Stud. Treat.*, 2019, 51–58 (atropine and scopolamine in medicine and ophthalmology)

U. A. Eze, G. I. Nathaniel and G. Pepple, *Journal of Advances in Medicine and Medical Research*, 2022, 34, 197–205 (ophthalmic uses of atropine)

S. A. Aldossary, *Biomedical & Pharmacology Journal*, 2022, 15, 691–697 (pharmacology of atropine)

V. Di Fazio, J. Vanhaebost, S. R. Wille, N. Samyn, G. Schmit and F. Bonbled, *Toxicologie Analytique et Clinique*, 2022, 34, S46 (murder by atropine eye-drops)

I. Jawaid, K. Saunders, C. J. Hammond, A. Dahlmann-Noor and M. A. Bullimore, *Eye,* 2024, 38, 434–441 (atropine and myopia)

7. CAPSAICIN

M. J. Caterina, M. A. Schumacher, M. Tominaga, T. A. Rosen, J. D. Levine and D. Julius, *Nature (London)*, 1997, 389, 816–824 (discovery of TRPV1 receptor)

J. Billing and P. W. Sherman, *Quart. Rev. Biol.*, 1998, 73, 3–49 (antimicrobial functions of spices)

M. J. Caterina, A. Leffler, A. B. Malmberg, W. J. Martin, J. Trafton, K. R. Petersen-Zeitz, M. Koltzenburg, A. I. Basbaum and D. Julius, *Science*, 2000, 288, 306–313 (impaired nociception and pain sensation in mice lacking the capsaicin receptor)

J. J. Tewksbury and G. P. Nabhan, *Nature*, 2001, 412, 203–204 (capsaicin and squirrel deterrent)

L. Collingham, *Curry: A Biography of a Dish*, London, Chatto and Windus, 2005

S. J. Conway, *Chem. Soc. Rev.*, 2008, 37, 1530–1545 (review of capsaicin and TRPV1)

J. J. Tewksbury, K. M. Reagan, N. J. Machnicki, T. A. Carlo, D. C. Haak, A. L. Calderon-Penaloza and D. J. Levey, *Proc. Natl. Acad. Sci. USA*, 2008, 105, 11808–11811 (evolutionary ecology of pungency in wild chillis)

C. T. Sen, *Curry: A Global History*, London, Reaktion Books, 2009

C. Uhlson, R. C. Murphy and K. M. Hargreaves, *J. Clin. Invest.,* 2010, 120, 1617–1626 (linoleic acid metabolites that activate TRPV1 and produce inflammation)

J. E. Mendelson, B. K. Tolliver, K. L. Delucchi, M. J. Baggott, K. Flower, C. W. Harris, G. P. Galloway and P. Berger, *Forensic Toxicol.*, 2010, 28, 33–37 (capsaicin increases the lethality of cocaine)

M. de L. Reyes-Escogido, E. G. Gonzalez-Mondragon and E. Vazquez-Tzompantzi, *Molecules*, 2011, 16, 1253–1270 (capsaicin review)

M. Liao, E. Cao, D. Julius and Y. Cheng, *Nature*, 2013, 504, 107–112; E. Cao, M. Cao, D. Julius and Y. Cheng, 2013, 504, 113–118 (TRPV1 ion channel structures)

L. Darré and C. Domene, *Mol. Pharmaceutics*, 2015, 12, 4454–4465 (binding of capsaicin to the TRPV1 ion channel)

K. Ohbuchi, Y. Mori, K. Ogawa, E. Warabi, M. Yamamoto and T. Hirokawa, *PLoS ONE*, 2016, 11, e0162543 (how vanilloids including capsaicin bind to TRPV1)

H. A. Anderson, *Chillies: A Global History*, London, Reaktion Books, 2016, reprint 2021

F. Yang and J. Zheng, *Protein & Cell*, 2017, 8, 169–177 (mechanism of TRPV1 channel activation by capsaicin)

Z. Shi, T. El-Obeid, M. Riley, M. Li, A. Page and J. Liu, *Nutrients*, 2019, 11, 1183 (high chilli intake associated with cognitive decline)

H. Ledford and E. Callaway, *Nature*, 2021, 598, 246 (capsaicin Nobel prize)

J. Hwang, Y. M. Choe, G.-H. Suh, B. C. Lee, I.-G. Choi, J. H. Lee, H. S. Kim, S. G. Kim, D. Yi and J. W. Kim, *Scientific Reports*, 2023, 13, 7942 (spicy food intake associated with Alzheimer-related cognitive decline in older adults with low physical activity)

T. Thornton, D. Mills and E. Bliss, *Nutrients*, 2023, 15, 1537 (capsaicin treatment can improve cognition in animals)

M. Musolino, M. D'Agostino, M. Zicarelli, M. Andreucci, G. Coppolino and D. Bolignano, *Int. J. Mol. Sci.*, 2024, 25, 791 (capsaicin may improve renal function)

D. Inyang, T. Saumtally, C. N. Nnadi, S. Devi and P.-W. So, *Int. J. Mol. Sci.*, 2023, 24, 10176 (the effects of capsaicin on Alzheimer's Disease reviewed)

8. CATHINONES AND KHAT

T. S. Emerson and J. E. Cisek, *Ann. Emerg. Med.*, 1993, 22, 1897; M. S. Goldstone, *JAMA*, 1993, 269, 2508 (methcathinone appears in USA)

K. Rushby, *Eating the Flowers of Paradise*, London, Constable, 1998 (khat)

A. Al-Motarreb, S. Briancon, N. Al-Jaber, B. Al-Adhi, F. Al-Jailani, M S. Salek and K. J. Broadley, *Br. J. Clin. Pharmacol.*, 2005, 59, 574–581 (khat chewing as a risk factor for acute myocardial infarction)

A. Stepens [sic], I. Logina, V. Liguts, P. Aldiņš, I. Ekšteina, A. Platkājis, I. Mārtiņsone, E. Tērauds, B. Rozentāle and M. Donaghy, *N. Engl. J. Med.*, 2008, 358, 1009–1017 (manganese causing a Parkinsonian syndrome in methcathinone users)

O. Roussel, M. Perrin, P. Herard, M. Chevance and P. Arpino, *Ann. Toxicol. Anal.*, 2009, 21, 169–177 (spectroscopic analysis of mephedrone)

N. B. Patel, *East African Med. J.*, 2009, 77, 329–332 (mechanism of action of cathinone)

W. M. Ali, M. Zubaid, A. Al-Motarreb, R. Singh, S. Al-Shereiqi, A. Shehab, W. Rashed, N. Q. Al-Sagheer, A. H. Saleh and J. Al Suwaidi, *Mayo Clin. Proc.*, 2010, 85, 974–980 (khat chewing associated with increased risk of stroke and death)

M. H. Chapman, M. Kajihara, G. Borges, J. O'Beirne, D. Patch, A. P. Dhillon, A. Crozier and M. Y. Morgan, *N. Engl. J. Med.*, 2010, 362, 1642–1644 (khat chewing and liver injury)

E. Y. Santali, A.-K. Cadogan, N. Nic Daeid, K. A. Savage and O. B. Sutcliffe, *Journal of Pharmaceutical and Biomedical Analysis*, 2011, 56, 246–255 (synth. and characterisation of mephedrone)

K. N. Cameron, R. Kolanos, E. Solis, R. A. Glennon and L. J. De Felice, *Br. J. Pharmacol.*, 2013, 168, 1750–1757; K. N. Cameron, R. Kolanos, R. Vekariya, L. J. De Felice and R. A. Glennon, *Psychopharmacology*, 2013, 227, 493–499 (mephedrone and MDPV act synergistically at the human dopamine transporter)

K. Cameron, R. Kolanos, R. Vekariya, L. De Felice and R. A. Glennon, *Psychopharmacology*, 2013, 227, 493–499 (Mephedrone and MDPV at the human dopamine transporter)

A. J. Eshleman, K. M. Wofrum, M. G. Hatfield, R. A. Johnson, K. V. Murphy and A. Janowsky, *Biochem. Pharmacol.*, 2013, 85, 1803–1815 (substituted methcathinones differ in transporter and receptor interactions)

R. A. Gregg and S. M. Rawls, *Life Sciences*, 2014, 97, 27–30 (designer cathinones reviewed)

K. Sellors, A. Jones and B. Chan, *Med. J. Aust.*, 2014, 201, 601–603 (strange behaviour and fatality associated with intravenous use of Flakka)

B. A. Froberg, M. Levine, M. C. Beuhler, B. S. Judge, P. W. Moore, K. M. Engebretsen, N. J. Mckeown, C. D. Rosenbaum, A. C. Young and D. E. Rusyniak, *J. Med. Toxicol.*, 2015, 11, 185–194 (acute MDPV toxicity)

K. Sikk and P. Taba, *International Review of Neurobiology*, 2015, 120, 257–271 (impure methcathinone leading to Parkinson's)

T. Nichols, P. Khondkar and S. Gibbons, *Phytochemistry Letters*, 2015, 13, 127–133 (mini-review of khat)

A. Yahya, Y. Rajeshwar, T. Eticha, G. Kahsay, D. Ali, H. Gebretsadik, T. Gebretsadik and Y. K. Janapati, *International Journal of Pharmacy and Pharmaceutical Research*, 2016, 8(1), 11–22 (khat chewing in Ethiopia)

C. Crespi, *Case Reports in Psychiatry*, 2016, Article ID 3460849 (Flakka-induced prolonged psychosis)

E. Papaseit, J. Moltó, R. Muga, M. Torrens, R. de la Torre and M. Farré, *Current Topics in Behavioral Neurosciences*, 2017, 32, 313–331 (clinical pharmacology of mephedrone)

A. Dsouza, L. Pereira and P. Levounis, *Current Psychiatry*, 2017, 16(9), e4–e5 ('Flakka': a low-cost, dangerous high)

K. Šíchová, N. Pinterová, M. Židková, R. R. Horsley, E. Lhotková, K. Štefková, Č. Vejmola, L. Uttl, M. Balíková, M. Kuchař and T. Páleníček, *Front. Psychiatry*, 2018, 8, 306 (effects of mephedrone in rats)

S. J. Simmons, J. M. Leyrer-Jackson, C. F. Oliver, C. Hicks, J. W. Muschamp, S. M. Rawls and M. F. Olive, *ACS Chem. Neurosci.*, 2018, 9, 2379–2394 (DARK Classics – cathinone-derived psychostimulants including MDPV)

T. O. Kolesnikova, S. L. Khatsko, K. A. Demin, V. A. Shevyrin and A. V. Kalueff, *ACS Chem. Neurosci.*, 2019, 10, 168–174 (DARK Classics in Chemical Neuroscience: 'Flakka')

B. Kalyanasundar, C. I. Perez, B. Arroyo, M. Gil Moreno and R. Gutierrez, *Front. Neurosci.*, 2020, 14, 572328 (cathine as appetite suppressant)

T. Heinonen, A. Korvenoja and E. Pekkonen, *Case Reports in Neurology*, 2021, 13, 131–134 (Flakka-induced stroke)

B. Silva et al., *Toxins*, 2022, 14, 71 (khat reviewed)

G. Bielecka-Papierz, E. Poleszak and A. Serefko, *Curr. Issues Pharm. Med. Sci.*, 2023, 36, 54–64 (mephedrone reviewed)

M. Bassi et al., *Neurotoxicology*, 2024, 103, 230–255 (MDPV and MDPHP compared)

N. Nadal-Gratacos, M. D. Pazos, D, Pubill, J. Camarasa, E. Escubedo, X. Berzosa and R. Lopez-Arnau, *ACS Pharmacol. Transl. Sci.*, 2024, 7, 2588–2603 (structure–activity relationships of synthetic cathinones; review)

9. CF_2Cl_2

F. Swarts, *Bull. Acad. Roy. Belg.*, 1892, 24, 309–320; 474–484 (the Swarts reaction)

T. Midgley Jr. and A. L. Henne, *Ind. Eng. Chem.*, 1930, 22, 542–545 ('organic fluorides as refrigerants')

R. J. Thompson, *Ind. Eng. Chem.*, 1932, 24, 620–623 ('Freon, a Refrigerant')

T. Midgley Jr., *Ind. Eng. Chem.*, 1937, 29, 241–244 (how CFCs originated)

A. B. Garrett, *J. Chem. Educ.*, 1962, 39, 1361–1362 (Midgley's account of the discovery of CF_2Cl_2)

J. E. Lovelock, *Nature*, 1971, 230, 379 (atmospheric fluorine compounds as indicators of air movements); J. E. Lovelock, R. J. Maggs and R. J. Wade, *Nature*, 1973, 241, 194–196 (halogenated hydrocarbons in and over the Atlantic)

J. E. Lovelock, *Journal of Chromatography A.*, 1974, 99, 3–12 (the electron capture detector)

P. A. Crutzen, *Can. J. Chem.*, 1974, 52, 1569–1581; M. J. Molina and F. S. Rowland, *Nature*, 1974, 249, 810–812 (chlorine atom-catalysed destruction of ozone)

N. N. Greenwood and A. Earnshaw, *Chemistry of the Elements*, Oxford, Pergamon, 1st edition, 1984, p. 323 (synthesis of CFCs)

J. C. Farman, B. C. Gardiner and J. D. Shanklin, *Nature*, 1985, 315, 207–210 (thinning of ozone layer over Antarctic)

J. K. Cockcroft and A. N. Fitch, *Zeitschrift für Kristallographie*, 1991, 197, 121–130 (the structure of solid CF_2Cl_2)

M. McFarland and J. Kaye, *Photochemistry and Photobiology*, 1992, 55, 911–929 (CFC review)

T. Midgley IV, *From the Periodic Table to Production: The Life of Thomas Midgley, Jr., the Inventor of Ethyl Gasoline and Freon Refrigerants*, Corona, CA, Stargazer Publ. Co., 2001

G. B. Kauffman, *Chem. Educator*, 2005, 10, 211–220 (the 75th anniversary of CFCs)

C. J. Giunta, *Bull. Hist. Chem.*, 2006, 31, 66–74 (Thomas Midgley Jr. and the Invention of CFC Refrigerants)

K.-H. Kim, Zang-H. Shon, H. T. Nguyen and E.-C. Jeon, *Atmospheric Environment*, 2011, 45, 1369–1382 (CFCs and halocarbon alternatives reviewed)

H. E. B. Viana and P. A. Porto, *J. Chem. Educ.*, 2013, 90, 1632–1638 (Thomas Midgley Jr. and CFCs)

S. A. Montzka et al., *Nature*, 2018, 557, 413–417 (increase in CFC-11 emissions)

L. M. Western et al., *Nature Geoscience*, 2023, 16, 309–313 (increase in CFC levels 2010–2020)

10. CHANEL NO. 5 AND 2-METHYLUNDECANAL

G. Darzens, *Comptes Rendus Hebdomadaires des séances de l'Académie des Sciences*, 1904, 139, 1214–1217 (synthesis of 2-methylundecanal)

R. W. Moncrieff, *The Chemical Senses*, London, Leonard Hill, 1967, pp. 407–411

S. Arctander, *Perfume and Flavor Chemicals II*, Montclair, NJ, 1969, compound 2132, Allured Publishing Corporation

A. Koyasako and R. A. Bernhard, *Journal of Food Science*, 1984, 48, 1807–1812 (2-methylundecanal in kumquat oil)

D. Enders and H. Dyker, *Liebigs Ann. Chem.*, 1990, 1107–1110 (synthesis of the enantiomers of 2-methylundecanal)

J. Emsley, *The Consumer's Good Chemical Guide*, Oxford, W. H. Freeman, 1994, pp. 1–23, esp. p. 2 and p. 302

P. Laszlo, *Bull. Hist. Chem.*, 1994, 15/16, 59–64 (Georges Darzens)

D. Veuillet-Gallot, *Le Guide du Parfum*, Tours, Éditions Hors Collection, 1996, pp. 59–61

C. Newman, *Perfume: The Art and Science of Scent*, Washington, D.C., National Geographic, 1998, pp. 131–138

N. M. de Barry, M. Turonnet and G. Vindry, *L'abcdaire du parfum*, Paris, Flammarion, 1999, pp. 38–39

S. Girard-Lagorge, *100 Parfums de légende*, Paris, Solar, 2000, pp. 44–45

J. Gibka and M. Glinski, *Flavour and Fragrance Journal*, 2006, 11, 480–483 (odours of aldehydes)

L. Turin, *The Secret of Scent*, London, Faber and Faber, 2006, pp. 52–56 (aldehydic scents)

C. Sell, ed., *The Chemistry of Fragrances: From Perfumer to Consumer*, Cambridge, Royal Society of Chemistry, 2nd edition, 2006

P. Kraft, C. Leard and P. Goutell, *From Rallet No. 1 to Chanel No. 5 via Mademoiselle Chanel No. 1*, Perfumer and Flavorist, October 2007, pp. 36–48

L. Turin and T. Sanchez, *Perfumes: The Guide*, New York, Viking Adult, 2008, pp. 259–261

T. J. Mazzeo, *The Secret of Chanel No. 5: The Biography of a Scent*, New York, HarperCollins, 2010

J. Picardie, *Coco Chanel. The Legend and the Life*, London, Harper Collins, 2010

S. A. Cotton, *Every Molecule Tells a Story*, Boca Raton, CRC Press, 2012, pp. 87–89 (smells of aldehydes)

A. Catalano, A. Mariconda, A. D'Amato, D. Iacopetta, J. Ceramella, M. Marra, C. Saturnino, M. S. Sinicropi and P. Longo, *Organics*, 2024, 5, 395–428 (review: aldehydes, including smells)

11. CHLOROMETHANE

J.-B. Dumas and E. M. Peligot, *Ann. Pharm.*, 1835, 15, 20 (original synth.)

J. E. Lovelock, *Nature*, 1975, 256, 193–1974 (detection of CH_3Cl in air over England)

E. C. Y. Inn et al., *Science*, 1982, 215, 665–667 (CH_3Cl from Mount St Helens)

X. Ni and L. P. Hager *Proc. Natl. Acad. Sci. USA*, 1998, 95, 12866–12871 (CH_3Cl from *Batis maritima* in salt marshes)

Y. Yokouchi, M. Ikeda, Y. Inuzuka and T. Yukawa, *Nature*, 2002, 416, 163–165 (CH_3Cl from tropical vegetation)

G. W. Gribble, *Chemosphere*, 2003, 52, 289–297 (diversity of naturally occurring organo-halogens)

J. T. G. Hamilton, W. C. McRoberts, F. Keppler, R. M. Kalin and D. B. Harper, *Science*, 2003, 301, 206–209 (chloride methylation by plant pectin)

M. A. Cochrane, *Nature*, 2003, 421, 913–919 (CH_3Cl from mass burning)

H. Anke and R. W. S. Weber, *Mycologist*, 2006, 20, 83–89 (CH_3Cl from wood rotting fungi)

F. Tassi et al., *J. Geophys. Res.*, 2012, 117, D17305 (CH_3Cl from Mount Etna)

E. C. Fayolle et al., *Nature Astronomy*, 2017, 1, 703–708 (CH_3Cl in interstellar space and in the coma of comet 67P/Churyumov–Gerasimenko (67P/C-G))

Y. Sailaukhanuly, Z. Sárossy, L. Carlsen and H. Egsgaard, *Chemosphere*, 2014, 111, 575–579 (formation of chloromethane by nucleophilic substitution of pectin)

G. W. Gribble, 'Naturally Occurring Organohalogen Compounds', in: *Progress in the Chemistry of Organic Natural Products*, Vol. 121, Switzerland, Springer, 2023, pp. 1–545

12. CINNAMALDEHYDE

J. Dumas and E. Peligot, *Ann. Chim. Phys.*, 1834, 57, 305–334 (discovery)

U. M. Senanayake, T. H. Lee and R. B. H. Wills, *J. Agric. Food Chem.*, 1978, 26, 822–824 (analysis of cinnamon bark oil)

G. Vernin et al., in: G. Charalambous (ed.), 'GC/MS Analysis of Cinnamon and Cassia Essential Oils', *Spices, Herbs and Edible Fungi*, Amsterdam, Elsevier, 1994, pp. 411–425 (analysis of cinnamon essential oils)

G. R. Mallavarapu, R. S. Chandrasekhara, B. R. Rajeswara Rao, P. N. Kaul and A. K. Battacharya, *Flav. Fragr. J.*, 1995, 10, 239–242 (eugenol in cinnamon leaf oil)

G. K. Jayaprakasha, L. Jaganmohan Rao and K. K. Sakariah, *Flav. Fragr. J.*, 1997, 12, 331–333 (cinnamyl acetate as the major compound in the oils of cinnamon fruits)

S.-S. Cheng, J.-Y. Liu, K.-H. Tsai, W.-J. Chen and S.-T. Chang, *J. Agric. Food Chem.*, 2004, 52, 4395–4400; W.-B. Ma, J.-T. Feng, Z.-L. Jiang and X. Zhang, *Journal of the American Mosquito Control Association,* 2014, 30, 199–203 (anti-mosquito activity)

N. K. Leela, 'Cinnamon and Cassia', in: V. A. Partasarathy, B. Chempakam and T. J. Zachariah (eds.), *Chemistry of Spices*, Wallingford, CAB International, 2008, pp. 124–145

F. Czerra, *Spices. A Global History*, London, Reaktion Books, 2009

R. Z. Behar, W. Luo, S. C. Lin, Y. Wang, J. Valle, J. F. Pankow and P. Talbot, *Tob. Control*, 2016, 25, ii94–ii102 (cinnamaldehyde found in 20/39 e-liquids analysed)

H. B. Bang, Y. H. Lee, S. C. Kim, C. K. Sung and K. J. Jeong, *Microb. Cell Factories*, 2016, 15, 1–12 (engineering *E. coli* for the production of cinnamaldehyde)

P. W. Clapp, E. A. Pawlak, J. T. Lackey, J. E. Keating, S. L. Reeber, G. L. Glish and I. Jaspers, *Am. J. Physiol. Lung Cell Mol. Physiol.*, 2017, 313, L278–L292 (cinnamaldehyde impair respiratory innate immune cell function)

P. W. Clapp, K. S. Lavrich, C. A. van Heusden, E. R. Lazarowski, J. L. Carson and I. Jaspers, *Am. J. Physiol. Lung Cell Mol. Physiol.*, 2019, 316, L470–L486 (cinnamaldehyde disrupts lung mitochondrial function)

A. A. Doyle and J. C. Stephens, *Fitoterapia*, 2019, 139, 104405 (cinnamaldehyde and its derivatives as antibacterial agents)

Z. Gan, J. Huang, J. Chen, M. F. Nisar and W. Qi, *Journal of Food Quality*, 2020, Article ID 8898692 (antifungal activities of cinnamaldehyde derivatives against *Penicillium digitatum*)

M. Hajinejad, M. Ghaddaripouri, M. Dabzadeh, F. Forouzanfar and S. Sahab-Negah, *BioMed Research International*, 2020, Article 1034325 (cinnamaldehyde and derivatives ameliorate neuroinflammatory pathways in neurodegenerative diseases)

B. Bhattacharya, V. Narain and M. Bondesson, *Aquatic Toxicology*, 2021, 240, 105995 (cinnamaldehyde in zebrafish)

S. Wang, O.-H. Kang and D.-Y. Kwon, *Int. J. Mol. Sci.*, 2021, 22, 2752 (cinnamaldehyde exhibits synergy with conventional antibiotic against MRSA)

P. J. Kuehl et al., *Inhal Toxicol.*, 2022, 34, 319–328 (thermal degradation of flavourings including cinnamaldehyde in e-liquids)

B. A. J. Jimenez, F. Awwad and I. Desgagné-Penix, *Antibiotics*, 2024, 13, 1095 (biosynthesis and antimicrobial properties of cinnamaldehyde)

N. O. F. Kassem et al., *Nicotine and Tobacco Research*, 2024, 26, 1445–1454 (toxicity of ingredients in e-cigarettes, including cinnamaldehyde)

13. CISPLATIN

M. Peyrone, *Ann. Chem. Pharm.*, 1844, 51, 1–29 (first synthesis of cis$[Pt(NH_3)_2Cl_2]$)
A. Werner, *Z. Anorg. Chem.*, 1893, 3, 267 (structure of cisplatin deduced)
B. Rosenberg, L. Van Camp and T. Krigas, *Nature*, 1965, 205, 698–699 (discovery of inhibited cell division)
B. Rosenberg, L. Van Camp, E. B. Grimley and A. J. Thomson, *J. Biol. Chem.*, 1967, 242, 1347–1352 (testing on mice with cancer)
B. Rosenberg, L. Van Camp, J. E. Trosko and V. H. Mansour, *Nature*, 1969, 222, 385–386 (testing on mice with cancer)
S. C. Dhara, *Indian J. Chem.,* 1970, 8, 193–194 (best synthesis of cisplatin)
M. J. Cleare and J. D. Hoeshcele, *Plat. Met. Rev.,* 1973, 17, 2–13; *Bioinorg. Chem.*, 1973, 2, 187–210 (structure-activity rules for platinum drugs)
R. Champion and J. Powell, *Champion's Story*, London, Victor Gollancz, 1981
P. M. Takahara, C. A. Frederick and S. J. Lippard, *J. Am. Chem. Soc.*, 1996, 118, 12309–12321 (crystal structure of cisplatin bound to Duplex DNA)
B. Lippert, ed., *Cisplatin: Chemistry and Biochemistry of a Leading Anticancer Drug*, Zurich, Verlag Helvetica Chimica Acta; Weinheim, Germany, Wiley-VCH, 1999
L. Armstrong, *It's Not about the Bike,* London, Yellow Jersey Press, 2000, esp. p. 86
J. Graham, M. Muhsin and P. Kirkpatrick, *Nature Reviews Drug Discovery*, 2004, 3, 11–12 (Oxaliplatin)
R. A. Alderden, M. D. Hall and T. W. Hambley, *J. Chem. Educ.*, 2006, 83, 728–734 (discovery and development of cisplatin)
L. Kelland, *Nature Reviews Cancer*, 2007, 7, 573–584 (resurgence of platinum-based cancer chemotherapy)
G. B. Kauffman, R. Pentimalli and M. D. Hall, *Platinum Metals Rev.*, 2010, 54, 250–256 (Michele Peyrone, discoverer of cisplatin)
T. C. Johnstone, G. Y. Park and S. J. Lippard, *Anticancer Res.*, 2014, 34, 471–476 ('understanding and improving platinum anticancer drugs')
G. F. de Sousa, S. R. Wlodarczyk and G. Monteiro, *Brazilian Journal of Pharmaceutical Sciences*, 2014, 50, 693–701 (carboplatin)
J. D. Hoeschele, *Anticancer Res.*, 2014, 34, 417–421; *Dalton Trans.*, 2016, 45, 10648–10650 (bio. of Barnett Rosenberg)
M. P. M. Marques, D. Gianolio, G. Cibin, J. Tomkinson, S. F. Parker, R. Valero, R. P. Lopes and L. A. E. Batista de Carvalho, *Phys. Chem. Chem. Phys.,* 2015, 17, 5155–5171 (a molecular view of cisplatin's mode of action)
T. C. Johnstone, K. Suntharalingam and S. J. Lippard, *Chem Rev.*, 2016, 116, 3436–3486 (later generations of platinum drugs)
R. Oun, Y. E. Moussa and N. J. Wheate, *Dalton Trans*, 2018, 47, 6645–6653 (the side effects of platinum-based chemo.)
F. A. Armstrong, R. Grinter, M. K. Johnson, N. E. Le Brun and M. T. Wilson, *Biogr. Mems Fell. R. Soc.*, 2022, 73, 471–497 (memoir of Andrew Thomson)
N. A. Yusoh, H. Ahmad, K. A. Vallis and M. R. Gill, *Medical Oncology*, 2025, 42, 262 (advances in platinum-based cancer therapy)

14. COCAINE

F. Wöhler, *Archiv der Pharmazie*, 1860, 152, 29–32 ('Ueber eine organische Base in der Coca')
A. Niemann, *Archiv der Pharmazie*, 1860, 153, 129–155, 291–308 ('ueber eine neue organische Base in den cocablättern')
R. J. Hrynchuk, R. J. Barton and B. E. Robertson, *Can. J. Chem.*, 1983, 61, 481–487 (cryst. struct.)

N. D. Volkow and A. C. Swann, eds., *Cocaine in the Brain*, New Brunswick, NJ, Rutgers University Press, 1990

A. C. Springfield, L. W. Cartmell, A. C. Aufderheided, J. Buikstra and J. Ho, *Forensic Science International*, 1993, 63, 269–275 (cocaine and metabolites in the hair of ancient Peruvian coca leaf chewers)

J. A. Nathanson, E. J. Hunnicutt, L. Kantham and C. Scavone, *Proc. Natl. Acad. Sci. USA*, 1993, 90, 9645–9648 (cocaine as a plant defence)

D. M. Perrine, *The Chemistry of Mind-Altering Drugs*, Washington, American Chemical Society, 1996, pp. 181–192

J. M. Moore, J. F. Casale and D. A. Cooper, *Journal of Chromatography A*, 1996, 756, 193–201 (truxillines in illegal cocaine)

A. Negrusz, J. L. Perry and C. M. Moore, *J. Forensic Sci.*, 1998, 43, 626–629 (detection of cocaine on U. S. banknotes)

R. Sleeman, I. F. A. Burton, J. F. Carter and D. J. Roberts, *Analyst,* 1999, 124, 103–108 (cocaine on banknotes)

J. R. Ehleringer, J. F. Casale, M. J. Lott and V. L. Ford, *Nature*, 2000, 408, 311 (isotopic fingerprint of cocaine from the region of growth)

D. Streatfeild, *Cocaine: An Unauthorized Biography*, London, Virgin, 2001

W. E. Brewer, R. C. Galipo, K. W. Sellers and S. L. Morgan, *Analytical Chemistry,* 2001, 73, 2371–2376 (cocaine and metabolites in hair)

M. Bowden, *Killing Pablo: The Hunt for the World's Richest, Most Powerful Criminal in History*, New York, Atlantic Books, 2001

S. Bencharit, C. L. Morton, Y. Xue, P. M. Potter and M. R. Redinbo, *Nature Structural Biology,* 2003, 10, 349–356 (hCE1 and metabolism of heroin and cocaine)

J. Calatayud and A. Gonzalez, *Anesthesiology,* 2003, 98, 1503–1508 (cocaine and anaesthesia)

A. Dronsfield, P. M. Ellis and K. Pooley, *Education in Chemistry,* 2007, 44, 183–186 (anaesthesia)

K. A. Ebejer, G. R. Lloyd, R. G. Brereton, J. F. Carter and R. Sleeman, *Forensic Science International*, 2007, 171, 165–170 (UK banknotes contaminated with cocaine)

R. L. Myers, *The 100 Most Important Chemical Compounds*, Westport, CT, Greenwood Press, 2007, pp. 88–91

P. Hazarika, S. M. Jickells, K. Wolff and D. A. Russell, *Angewandte Chemie, International Ed.*, 2008, 47, 10167–10170 (cocaine and metabolites in fingerprints)

T. D. Dillehay, J. Rossen, D. Ugent, A. Karathanasis, V. Vásquez and P. J. Netherly, *Antiquity*, 2010, 84, 939–953 (early Holocene coca chewing in northern Peru)

J. E. Mendelson et al., *Forensic Toxicol.*, 2010, 28, 33–37 (capsaicin increases toxicity of cocaine)

J. Kinyua and T. A. Anderson, *J. Forensic Sci.*, 2012, 57, 1349–1353 (benzoylecgonine in wastewater)

M. J. Bailey et al., *Analyst*, 2015, 140, 6254–6259 (detecting cocaine and metabolites in fingerprints by surface mass spec.)

S. Castiglioni, A. Borsotti, F. Riva and E. Zuccato, *Drug and Alcohol Review*, 2016, 35, 128–132 (detecting drugs in wastewater in Milan)

A. J. Johnston, S. Busch, L. C. Pardo, S. K. Callear, P. C. Biggin and S. E. McLain, *Phys. Chem. Chem. Phys.,* 2016, 18, 991–999 (structure of cocaine in solution)

J. R. Mallette, J. F. Casale, J. Jordan, D. R. Morello and P. M. Beyer, *Scientific Reports*, 2016, 6, 23520 (geolocation of cocaine from Bolivia, Colombia, and Peru)

E. A. Griffin et al., *Sci. Adv.*, 2017, 3, e1701682 (prior exposure to alcohol enhances vulnerability to cocaine)

G. Troiano, I. Mercurio, M. Golfera, N, Nante, P. Melai, M. Lancia and M. Bacci, *European Journal of Public Health*, 2017, 27, 1097–1101 (cocaine contamination of banknotes)

L. R. Drake and P. J. H. Scott, *ACS Chem. Neurosci.*, 2018, 9, 2358–2372 (DARK Classics in Chemical Neuroscience: Cocaine)

B. G. Chavez et al., *PNAS*, 2022, 119, e2215372119 (cocaine biosynthesis)

J. Pergolizzi, F. Breve, P. Magnusson, J. A. K. LeQuang and G. Varrassi, *Cureus*, 2022, 14, e22498 (cocaethylene: when cocaine and alcohol are taken together)

Y.-J. Wang, J.-P. Huang, T. Tian, Y. Yan, Y. Chen, J. Yang, J. Chen, Y.-C. Gu and S.-X. Huang, *J. Am. Chem. Soc.*, 2022, 144, 22000–22007 (cocaine biosynthesis)

V. de Luca, *PNAS*, 2023, 120, e2218838120 (how plants produce cocaine)

S. Shastry, O. Manoochehri, L. D. Richardson and A. F. Manini, *Acad. Emerg. Med.*, 2023, 30, 82–88 (cocaethylene cardiotoxicity in ED patients)

J. van Amsterdam, F. Gresnigt and W. van den Brink, *J. Clin. Med.*, 2024, 13, 1475 (cardiovascular risks of simultaneous use of alcohol and cocaine)

N. Petrovičová et al., *Toxics*, 2025, 13, 242 (banknotes as a source of drug and pharmaceutical contamination)

15. CODEINE, OXYCODONE AND DESOMORPINE

M. Freund and E. Speyer, *J. Prakt. Chem.*, 1916, 94, 135–178 (original synth. of oxycodone)

E. Mosettig, F. L. Cohen and L. F. Small, *J. Am. Chem. Soc.,* 1932, 54, 793–801; L. F. Small and D. E. Morris, *J. Am. Chem. Soc.*, 1933, 55, 2874–2885; N. B. Eddy and H. A. Howes, *J. Pharmacol. Exp. Ther.*, 1935, 55, 257–267 (synthesis of desomorphine)

P. A. Janssen, *Brit. J. Anaesth.*, 1962, 34, 260–268; L. J. Sargent and E. L. May, *J. Med. Chem.*, 1970, 13, 1061–1063 (potency of desomorphine compared with morphine)

D. V. Canfield, J. Barrick and B. C. Giessen, *Acta Cryst.*, 1987, C43, 977–979 (codeine cryst. struct.)

F. B. Ross and M. T. Smith, *Pain*, 1997, 73, 151–157; E. Kalso, *Pain,* 2007, 132, 227–228; M. T. Smith, *Curr. Opin. Anaesthesiol.*, 2008, 21, 596–601 (how codeine and oxycodone work)

B. Meier, *Pain Killer: A "Wonder" Drug's Trail of Addiction and Death*, Emmaus, PA, Rodale Press, 2003 (Oxycontin)

D. Carise, K. L. Dugosh, A. T. McLellan, A. Camilleri, G. E. Woody and K. G. Lynch, *Am. J. Psychiatry*, 2007, 164, 1750–1756 (prescription OxyContin abuse)

J. Riley, E. Eisenberg, G. Müller-Schwefe, A. M. Drewes and L. Arendt-Nielsen, *Current Medical Research and Opinion*, 2008, 24, 175–192 (oxycodone in pain management)

R. G. Homan, B. Hudson and R. C. Rowe, *Popular Medicines*, London, Pharmaceutical Press, 2008 (Dr Collis Browne's Chlorodyne)

P.-A. Chouvy, *Opium: Uncovering the Politics of the Poppy,* London, I. B. Tauris & Co., 2009

J. M. Hagel and P. J. Facchini, *Nature Chemical Biology*, 2010, 6, 273–275 (biosynth of codeine and morphine)

T. Dormandy, *Opium. Reality's Dark Dream*, New Haven and London, Yale University Press, 2012

T. J. Cicero, M. S. Ellis and H. L. Surratt, *The New England Journal of Medicine*, 2012, 367, 187–189 (abuse-deterrent formulation of OxyContin)

G. B. Kok and P. J. Scammells, *RSC Advances*, 2012, 2, 11318–11325 (improved synth. of oxycodone)

S. Srimurugan, C.-J. Su, H.-C. Shu, K. Murugan and C. Chen, *Monatsh. Chem.*, 2012, 143, 171–174 (improved synth. of desomorphine)

J. V. Pergolizzi, R. Taylor and R. B. Raffa, *Adv. Ther.*, 2015, 32, 485–495 (abuse-resistance of oxycontin)

E. A. Alves et al., *Forensic Sci. Int.*, 2015, 257, 76–82 (the harmful chemistry behind "krokodil": street-like synthesis and product analysis)

J. F. Neves, E. A. Alves, J. X. Soares, S. M. Cravo, A. M. S. Silva, A. D. P. Netto, F. Carvalho, R. J. Dinis-Oliveira and C. M. Afonso, *Data Analysis*, 2016, 6, 83–88 (data analysis of "krokodil" samples obtained by street-like synthesis)

D. H. Â. Florez, A. M. dos Santos Moreira, P. R. da Silva, R. Brandão, M. M. C. Borges, F. J. M. de Santana and K. B. Borges, *Drug Alcohol Depend.*, 2017, 173, 59–68 (desomorphine)

L. Inglis, *Milk of Paradise*, London, Macmillan, 2018 (opium)

C. C. Presley and C. W. Lindsley, *ACS Chem Neurosci.*, 2018, 9, 2503–2518 (DARK Classics: Opium, a Historical Perspective)

P. Alam, S. Borkokoty, M. K. Siddiqi, A. Ehtram, N. Majid, M. Uddin and R. H. Khan, *ACS Chem. Neurosci.*, 2019, 10, 182–189 (DARK Classics in Chemical Neuroscience: Opium, a Friend or Foe)

P. R. Keefe, *Empire of Pain*, London, Picador, 2021 (Purdue Pharma and Oxycontin)

E. A. Alves, *ACS Chem. Neurosci*, 2020, 11, 3968–3978 (DARK Classics in Chemical Neuroscience: Krokodil)

H. Hussain, S. Angly, G. Michel, E. Garcia and A. Fadel, *Eur. J. Case Rep. Intern. Med.*, 2023, 10, 004181 (Krokodil-associated cutaneous ulceration, cardiac arrhythmia and thrombi)

16. DIMETHYLMERCURY

W. C. Zeise, *Annalen der Pharmacie*, 1834, 11(1), 1–10 ('Ueber das Mercaptan')

G. B. Buckton, *Liebigs Ann. Chem*, 1858, 108, 103; 1859, 109, 219 (synthesis of Me_2Hg)

H. Gilman and R. E. Brown, *J. Amer. Chem. Soc.*, 1930, 52, 3314–3317 (synthesis of Me_2Hg)

D. Hunter, R. R. Bomford and D. S. Russell, *Quart. J. Med.*, 1940, 9, 193–213 (poisoning by methylmercury compounds)

L. J. Goldwater, *Mercury; A History of Quicksilver*, Baltimore, York Press, 1972

K. Kashiwabara, S. Konaka, T. Iijima and M. Kimura, *Bull. Chem. Soc. Japan*, 1973, 46, 407–409 (structure by electron diffraction)

J. Pezderová, A. Jirásek, M. Mráz and J. Pechan, *Int. Arch. Arbeitsmed.*, 1974, 33, 323–328 (post-mortem and clinical signs of dimethylmercury poisoning)

W. E. Smith and A. M. Smith, *Minamata*, New York, Holt, Rinehart and Winston, 1975

N. J. Taylor, Y. S. Wong, P. C. Chieh and A. J. Carty, *JCS Dalton*, 1975, 438–442 (synth. and structure of CH_3Hg(cysteinate))

C. U. Eccles and Z. Aman, eds., *The Toxicity of Methylmercury*, Baltimore, The John Hopkins University Press, 1987

R. J. Gillespie and I. Hargittai, *The VSEPR Model of Molecular Geometry*, Boston, Allyn and Bacon, 1991, p. 166

J. E. Macintyre, ed., *Dictionary of Organometallic Compounds*, Chapman and Hall, London, 2nd edition, 1995, Compound Hg–00076

M. Harada, *Crit. Rev. Toxicol.*, 1995, 25, 1–24 (Minamata disease)

S. Budavari, ed., *The Merck Index*, Merck and Co Ltd, 12th edition, 1996, Compound 3299 (p. 550)

N. Hachiya, *Japan Medical Association Journal*, 2006, 49, 112–118 (history of Minamata disease)

M. Blayney, J. Winn and D. Nierenberg, *Chem. Eng. News*, 12 May 1997, 75(19), 7; D. W. Nierenberg et al., *N. Engl. J. Med.*, 1998, 338, 1672–1676 (Wetterhahn case)

R. W. Siegler, D. W. Nierenberg and W. F. Hickey, *Human Pathology*, 1999, 30, 720–723 (fatal poisoning from liquid dimethylmercury)

H. H. Harris, I. J. Pickering and G. N. George, *Science*, 2003, 301, 1203 (CH_3Hg(cysteinate) in fish)

J. Timbrell, *The Poison Paradox*, Oxford, OUP, 2005, esp. pp. 110–118 (Minamata poisoning)

K. Eto, M. Marumoto and M. Takeya, *Neuropathology*, 2010, 30, 471–479 (pathology of Minmata disease)

G. N. George, S. P. Singh, G. J. Myers, G. E. Watson and I. J. Pickering, *J. Biol. Inorg. Chem.*, 2010, 15, 709–715 (CH_3Hg(cysteinate) in human hair)

M. Korbas, J. L. O'Donoghue, G. E. Watson, I. J. Pickering, S. P. Singh, G. J. Myers, T. W. Clarkson and G. N. George, *ACS Chem. NeuroSci.*, 2010, 1, 810–818 (XAS of methylmercury cysteinate in human brain)

S. Jonsson, N. M. Mazrui and R. P. Mason, *Scientific Reports*, 2016, 6, 27958 (mercury in the oceans)

J. West, D. Babi, A. Azaroff and S. Jonsson, *Limnol. Oceanogr. Methods*, 2023, 21, 837–846 (dimethylmercury in natural waters)

H. M. Adams, X. Cui, C. H. Lamborg and A. T. Schartup, *Environ. Sci. Technol.*, 2024, 58, 10591–10600 (dimethylmercury as a source of monomethylmercury in the CCS)

Y. He, H. Inman, D. C. Kadko, M. P. Stephens, D. E. Hammond, W. M. Landing and R. P. Mason, *Sci. Adv.*, 2025, 11, eadr3805 (methylmercury in aerosol and Arctic rain links with air-sea exchange of dimethylmercury)

17. DIMETHYLSULFIDE

D. D. Bills and T. W. Keenan, *J. Agric. Food Chem.*, 1968, 16, 643–645 (Me_2S in sweet corn)

J. Lovelock, R. A. Maggs and R. A. Rasmussen, *Nature*, 1972, 237, 452–453; J. Lovelock, *Philos. Trans. R. Soc. Lond. B*, 1997, 352, 143–147 (Me_2S and climate)

T. Iijima, S. Tsuchiya and M. Kimura, *Bull. Chem. Soc. Japan*, 1977, 50, 2564–2567 (structure of Me_2S)

T. Talou, M. Delmas and A. Gaset, *J. Sci. Food Agric.,* 1989, 48, 57–62 (headspace volatiles from entire black truffle (*Tuber melanosporum*))

T. Talou, A. Gaset, M. Delmas, M. Kulifaj and C. Montant, *Mycol. Res.*, 1990, 94, 277–278 (dogs and pigs detect Me_2S in black truffles)

F. L. Suarez, J. Springfield and M. D. Levitt, *Gut*, 1998, 43, 100–104 (Me_2S in human flatulence)

J. Hansen, *Appl. Environ. Microbiol.*, 1999, 65, 3915–3919 (Me_2S in brewing)

A. S. Henriksson, M. Sarnthein, G. Eglinton and J. Poynter, *Geology*, 2000, 28, 499–502 (Me_2S over 200 000 years in equatorial Atlantic)

P. L. H. McSweeney and M. J. Sousa, *Lait*, 2000, 80, 293–324 (Me_2S in cheese)

M. C. Stensmeyr, I. Urru, I. Collu, M. Celander, B. S. Hansson and A. M. Angioy, *Nature*, 2002, 420, 625–626 (Me_2S in arum lily)

J. R. Pawlik, G. McFall and S. Zea, *J. Chem. Ecol.*, 2002, 28, 1103–1115 (Me_2S in sponges)

L. Pripis-Nicolau, G. Revel, A. Bertrand and A. Lonvaud-Funel, *J. Appl. Microbiol*, 2004, 96, 1176–1184 (Me_2S in plants and vegetables)

T. G. Chasteen and R. Bentley, *J. Chem. Educ.*, 2004, 81, 1524–1528 (Me_2S in the sulfur cycle)

R. E. March, D. S. Richards and R. W. Ryan, *Int. J. Mass Spectrometry*, 2006, 249–250, 60–67 (Me_2S in black truffles)

E. Aprea, F. Biasioli, S. Carlin, G. Versini, T. D. Märk and F. Gasperi, *Rapid Commun. Mass Spectrom.*, 2007, 21, 2564–2572 (Me_2S in white truffles)

J. D. Todd et al., *Science*, 2007, 315, 666–669; J. D. Todd et al., *Environmental Microbiology*, 2010, 12, 327–343 (genes that make Me_2S in bacteria and the sulfur cycle)

G. B. Cunningham, V. Strauss and P. G. Ryan, *The Journal of Experimental Biology*, 2008, 211, 3123–3127 (African penguins detect Me_2S)

L. Culleré, V. Ferreira, B. Chevret, M. E. Venturini, A. C. Sánchez-Gimeno and D. Blanco, *Food Chemistry*, 2010, 122, 300–306 (Me_2S in black truffles and summer truffles)

F. Martin et al., *Nature*, 2010, 464, 1033–1038 (genome of Périgord black truffle and Me_2S)
R. Splivallo, S. Ottonello, A. Mello and P. Karlovsky, *New Phytologist*, 2011, 189, 688–699 (truffle volatiles review)
Z. Nowak, *Truffle: A Global History*, London, Reaktion Books, 2013
C. N. Harvey-Woodworth, *British Dental Journal*, 2013, 214, E20 (Me_2S in halitosis)
N. M. Levine, *Science*, 2016, 354, 418–419; K. B. Ksionzek, O. J. Lechtenfeld, S. L. McCallister, P. Schmitt-Kopplin, J. K. Geuer, W. Geibert and B. P. Koch, *Science*, 2016, 354, 456–459 (marine sulphur cycle)
P. C. Schmidberger and P. Schieberle, *J. Agric. Food Chem.*, 2017, 65, 9287–9296 (Me_2S in aroma of White Alba truffle and Burgundy truffle)
K. Thume, B. Gebser, L. Chen, N. Meyer, D. J. Kieber and G. Pohnert, *Nature*, 2018, 563, 412–415 (marine sulphur cycle)

18. FENTANYL AND ITS DERIVATIVES

J. C. Haigh, L. J. Lee and R. E. Schweinsburg, *J. Wildl. Dis.*, 1983, 19, 140–144 (immobilising polar bears)
T. H. Stanley, T. D. Egan and H. Van Aken, *Anesthesia & Analgesia*, 2008, 106, 451–462 (tribute to Paul A. J. Janssen)
A. V. George, J. J. Lu, M. V. Pisano, J. Metz and T. B. Erickson, *Am. J. Emerg. Med.*, 2010, 28, 530–532 (carfentanil, the vet and the elk)
J. R. Riches, R. W. Read, R. M. Black, N. J. Cooper and C. M. Timperley, *J. Anal. Toxicol.*, 2012, 36, 647–656 (analysis shows carfentanil used in Moscow theatre siege)
J. J. Li, *Blockbuster Drugs*, New York, OUP, 2014, 146–152 (also *Duragesic*)
T. H. Stanley, *The Journal of Pain*, 2014, 15, 1215–1224 ('the fentanyl story')
J. Mounteney, I. Giraudon, G. Denissov and P. Griffiths, *International Journal of Drug Policy*, 2015, 26, 626–631 (fentanyl deaths rise in Europe)
K. McLaughlin, *Science*, 2017, 355, 1364–1366 (deadly opioid chemistry)
K. Megget, *Chem. Ind.*, 2018, (10), 19–21 (fentanyl drug crisis)
N. Misailidi, I. Papoutsis, P. Nikolaou, A. Dona, C. Spiliopoulou and S. Athanaselis, *Forensic Toxicol.*, 2018, 36, 12–32 (review of ocfentanil and carfentanil)
S. M. Burns, C. W. Cunningham and S. L. Mercer, *ACS Chem. Neurosci.*, 2018, 9, 2428–2437 (DARK classics in chemical neuroscience: fentanyl)
D. Ciccarone, *Int. J. Drug Policy*, 2019, 71, 183–188 (the U. S. triple wave epidemic and its suppliers)
B. Westhoff, *Fentanyl, Inc.*, London, Grove Atlantic, 2019
J. L. S. Leen and D. N. Juurlink, *Can. J. Anesth.*, 2019, 66, 414–421 (carfentail review)
A. E. Ringuette, M. Spock, C. W. Lindsley and A. M. Bender, *ACS Chem. Neurosci.*, 2020, 11, 3955–3967 (DARK Classics in Chemical Neuroscience: Carfentanil)
Q. N. Vo, P. Mahinthichaichan, J. Shen and C. R. Ellis, *Nature Communications*, 2021, 12, 984 (how μ-opioid receptor recognizes fentanyl)
D. Ciccarone, *Curr. Opin. Psychiatry*, 2021, 34, 344–350 (the rise of illicit fentanyls, stimulants and the fourth wave of the opioid overdose crisis)
P. Skolnick, *Pharmacology & Therapeutics*, 2022, 233, 108019 (US overdose deaths)
K. Xiao, Q. Zheng and L. Bao, *BMC Anesthesiology*, 2022, 22, 268 (fentanyl activates ovarian cancer)

19. HEXANAL AND GREEN GRASS

A. Hatanaka, *Phytochemistry*, 1993, 34, 1201–1218 (green odour reviewed)

W. Kirstine, I. Galbally, Y. Ye and M. Hooper, *J. Geophys. Res.*, 1998, 103, 10605–10619 (enhanced emissions of (Z)-3-hexenal,(E)-2-hexenal,(Z)-2-hexen-1-ol, (Z)-3-hexen-1-ol, and (Z)-3-hexenyl acetate from freshly mown pasture)
R. Fall, T. Karl, A. Hansel, A. Jordan and W. Lindinger, *J. Geophys. Res.*, 1999, 104, 15963–15974 (VOCs emitted after leaf wounding)
G. Bisignano, M. G. Lagana, D. Trombetta, S. Arena, A. Nostro, N. Uccella, G. Mazzanti and A. Saija, *FEMS Microbiology Letters*, 2001, 198, 9–13 ((E)-2-hexenal, not hexanal has antibacterial activity)
T. Karl, R. Fall, A. Jordan and W. Lindinger, *Environ. Sci. Technol.*, 2001, 35, 2926–2931 ((Z)-3-hexenal, (E)-2-hexenal and other C_6 compounds formed in mowing lawn)
T. Sasabe et al., *Chem. Senses*, 2003, 28, 565–572 (PET study of activation of cerebral blood flow in primary olfactory cortex of monkeys by green odour)
Y. Watanabe, T. Sasabe, K. Yamaguti, M. Kobayashi, S. Yamamoto, H. Kuratsune, K. Sano, A. Hatanaka, H. Tsukada and H. Onoe, *Chem. Senses*, 2005, 30, Suppl 1, i268–i269 (anti-fatigue effects of green odour)
I. Kubo, K. Fujita, A. Kubo, K. Nihei and T. Ogura, *J. Agric. Food Chem.*, 2004, 52, 3329–3332 (Antibacterial activity of (*E*)-2-alkenals from cinnamon against *Salmonella choleraesuis*)
K. Matsui, *Current Opinion in Plant Biology*, 2006, 9, 274–280 (GLVs reviewed)
M. Steinhaus, J. Bogen and P. Schieberle, *Developments in Food Science*, 2006, 43, 189–192 (hexenal key aroma compound in apple juice)
T. Oka, S. Hayashida, Y. Kaneda, M. Takenaga, Y. Tamagawa, S. Tsuji and A. Hatanaka, *BioPsychoSocial Medicine*, 2008, 2, 2 (green odour has an anti-stress effect in humans, reducing blood pressure etc.)
S. Allmann and I. T. Baldwin, *Science*, 2010, 329, 1075–1078 (insects betray themselves in nature to predators by rapid isomerization of GLVs)
Y. Nikaido, S. Miyata and T. Nakashima, *Physiol. Behav.*, 2011, 103, 547–556 (a mixture of cis-3-hexenol and trans-2-hexenal reduce stress in rats)
M. Wikelski, M. Quetting, Y. Cheng, W. Fiedler, A. Flack, A. Gagliardo, R. Salas, N. Zannoni and J. Williams, *Scientific Reports*, 2021, 11, 12912 (GLV smell attracts white storks to freshly cut meadows)
F. Lucht, M. Rinken and S. Haberlandt, *International Journal of Food Science and Technology,* 2021, 56, 909–918 (hexanal, (Z)-2-hexenal and hexyl acetate in apple cultivars and their juices)

20. IVERMECTIN

R. W. Burg et al., *Antimicrob. Agents Chemother.*, 1979, 15, 361–367 (production of avermectins)
T. W. Miller et al., *Antimicrob. Agents Chemother.*, 1979, 15, 372–378 (efficacy of avermectins)
J. C. Chabala, H. Mrozik, R. L. Tolman, P. Eskola, A. Lusi, L. H. Peterson, M. F. Woods, M. H. Fisher and W. C. Campbell, *J. Med. Chem.*, 1980, 23, 1134–1136 (synth. of ivermectin)
W. C. Campbell, M. H. Fisher, E. O. Stapley, G. Albers-Schönberg and T. A. Jacob, *Science*, 1983, 221, 823–828 (ivermectin as a new antiparasitic agent)
M. H. Fisher, *Pure Appl. Chem.*, 1990, 62, 1231–1240 (avermectin review)
K. Collins, *Perspect. Biol. Med.*, 2004, 47, 100–109 (history of the Mectizan donation program)
S. Ōmura and A. Crump, *Nat. Rev. Microbiol.*, 2004, 2, 984–989 (the ivermectin story)
Y. J. Yoon, E.-S. Kim, Y.-S. Hwang and C.-Y. Choi, *Appl. Microbiol. Biotechnol.*, 2004, 63, 626–634 (biosynthesis)
A. Crump and K. Otoguro, *Trends in Parasitology*, 2005, 21, 126–132 (Ōmura's work)
T. G. Geary, *Trends in Parasitology*, 2005, 21, 530–532 (20 years of ivermectin)

A. D. Hopkins, *Eye*, 2005, 19, 1057–1066 (improvements upon ivermectin treatment)
S. Ōmura, Int. *J. Antimicrob. Ag.*, 2008, 31, 91–98 (25 years of Ivermectin)
A. Crump and S. Ōmura, *Proc. Jpn. Acad., Ser. B.*, 2011, 87, 13–28 (ivermectin review)
J. M. Sheele, J. F. Anderson, T. D. Tran, Y. A. Teng, P. A. Byers, B. S. Ravi and D. E. Sonenshine, *J. Emerg. Medicine*, 2013, 45, 433–440 (ivermectin and bed bugs)
X. Huang, H. Chen and P. L. Shaffer, *Structure*, 2017, 25, 945–950 (how ivermectin binds to receptors)
D. S. Ashour, *International Journal of Antimicrobial Agents*, 2019, 54, 134–142 (review of ivermectin treatment of parasitic infections)
S. Reardon, *Nature*, 2021, 596, 173–174 (on flawed claims for Ivermectin vs. COVID)
E. Lopez-Medina et al., *JAMA*, 2021, 325, 1426–1435 (results of clinical trial vs. COVID-19 in Colombia)
M. Popp, S. Reis, S. Schießer, R. Hausinger, M. Stegemann, M.-I. Metzendorf, P. Kranke, P. Meybohm, N. Skoetz and S. Weibel, *Cochrane Database of Systematic Reviews*, 2022, (6), Art. No.: CD015017 (Ivermectin for preventing and treating COVID-19)
K. U. Abbas, S. Muhammad and S. F. Ding, *Indian J. Pharm. Sci.*, 2022, 84(1) Spl. Issue 87–91 (Ivermectin versus mild COVID-19)
C. Chaccour et al., *N. Engl J. Med.*, 2025, 393, 362–375 (Ivermectin and malaria)

21. LIMONENE

L. Friedman and J. G. Miller, *Science*, 1971, 172, 1044–1046 (likely origin of incorrect statement on smells of limonene isomers)
A. F. Thomas and Y. Bessière, *Nat. Prod. Rep.*, 1989, 6, 292–309 (limonene reviewed)
M. H. Boelens, H. Boelens and L. H. van Gemert, *Perfum. Flavor.*, 1993, 18, 2 (query of smells of limonene isomers)
A. Hinterholzer and P. Schieberle, *Flavour Fragr. J.*, 1998, 13, 49–55 (key aroma compounds in hand-squeezed Valencia late oranges juice)
C. Sell, *Chem. Biodiversity*, 2004, 1, 1899–1920 (correct statement on isomers of limonene)
D. C. Hyatt, B. Youn, Y. Zhao, B. Santhamma, R. M. Coates, R. B. Croteau and C.-H. Kang, *PNAS*, 2007, 104, 5360–5365 (structure of (-)-limonene synthase)
P. Laszlo, *Citrus. A History*, Chicago, University of Chicago Press, 2007
P. Erasto and A. M. Viljoen, *Natural Product Communications*, 2008, 3, 1193–1202 (limonene reviewed)
M. Averbeck and P. H. Schieberle, *Eur. Food Res. Technol.*, 2009, 229, 611–622 (key aroma compounds in freshly reconstituted orange juice from concentrate)
R. Ciriminna, M. Lomeli-Rodriguez, P. D. Cara, J. A. Lopez-Sanchez and M. Pagliaro, *Chem. Commun.*, 2014, 50, 15288–15296 (limonene and the bioeconomy)
T. M. Attard, B. Watterson, V. L. Budarin, J. H. Clark and A. J. Hunt, *New J. Chem.*, 2014, 38, 2278–2283 (microwave assisted extraction of (D)-limonene from orange waste)
R. Fernández del Río, M. E. O'Hara, A. Holt, P. Pemberton, T. Shah, T. Whitehouse and C. A. Mayhew, *EBioMedicine*, 2015, 2, 1243–1250 (volatile biomarkers in breath associated with liver cirrhosis)
B. R. Morehouse, R. P. Kumar, J. O. Matos, S. N. Olsen, S. Entova and D. D. Oprian, *Biochemistry*, 2017, 56, 1706–1715 (structure of (+)-limonene synthase)
L. Kvittingen, B. J. Sjursnes and R. Schmid, *J. Chem. Educ.*, 2021, 98, 3600–3607 ('Limonene in Citrus: A String of Unchecked Literature Citings?')
Y. Han, W. Chen and Z. Sun, *Food Saf.*, 2021, 12918, 1–14 (antimicrobial activity of limonene against *Staphylococcus aureus*)
A. D. Ribeiro et al., *Bol. Latinoam. Caribe. Plant Med. Aromat.*, 2023, 22, 581–593 (antimicrobial activity of limonene reviewed)

D. J. Mabberley, *Citrus. A World History*, London, Thames and Hudson, 2024

22. LSD

Drs. Gabbai, Lisbonne and Pourquier, *BMJ*, 1951, 2, 650–651 ('Ergot poisoning at Pont St Esprit')
D. E. Smith, *California Medicine*, 1969, 110, 472–476 (LSD and Haight-Ashbury)
J. G. Fuller, *The Day of St Anthony's Fire*, London, Hutchinson, 1969 (Pont St. Esprit)
L. R. Caporael, *Science*, 1976, 192, 21–26 (ergotism and Salem)
J. Stevens, *Storming Heaven: LSD and the American Dream*, London, Heinemann, 1988
M. K. Matossian, *Poisons of the Past: Molds, Epidemics, and History*, New Haven, Yale University Press, 1989 (ergotism and history)
P. W. J. van Dongen and A. N. J. A. de Groot, *European Journal of Obstetrics & Gynecology*, 1995, 60, 109–116 (ergot alkaloids)
D. Black, *Acid: The Secret History of LSD*, London, Vision Paperbacks, 1998
M. A. Lee, *Acid Dreams: The Complete Social History of LSD*, New York, Grove Press, 2000
J. González-Maeso et al., *Neuron*, 2007, 53, 439–452 (LSD and signalling pathways)
T. Passie, J. H. Halpern, D. O. Stichtenoth, H. M. Emrich and A. Hintzen, *CNS Neuroscience & Therapeutics*, 2008, 14, 295–314 (pharmacology of LSD reviewed)
A. Roberts, *Albion Dreaming: A Popular History of LSD in Britain*, London, Marshall Cavendish, 2008
M. R. Lee, *J. R. Coll. Physicians Edinb.*, 2009, 39, 179–184 (history of ergot of rye (*Claviceps purpurea*) to 1900)
H. P. Albarelli Jr., *A Terrible Mistake*, Walterville, OR, Trine Day, 2009, pp. 350–358 for Pont St. Esprit
A. Hofmann, *LSD My Problem Child*, Oxford, OUP, 2013 (memoirs)
M. B. Liester, *Current Drug Abuse Reviews*, 2014, 7, 146–156 (review of LSD in treating addictions)
D. E. Nichols, *Pharmacol. Rev.*, 2016, 68, 264–355 (psychedelics reviewed)
D. Wacker et al., *Cell*, 2017, 168, 377–389 (cryst. struct. of an LSD-Bound human serotonin receptor)
D. E. Nichols and C. S. Grob, *Forensic Sci. Int.*, 2018, 284, 141–145 ('Is LSD Toxic?' LSD fatalities)
D. E. Nichols, *ACS Chem. Neurosci.*, 2018, 9, 2331–2343 (Dark Classics: Lysergic Acid Diethylamide (LSD))
R. Robison et al., *JAMA*, 2025, 334, 1358–1372 (clinical trials of MM120 (Lysergide) in treatment of Generalized Anxiety Disorder)

23. MEFLOQUINE

D. Whyte, *British Medical Journal,* 1984, 285, 1776–1779 (mepacrine and the Chindits)
L. J. Bruce-Chwatt, *British Medical Journal*, 1984, 288, 796 (mepacrine and Wingate's toxic psychosis)
K. J. Palmer, S. M. Holliday and R. N. Brogden, *Drugs*, 1993, 45, 430–475 (review of mefloquine)
G. D. Shanks, *Military Medicine*, 1994, 159, 275–281 (the rise and fall of mefloquine as an antimalarial drug in South East Asia)
M. Foley and L. Tilley, *Int. J. Parasitol.*, 1997, 27, 231–240; *Pharmacol. Ther.*, 1998, 79, 55–87 (quinoline antimalarials review)
A. Robert, F. Benoit-Vical, O. Dechy-Cabaret and B. Meunier, *Pure and Applied Chemistry,* 2001, 73, 1173–1188 (malaria drugs)

G. B. Stefano, N. Pilonis, R. Ptacek and R. M. Kream, *Med. Sci. Monit.*, 2017, 23, 2890–2896 (opium and morphine history)

L. Inglis, *Milk of Paradise*, London, Macmillan, 2018 (opium)

D. Ciccarone, *Int. J. Drug Policy*, 2019, 71, 183–188 (the triple wave epidemic and its suppliers)

C. C. Presley and C. W. Lindsley, *ACS Chem. Neurosci.*, 2018, 9, 2503–2518 (DARK Classics in Chemical Neuroscience: Opium, a Historical Perspective)

P. Alam, S. Borkokoty, M. K. Siddiqi, A. Ehtram, N. Majid, M. Uddin and R. H. Khan, *ACS Chem. Neurosci.*, 2019, 10, 182–189 (DARK Classics in Chemical Neuroscience: Opium, a Friend or Foe)

J. Mella-Raipán, J. Romero-Parra and G. Recabarren-Gajardo, *ACS Chem. Neurosci.*, 2020, 11, 3905–3927 (DARK Classics in Chemical Neuroscience: Heroin and Desomorphine)

S. N. Michaleas, K. Laios, G. Tsoucalas and G. Androutsos, *Toxicology Reports*, 2021, 8, 411–414 (Paracelsus)

M. S. Milella, G. D'Ottavio, S. De Pirro, M. Barra, D. Caprioli and A. Badiani, *Translational Psychiatry*, 2023, 13, 120 (heroin and its metabolites)

27. MUSCONE AND SYNTHETIC MUSKS

German patent 47 599 in 1888; also see A. Baur, *Ber. Dtsch. Chem. Ges.*, 1891, 24, 2832–2836 (Musk Baur)

A. Baur, *Ber. Dtsch. Chem. Ges.*, 1894, 27, 1614–1619 (Musk Ambrette)

A. Baur-Thurgau, *Ber. Dtsch. Chem. Ges.*, 1898, 31, 1344–1349 (Musk Ketone)

L. Ruzicka, M. Stoll and H. Schinz, *Helv. Chim. Acta*, 1926, 9, 249–264; L. Ruzicka, *Helv. Chim. Acta*, 1926, 9, 715–729, 1008–1017 (muscone)

R. W. Moncrieff, *The Chemical Senses*, Leonard Hill, London, 1967, pp. 418–421, 436–443, 630–639

M. G. J. Beets, in: E. T. Theimer (ed.), 'Odor and Stimulant Structure', *Fragrance Chemistry*, Academic Press, 1982, pp. 77–122

B. D. Mookherjee and R. A. Wilson, in: E. T. Theimer (ed.), 'The Chemistry and Fragrance of Natural Musk Compounds', *Fragrance Chemistry*, Academic Press, 1982, pp. 433–494

E. G. Nash, E. J. Nienhouse, T. A. Silhavy, D. E. Humbert and M. J. Mish, *J. Chem. Educ.*, 1970, 47, 705–706 (syntheses of Musk Ketone and Musk Xylene)

D. P. Anonis, *Perfumer and Flavorist*, Jan–Feb 1997, 22, 43–47 (civetone)

R. Fenn, *Perfumer & Flavorist*, 1999, 24, 17 (musks)

J. Louie, C. W. Bielawski and R. H. Grubbs, *J. Am. Chem. Soc.*, 2001, 113, 11312 (muscone synth.)

C. Fehr and N. Chaptal-Gradoz, *Helv. Chim Acta*, 2002, 85, 533–543; C. Fehr, N. Chaptal-Gradoz and J. Galindo, *Chem. Eur. J.*, 2002, 8, 853–858 (Vulcanolide)

M. Eh [sic], *Chemistry & Biodiversity*, 2004, 1, 1975–1984 (new alicycllc musks)

H. Matsuda, S. Watanabe and H. Yamamoto, *Chemistry & Biodiversity*, 2004, 1, 1985–1991 (new macrocyclic musks)

P. Kraft, *Aroma Chemicals IV: Musks*, Chapter 7 of D. J. Rowe, ed., *Chemistry and Technology of Flavors and Fragrances*, Blackwell and CRC Press, 2005, pp. 143–168

C. Sell, 'Musks', in: *The Chemistry of Fragrances*, Royal Society of Chemistry, 2006, pp. 95–106

O. R. P. David, *Eur. J. Org. Chem.*, 2017, (1), 4–13 (definitive review on nitromusks)

O. R. P. David, *Chem. Eur. J.*, 2020, 26, 7537–7555 (chemical history of polycyclic musks)

K. Yoshikawa, J. Deguchi, J. Hu, H.-Y. Lu and H. Matsunami, *Current Biology*, 2022, 32, 5172–5179 (OR5A2 receptor responds to four types of musks)

N. Sato-Akuhara, C. Trimmer, A. Keller, Y. Niimura, M. Shirasu, J. D. Mainland and K. Touhara, *Chemical Senses*, 2023, 48, 1–9 (OR5AN1 and musks)

R. Emter, C. Merillat, S. Dossenbach and A. Natsch, *Chemical Senses*, 2024, 49, bjae015 (the trilogy of human musk receptors)

R. S. Pawar, E. Grundel, A. R. Fardin-Kia and J. I. Rader, *Journal of Pharmaceutical and Biomedical Analysis*, 2014, 88, 457–466 (amphetamines not present in *Acacia rigidula*)
N. Ohler, *Blitzed. Drugs in Nazi Germany*, London, Allen Lane, 2016
L. Kamieński, *Shooting Up. A History of Drugs in Warfare*, London, Hurst, 2016
S. Moeller, H. B. Huttner, T. Struffert and H. H. Müller, *Alcoholism and Drug Addiction*, 2016, 29, 39–41 (irreversible brain damage caused by methamphetamine)
T. J. Abbruscato and P. C. Trippier, *ACS Chem. Neurosci.*, 2018, 9, 2373–2378 (DARK Classics: Methamphetamine)
D. Trumbore and D. J. Nelson, *The Science of Breaking Bad*, Cambridge, MA, MIT Press, 2019
A. N. Edinoff, S. E. Kaufman, K. M. Green, D. A. Provenzano, J. Lawson, E. M. Cornett, K. S. Murnane, A. M. Kaye and A. D. Kaye, *Health Psychology Research*, 2022, 10, 38161 (adverse effects of methamphetamine use)
A. A. Guerin, T. Bridson, H. M. Plapp and G. Bedi, *Neuroscience and Biobehavioral Reviews*, 2023, 153, 105380 (young methamphetamine users)

26. MORPHINE AND HEROIN

A. Conan-Doyle, 'The Man with the Twisted Lip', *Strand Magazine*, 1891; *The Adventures of Sherlock Holmes*, 1892
J. Kaplan, *The Hardest Drug*, Chicago, The University of Chicago Press, 1983
L. D. Kapoor, *Opium Poppy. Botany, Chemistry and Pharmacology*, Binghamton, The Haworth Press, 1995
D. M. Perrine, *The Chemistry of Mind-Altering Drugs*, Washington, ACS, 1996, pp. 43–66
M. Booth, *Opium. A History*, London, Simon and Schuster, 1996
B. Hodgson, *Opium: A Portrait of the Heavenly Demon*, London, Souvenir Press, 2000
B. Hodgson, *In the Arms of Morpheus: The Tragic History of Laudanum, Morphine, and Patent Medicines*, Buffalo, NY, Firefly Books, 2001
R. Davenport-Hines, *The Pursuit of Oblivion. A Global History of Narcotics 1500–2000*, London, Weidenfeld and Nicolson, 2001
P. R. Blakemore and J. D. White, *Chem. Commun.*, 2002, 1159–1168 (morphine, the Proteus of organic molecules)
A. M. Trescot, S. Datta, M. Lee and H. Hansen, *Pain Physician*, 2008, 11, S133–S153 (opiates and the mu-opioid receptor)
R. G. Homan, B. Hudson and R. C. Rowe, *Popular Medicines*, London, Pharmaceutical Press, 2008 (Dr Collis Browne's Chlorodyne)
D. Ciccarone, *Int. J. Drug Policy*, 2009, 20, 277–282 (heroin and America)
P-A. Chouvy, *Opium: Uncovering the Politics of the Poppy*, London, I. B. Tauris, 2009.
T. Dormandy, *Opium. Reality's Dark Dream*, New Haven and London, Yale University Press, 2012.
K. L. Dorrington and W. Poole, *British Journal of Anaesthesia*, 2013, 110, 7–12 (Wren and opium anaesthetic)
A. Onoyovwe, J. M. Hagel, X. Chen, M. F. Khan, D. C. Schriemer and P. J. Facchini, *The Plant Cell*, 2013, 25, 4110–4122 (morphine biosynthesis)
A. Gottås, E. L. Øiestad, F. Boix, V. Vindenes, Å. Ripel, C. H. Thaulow and J. Mørland, *British Journal of Pharmacology*, 2013, 170, 546–556; D. Perekopskiy and E. A. Kiyatkin, *ACS Chem. Neurosci.*, 2019, 10, 8, 3409–3414 (6-monoacetylmorphine as a heroin metabolite)
J. T. Jones, M. Jones, B. Jones, K. Sulaiman, C. Plate and Douglas Lewis, *Ther. Drug. Monit.*, 2015, 37, 45–52 (detection of codeine, morphine, 6-monoacetylmorphine)

28. NYLON

R. Adams, *Biographical Memoir of Wallace Hume Carothers 1896–1937*, National Academy of Sciences of the USA, 1939, Vol XX, No. 12, pp. 293–309

P. W. Morgan and S. L. Kwolek, *Journal of Chemical Education*, 1959, 36, 182–184 (the nylon rope trick)

H. F. Mark, *Giant Molecules*, Amsterdam, Time-International Nederland, 1968, pp. 105–109

S. Fenichell, *Plastic. The Making of a Synthetic Century*, New York, HarperCollins, 1996, pp. 134–174

M. Hermes, *Enough for One Lifetime, Wallace Carothers the Inventor of Nylon*, York, PA, Chemical Heritage Foundation, 1996

S. Handley, *Nylon: the Manmade Fashion Revolution*, London, Bloomsbury Publishing, 1999

S. B. McGrayne, *Prometheans in the Lab: Chemistry and the Making of the Modern World*, New York, McGraw-Hill, 2001, pp. 106–147

P. Le Couteur and J. Burreson, *Napoleon's Buttons: How 17 Molecules Changed History*, New York, Tarcher/Putnam, 2003, pp. 105–123

R. J. Giguere, ed., *Molecules that Matter*, Philadelphia, Chemical Heritage Foundation, 2008, pp. 81–96

W. Gratzer, *Giant Molecules: From Nylon to Nanotubes*, Oxford, 2009, pp. 126–127, 157–158

29. OCTANAL

M. H. Boelens, *Perfumer and Flavorist*, 1991, 16, 17–34 (review of citrus oil composition)

H. Zhao, L. Ivic, J. M. Otaki, M. Hashimoto, K. Mikoshiba and S. Firestein, *Science*, 1998, 279, 237–242 (rat OR-17)

D. Rowe and B. Tangel, *Perfumer and Flavorist*, 1999, 24, 36–44 (citrus flavour)

A. Buettner and P. Schieberle, *J. Agric. Food Chem.*, 2001, 49, 2387–2394 ('hand-squeezed' Valencia orange juice)

H. D. Douglas, J. E. Co, T. H. Jones and W. E. Conner, *Naturwiss.*, 2001, 88, 330–332; *J. Chem. Ecol.*, 2004, 30, 1921–1935 (auklet phenomenon and composition of emissions)

F. Xu, N. Liu, I. Kida, D. L. Rothman, F. Hyder and G. M. Shepherd, *PNAS*, 2003, 100, 11029–11034 (aldehydes and mouse OR-17 receptor)

A. Högnadóttir and R. L. Rouseff, *Journal of Chromatography A*, 2003, 998, 201–211 (analysis of orange essence oil)

J. C. Hagelin, I. L. Jones and L. E. L. Rasmussen, *Proc. R. Soc. Lond. B*, 2003, 270, 1323–1329 (auklet emission phenomenon)

H. D. Douglas, J. E. Co, T. H. Jones, W. E. Conner and J. F. Day, *J. Med. Entomol.*, 2005, 42, 647–651 (auklet mixture repels mosquitoes)

H. D. Douglas, *J. Chem. Ecol.*, 2006, 32, 2559–2567 (auklet emissions and ticks)

M. Ghaninia, M. Larsson, B. S. Hansson and R. Ignell, *J. Exp. Biol.*, 2008, 211, 3020–3027 (octanal and mosquitoes)

R. Perez-Cacho and R. L. Rouseff, *J. Agric. Food Chem.*, 2008, 56, 9785–9796 (processing and storage effects on orange juice aroma)

R. Perez-Cacho and R. L. Rouseff, *Crit. Rev. Food Sci.*, 2008, 48, 681–695 (fresh squeezed orange juice odour reviewed)

Z. Peterlin, Y. Li, G. X. Sun, R. Shah, S. Firestein and K. Ryan, *Chemistry & Biology*, 2008, 15, 1317–1327 (rat OR-17, aldehyde chain size and response)

B. Lawrence, *Perfumer and Flavorist*, 2009, 34, 48–57 (essential oils)

M. Averbeck and P. Schieberle, *Eur. Food Res. Technol.*, 2009, 229, 611–622 (reconstituted orange juice)

M. D. Kurland, M. B. Newcomer, Z. Peterlin, K. Ryan, S. Firestein and V. S. Batista, *Biochemistry*, 2010, 49, 6302–6304 (rat OR-17)

J. G. Logan, N. M. Stanczyk, A. Hassanali, J. Kemei, A. E. G. Santana, K. A. L. Ribeiro, J. A. Pickett and A. J. Mordue (Luntz), *Malaria J.*, 2010, 9, 239 (octanal and mosquitoes)
X. Li, J.-N. Ren, G. Fan and S.-Y. Pan, *J. Food Sci. Technol.*, 2018, 55, 4530–4543 (freshly-squeezed orange juice)
M. T. Liu, J. Ho, J. Liue, R. Purakaita, U. N. Morzane, L. Ahmede, Victor S. Batistae, H. Matsunami and K. Ryan, *Org. Biomol. Chem.*, 2018, 16, 2541–2548; M. T. Liu, M. Na, Y. Li, M. Biscoe and K. Ryan, *ChemEur. J.*, 2020, 26, 11462–11469 (conformation and sensing of aldehydes)
P. J. Weldon, *Ibis*, 2023, 165, 1054–1067 (non-host odours in avian defence)
I. L. Jones and B. E. Lyon, *Current Biology*, 2023, 33, R1173–R1175 (auklets)

30. OCTANE

P. W. Atkins, *Molecules*, New York Scientific American Library, 1987, pp. 37–38
A. Perdih and F. Perdih, *Acta Chim. Slov.*, 2006, 53, 306–315; A. Demirbas, M. A. Balubaid, A. M. Basahel, W. Ahmad and M. H. Sheikh, *Petroleum Science and Technology*, 2015, 33, 1190–1197 (lists of RON values for hydrocarbons)
G. Bailey, 'The Narrow Margin of Criticality: The Question of the Supply of 100-Octane Fuel in the Battle of Britain', *The English Historical Review*, 2008, CXXIII, 394–411
M. E. Bowden and N. Gussman, 'Isooctane', in: *Molecules that Matter*, Philadelphia, Chemical Heritage Foundation, 2008, pp. 32–49
B. M. Goortani, A. Gaurav, A. Deshpande, F. T. T. Ng and G. L. Rempel, *Industrial & Engineering Chemistry Research*, 2015, 54, 3570–3581 (production of isooctane from isobutene)
D. Splitter, A. Pawlowski and R. Wagner, 'A Historical Analysis of the Co-evolution of Gasoline Octane Number and Spark-Ignition Engines', *Frontiers in Mechanical Engineering*, 2016, 1, 16

31. PARACETAMOL

H. N. Morse, *Berichte der deutschen chemischen Gesellschaft*, 1878, 11, 232–233 (original synthesis of phenacetin)
A. Cahn and P. Hepp, *Centralblatt für klinische Medizin*, 1886, 7, 561–564 (original synthesis of acetanilide, aka Antifebrin)
D. Lester, L. A. Greenberg and R. P. Carroll, *J. Pharmacol. Exp. Ther.*, 1947, 90, 68–75; B. B. Brodie and J. Axelrod, *J. Pharmacol. Exp. Ther.*, 1948, 94, 22–28; idem, 29–38; F. B. Flinn and B. B. Brodie, *J. Pharmacol. Exp. Ther.*, 1948, 94, 76–77 (metabolism of acetanilide in the body)
B. B. Brodie and J. Axelrod, *J. Pharmacol. Exp. Ther.*, 1948, 94, 58–67 (metabolism of phenacetin)
G. Margetts, *J. Int. Med. Res.*, 1976, 4, Suppl 4, 55–70 (phenacetin, paracetamol and the liver)
M. Beck, S. Monroe, L. R. Prout, M. Hager and R. LaBreque, *Newsweek*, October 11, 1982, p. 32 (the Tylenol murders)
V. F. Fairbanks, *Mayo Clin. Proc.*, 1994, 69, 889–892 (coal-tar derived chemicals)
A. L. Jones, P. C. Hayes, A. T. Proudfoot, J. A. Vale and L. F. Prescott, *Brit. Med. J. (Clin. Res. Ed.)*, 315, 1997, 301–303 (Should methionine be added to every paracetamol tablet?)
J. J. Johnston, P. J. Savarie, T. M. Primus, J. D. Eisemann, J. C. Hurley and D. J. Kohler, *Environ. Sci. Technol.*, 2002, 36, 3827–3833 (Paracetamol and brown snakes)
K. D. Rainsford, ed., *Aspirin and Related Drugs*, London and New York, Taylor and Francis, 2004, esp. pp. 181–213 (paracetamol)
D. Bartlett, *J. Emerg. Nurs.*, 2004, 30, 281–283 (Treatment for overdoses)
J. Timbrell, *The Poison Paradox*, Oxford, OUP, 2005, pp. 49–56 (paracetamol toxicity)

M. Mahachandra, Yassierli and E. D. Garnaby, *Procedia Manufacturing*, 2015, 4, 471–477 (peppermint in a fragrance to maintain drivers' alertness.)
A. Kumar, U. Baitha, P. Aggarwal and N. Jamshed, *Int. J. App. Basic Med. Res.*, 2016, 6, 137–139 (fatal case of menthol poisoning)
M. M. Diver, Y. Cheng and D. Julius, *Science*, 2019, 365, 1434–1440 (TRPM8 inhibition and desensitization)
L. Xu, Y. Han, X. Chen, A. Aierken, H. Wen, W. Zheng, H. Wang, X. Lu, Z. Zhao, C. Ma, P. Liang, W. Yang, S. Yang and F. Yang, *Nat. Commun.*, 2020, 11, 3790 (menthol molecule binds to the TRPM8 ion channel)
K. Wailoo, *Pushing Cool: Big Tobacco, Racial Marketing, and the Untold Story of the Menthol Cigarette*, University of Chicago Press, 2021
D. Mendez and T. T. T. Le, *Tob. Control*, 2022, 31, 569–571 (menthol smoking and the African American population 1980–2018)
X. Chen, L. Xu, H. Zhang, H. Wen and F. Yang, *Front. Pharmacol.*, 2022, 13, 898670 (activation of TRPM8 by the stereoisomers of menthol)

25. METHAMPHETAMINE

A. Ogata, *J. Pharm. Soc. Jpn.*, 1919, 445, 193–216; *Chem. Abs.*, 1919, 13, 1709 (synth)
D. E. Smith and C. M. Fischer, *Clinical Toxicology*, 1970, 3, 117–124 (speed and Haight-Ashbury)
H. L. Jin and T. E. Beesley, *Chromatographia*, 1994, 38, 595–598 (methamphetamine isomers)
B. A. Clement, C. M. Goff and T. D. A. Forbes, *Phytochemistry*, 1998, 38, 266–279 (claims of amphetamine and methamphetamine presence in *Acacia rigidula* shrub)
A. Shulgin and A. Shulgin, *Pihkal: A Chemical Love Story*, Berkeley, CA, Transform Press, 1995
D. R. Mottram, ed., *Drugs in Sport*, London, E and F. N. Spon, 2nd edition, 1996, esp. p. 19 and pp. 86–94
J. Cash, *Cash: The Autobiography*, San Francisco, Harper Collins, 1997
W. Fotheringham, *Put Me Back on My Bike: In Search of Tom Simpson*, London, Yellow Jersey Press, 2002
A. R. Green, A. O. Mechan, J. M. Elliott, E. O'Shea and M. Colado, *Pharmacol. Rev.*, 2003, 55, 463–508 (MDMA pharmacology)
P. M. Thompson et al., *J. Neurosci.*, 2004, 24, 6028–6036 (structural abnormalities in the brains of methamphetamine users)
A. Ross, *Unfinished Business: Alain Baxter*, Stockport, Dewi Lewis Media, 2005
M. F. Holley, *Crystal Meth*, Mustang, OK, Tate Publishing, 2005
L. Iversen, *Speed, Ecstasy, Ritalin*, Oxford, OUP, 2006 (the best single source)
C. Guillot and D. Greenway, *J. Psychopharmacol.*, 2006, 20, 411–416 (recreational use of methamphetamine linked to depression)
N. Reding, *Methland. The Death and Life of an American Small Town*, New York, Bloomsbury, 2009
F. Roussotte, L. Soderberg and E. Sowell, *Neuropsychol. Rev.*, 2010, 20, 376–397 (effect of meth exposure on foetus)
C. Morrison, F. J. Smith, T. Tomaszewski, K. Stawiarska and M. Biziuk, *Chirality*, 2011, 23, 519–522 (separating methamphetamine isomers)
R. J. Defalque and A. J. Wright, *Bulletin of Anesthesia History*, 2011, 29, 21–24 ('Methamphetamine for Hitler's Germany: 1937 to 1945')
J. M. Hagel, R. Krizevski, F. Marsolais, E. Lewinsohn and P. J. Facchini, *Trends in Plant Science*, 2012, 17, 404–412 (biosynthesis of amphetamine analogues in plants)

A. L. Baggish and D. R. Hill, *Antimicrobial Agents and Chemotherapy*, 2002, 46, 1163–1173 (atovaquone reviewed)

A. R. Croft, *J. Roy. Soc. Med.*, 2007, 100(4), 170–174 (a lesson learnt: the rise and fall of Lariam)

S. Toovey, *Travel Med. Infect. Dis.*, 2009, 7, 2–6 (mefloquine neurotoxicity reviewed)

C. Schneider, M. Adamcova, S. S. Jick, P. Schlagenhauf, M. K. Miller, H.-G. Rhein and C. R. Meier, *Travel Med. Infect. Dis.*, 2013, 11, 71–80 (antimalarial drugs risk neuropsychiatric disorders)

E. C. Ritchie, J. Block and R. L. Nevin, *J. Am. Acad. Psychiatry Law*, 2013, 41, 224–235 (psychiatric side effects of mefloquine: applications to forensic psychiatry)

G. L. Nixon, D. M. Moss, A. E. Shone, D. G. Lalloo, N. Fisher, P. M. O'Neill, S. A. Ward and G. A. Biagini, *J. Antimicrob. Chemother.*, 2013, 68, 977–985 (antimalarial pharmacology and therapeutics of atovaquone)

F. Gobbi, A. Rossanese, D. Buonfrate, A. Angheben, C. Postiglione and Z. Bisoff, *World J. Clin. Cases*, 2014, 2, 12–15 (epilepsy triggered by mefloquine)

A. Ringqvist, P. Bech, B. Glenthøj and E. Petersen, *Travel Med. Infect. Dis.*, 2015, 13, 80–88 (acute and long-term psychiatric side effects of mefloquine)

R. L. Nevin, 'Mefloquine and Posttraumatic Stress Disorder', in: E. C. Ritchie (ed.), *Textbook of Military Medicine. Forensic and Ethical Issues in Military Behavioral Health*, Washington, DC, Borden Institute, 2015, pp. 277–296

J. Livezey, T. Oliver and L. Cantilena, *Drug Saf. – Case Rep.*, 2016, 3, 7 (prolonged neuropsychiatric symptoms in a military service member exposed to Mefloquine)

R. L. Nevin, *Pharmacol. Res. Perspect.*, 2017, 5, e00328 ('A serious nightmare: psychiatric and neurologic adverse reactions to mefloquine are serious adverse reactions')

J. Manaranche et al., *J. Antimicrob. Chemother.,* 2024, 79, 2877–2886 (ganaplacide/lumefantrine versus artemisinin resistant *Plasmodium falciparum*)

R. Sangana et al., *Journal of Clinical Pharmacology*, 2025, 65, 179–189 (trial of ganaplacide and lumefantrine against *Plasmodium falciparum* Malaria)

24. MENTHOL

K. Tani, T. Yamagata, S. Akutagawa, H. Kumobayashi, T. Taketomi, H. Takaya, A. Miyashita, R. Noyori and S. Otsuka, *J. Am. Chem. Soc.*, 1984, 106, 5208–5217 (menthol synthesis)

B. Raudenbush, N. Corley and W. Eppich, *J. Sport and Exercise Psychology*, 2001, 23, 156–160 (peppermint and sport)

D. D. McKemy, W. M. Neuhausser and D. Julius, *Nature*, 2002, 416, 52–58 (TRPM8 receptor for menthol)

R. Noyori, Angew. *Chem. Int. Ed.,* 2002, 41, 2008–2022 (Nobel lecture)

A. M. Peier, A. Moqrich, A. C. Hergarden, A. J. Reeve, D. A. Andersson, G. M. Story, T. J. Earley, I. Dragoni, P. McIntyre, S. Bevan and A. Patapoutian, *Cell*, 2002, 108, 705–715 (TRPM8 receptor senses cold stimuli and menthol)

S. Barker, P. Grayhem, J. Koon, J. Perkins, A. Whalen and B. Raudenbush, *Perceptual and Motor Skills*, 2003, 97, 1007–1010 (peppermint odour enhances clerical performance)

R. B. Croteau, E. M. Davis, K. L. Ringer and M. R. Wildung, *Naturwissenschaften*, 2005, 92, 562–577; E. M. Davis, K. L. Ringer, M. E. McConkey and R. B. Croteau, *Plant Physiology*, 2005, 137, 873–881 (biosynthesis)

M. Moss, S. Hewitt, L. Moss and K. Wesnes, *Intern. J. Neuroscience*, 2008, 118, 59–77 (peppermint enhances accuracy of memory)

G. P. P. Kamatou, I. Vermaak, A. M. Viljoen and B. M. Lawrence, *Phytochemistry*, 2013, 96, 15–25 (review)

R. J. Wickham, *Yale Journal of Biology and Medicine*, 2015, 88, 279–287 (addictive properties of menthol cigarettes)

E. D. Högestätt, B. A. G. Jönsson, A. Ermund, D. A. Andersson, H. Björk, J. P. Alexander, B. F. Cravatt, A. I. Basbaum and P. M. Zygmunt, *J. Biol. Chem.*, 2005, 280, 31405–31412 (AM404)

A. Ottani, S. Leone, M. Sandrini, A. Ferrari and A. Bertolini, *Euro. J. Pharmacol.*, 2006, 531, 80–281 (Paracetamol and cannabinoid CB1 receptors)

A. Bertolini, A. Ferrari, A. Ottani, S. Guerzoni, R. Tacchi and S. Leone, *CNS Drug Reviews*, 2006, 12, 250–275 (paracetamol review)

G. B. Steventon and A. Hutt, in: R. H. Waring, G. B. Steventon and S. C. Mitchell (eds.), 'Steventon and Hutt - "Paracetamol"', *Molecules of Death*, London, Imperial College Press, 2nd edition, 2007, pp. 253–263

F. Tennant, 'Howard Hughes and Pseudoaddiction', *Practical Pain Management*, July/August 2007, 12–29

B. Anderson, *J. Pediatric Anesthesia*, 2008, 18, 915–921 (how paracetamol acts)

D. A. Andersson et al., *Nat. Commun.*, 2011, 2, 551 (paracetamol metabolites activate receptors in spinal cord)

A. W. Jones, *Drug Test. Analysis,* 2011, 3, 337–344 ('Early drug discovery and the rise of pharmaceutical chemistry')

G. G. Graham, M. J. Davies, R. O. Day, A. Mohamudally and K. F. Scott, *Inflammopharmacol.*, 2013, 21, 201–232 (paracetamol review)

M. Jóźwiak-Bębenista and J. Z. Nowak, *Acta Poloniae Pharmaceutica – Drug Research*, 2014, 71, 11–23 (paracetamol: mechanism of action, applications and safety concerns)

C. I. Ghanem, J. P. María, J. E. Manautou and A. D. Mottino, *Pharmacol Res.*, 2016, 109, 119–131 (acetaminophen: from liver to brain)

J. C. McCrae, E. E. Morrison, I. M. MacIntyre, J. W. Dear and D. J. Webb, *Br. J. Clin. Pharmacol*, 2018, 84, 2218–2230 (long-term adverse effects of paracetamol reviewed)

T. Mathies and R. E. Mauldin, *Sci. Rep.*, 2020, 10, 845 ('lethal methemoglobinemia in the invasive brown treesnake after acetaminophen ingestion')

M. G. Nafus, S. R. Siers, B. A. Levine, Z. C. Quiogue and A. A. Y. Adams, *J. Wildl. Manag.*, 2022, 86, e22136 (evaluation of snake suppression programme on Guam)

N. S. Abraham, S. Mishra and S. Bhatnagar, *Palliat. Med. Pract.*, 2024, 18(2), 109–112 (acetaminophen-induced liver Injury)

J. Sheikh et al., *BMJ*, 2025, 391, e088141 (umbrella review of systematic reviews on possible link between paracetamol (Tylenol) and autism)

32. PENICILLIN

A. Fleming, *British Journal of Experimental Pathology*, 1929, 10, 226–236 (antibacterial activity of penicillin cultures)

E. P. Abraham, E. Chain, C. M. Fletcher, W. H. Florey, A. D. Gardner, N. G. Heatley and M. A. Jennings, *The Lancet*, 1941, 238, 177–189 (early cases healed by penicillin)

A. Fleming, *The Lancet*, 1943, 242, 434–38 ('streptococcal meningitis treated with penicillin')

G. T. Stewart and R. J. Holt, *BMJ*, 1963, 1, 308–311 ('Evolution of Natural Resistance to the newer penicillins'- MRSA)

S. Selwyn, *The Beta-lactam Antibiotics: Penicillins and Cephalosporins in Perspective*, London, Hodder, 1980

J. C. Sheehan, *Enchanted Ring: Untold Story of Penicillin*, Cambridge, MA, MIT Press, 1982

C. Fletcher, *British Medical Journal*, 1984, 289, 1721–1723 (re first clinical use of penicillin)

S. W. Queener, *Antimicrobial Agents and Chemotherapy,* 1990, 34, 943–948 (penicillin and cephalosporin biosynthesis)

P. L. Roach, I. J. Clifton, C. M. H. Hensgens, N. Shibata, C. J. Schofield, J. Hajdu and J. E. Baldwin, *Nature,* 1997, 387, 827–830 (isopenicillin *N* synthase and penicillin formation)

J. Emsley, *Molecules at an Exhibition*, Oxford, OUP, 1998, pp. 72–77

J. Mann, *The Elusive Magic Bullet*, Oxford, Oxford University Press, 1999, pp. 39–60

F. W. Diggins, *British Journal of Biomedical Science,* 1999, 56, 83–93; F. W. Diggins, *Biologist (London)*, 2000, 47, 115–119 (the discovery of penicillin)

G. Dodson, *Biogr. Mems. Fell. R. Soc. Lond.*, 2002, 48, 179–219 (memoir of Dorothy Hodgkin, see pp. 198–200 for penicillin)

M. C. Enright, D. A. Robinson, G. Randle, E. J. Feil, H. Grundmann and B. G. Spratt, *PNAS*, 2002, 99, 7687–7692 (evolutionary history of MRSA)

V. P. Sandanayaka and A. S. Prashad, *Current Medicinal Chemistry,* 2002, 9, 1145–1165 (resistance to beta-lactam antibiotics)

I. Trehan, F. Morandi, L. C. Blaszczak and B. K. Shoichet, *Chemistry & Biology,* 2002, 9, 971–980 (structure of amoxycillin bound to β-lactamase)

E. Lax, *The Mould in Dr Florey's Coat: How Penicillin Began the Age of Miracle Cures*, London, Little, Brown, 2004

K. Brown, *Penicillin Man*, Stroud, Sutton, 2004 (biography of Fleming)

R. Bud, *Penicillin: Triumph and Tragedy*, Oxford, Oxford University Press, 2007

K. C. Nicolaou and T. Montagnon, *Molecules that Changed the World*, Weinheim, Wiley-VCH, 2008, pp. 97–106 (penicillin)

R. J. Giguere, 'Penicillin G', in: *Molecules that Matter*, Philadelphia, Chemical Heritage Foundation, 2008, pp. 51–64

G. L. Patrick, *An Introduction to Medicinal Chemistry*, Oxford, Oxford University Press, 4th edition, 2009, pp. 433–448

P. W. Smith, K. Watkins and A. Hewlett, *Am. J. Infect. Control*, 2012, 40, 35–42 (infection control through the ages)

M. Lobanovska and G. Pilla, *Yale Journal of Biology and Medicine*, 2017, 90, 135–145 (penicillin's discovery and antibiotic resistance)

B. Ribeiro da Cunha, L. P. Fonseca and C. R. C. Calado, *Antibiotics*, 2019, 8, 45 (antibiotic discovery)

B. Short, *Journal of Military and Veterans' Health*, 2021, 29, 36–42 (production of natural penicillin and the search for synthetic penicillin in WW2)

D. Carcione, C. Siracusa, A. Sulejmani, V. Leoni and J. Intra, *Antibiotics*, 2021, 10, 995 (beta-lactamase inhibitors)

G. Muteeb, M. T. Rehman, M. Shahwan and M. Aatif, *Pharmaceuticals*, 2023, 16, 1615 (antibiotics and antibiotic resistance)

33. POLYTHENE

E. Bamberger and F. Tschimer, *Berichte der Deutschen Chemischen Gesellschaft*, 1900, 33, 955–959 (first synthesis)

R. O. Gibson, *The Discovery of Polythene*, London, Royal Institute of Chemistry Lecture Series 1964, Number 1

H. F. Mark et al., *Giant Molecules*, Amsterdam, Time-life Books, 1967

F. M. McMillan, *The Chain Straighteners*, London, Macmillan, 1979, esp. pp. 11–14 and 56–72

S. Fenichell, *Plastic. The Making of a Synthetic Century*, New York, HarperCollins, 1996, esp. pp. 199–203 and 230–234

R. J. Giguere, ed., *Molecules that Matter*, Philadelphia, Chemical Heritage Foundation, 2008, pp. 64–79

34. QUININE

A. Strecker, *Annalen*, 1854, 91, 155 (quinine)

P. Rabe and K. Kindler, *Chemische Berichte*, 1918, 51, 466–467 (quinine synth.)
R. B. Woodward and W. E. Doering, *J. Amer. Chem. Soc.*, 1945, 67, 849 (synthesis of quinine)
D. Whyte, *British Medical Journal,* 1984, 285, 1776–1779; L. J. Bruce-Chwatt, *British Medical Journal*, 1984, 288, 796 (mepacrine, the Chindits and Wingate's toxic psychosis)
H. Hobhouse, *Seeds of Change: Six Plants that Transformed Mankind*, London, Sidgwick and Jackson, 1985, pp. 1–40 (quinine)
J. N. D. Kelly, *The Oxford Dictionary of Popes*, Oxford, OUP, 1986, pp. 280–281 (Pope Urban VIII)
R. S. Desowitz, *The Malaria Capers*, New York and London, W. W. Norton, 1991
M. J. Dobson, *Journal of the Royal Society of Medicine*, 1989, 82, 3–7; M. J. Dobson, *Parassitologia*, 1994, 36, 35–60 (malaria in England)
T. W. Keeble, *J. R. Soc. Med.*, 1997, 90, 285–290 (Robert Talbor)
P. Reiter, *Emerging Infectious Diseases*, 2000, 6(1), 1–11 (malaria in England in the Little Ice Age)
G. Stork, D. Niu, A. Fujimoto, E. R. Koft, J. M. Balkovec, J. R. Tata and G. R. Dake, *J. Am. Chem. Soc.,* 2001, 123, 3239–3242 (first stereoselective total synthesis of quinine)
M. Honigsbaum, *The Fever Trail: In Search of the Cure for Malaria*, Basingstoke, Macmillan, 2001
A. Spielman and M. D'Antonio, *Mosquito: The Story of Man's Deadliest Foe*, London, Faber and Faber, 2001
P. Le Couteur and J. Burreson, *Napoleon's Buttons: How 17 Molecules Changed History*, New York, Tarcher, 2003, pp. 332–341 (quinine)
F. Rocco, *The Miraculous Fever-tree: Malaria, Medicine and the Cure that Changed the World*, London, HarperCollins, 2003 (quinine)
T. Chin and P. D. Welsby, *Postgrad. Med. J.*, 2004, 80, 663–666 (malaria in the UK: past, present, and future)
T. S. Kaufman and E. A. Rúveda, *Angew. Chem. Int. Ed. Engl.*, 2005, 44, 854–885 ('The Quest for Quinine')
R. M. Packard, *The Making of a Tropical Disease*, Baltimore, MD, The John Hopkins University Press, 2007 ('a short history of malaria')
J. I. Seeman, *Angewandte Chemie, International Ed.*, 2007, 46, 1378; A. C. Smith and R. M. Williams, *Angewandte Chemie, International Ed.*, 2008, 47, 1736 (Woodward–Doering synthesis of quinine)
K. C. Nicolaou and T. Montagnon, *Molecules that Changed the World*, Weinheim, Wiley-VCH, 2008, pp. 57–66 (quinine: history and chemistry)
S. W. Lindsay, D. G. Hole, R. A. Hutchinson, S. A. Richards and S. G. Willis, *Malaria Journal*, 2010, 9, 70 (risk of malaria returning to the UK)
I. W. Sherman, *Magic Bullets to Conquer Malaria*, Washington, American Society for Microbiology, 2011, esp. pp. 23–51
H. Gay and W. P. Griffth, *The Chemistry Department at Imperial College. A History, 1845–2000*, New Jersey, World Scientific, 2016, pp. 40–42 (Perkin, quinine and mauve)
F. Trenti, K. Yamamoto, B. Hong, C. Paetz, Y. Nakamura and S. E. O'Connor, *Org. Lett.*, 2021, 23, 1793–1797 (quinine biosynthesis)
A. Jovanović and D. Krajnović, *Acta Medica Medianae*, 2023, 62, 61–69 (quinine history)
C. Bisaccia, L. S. De Santo and N. G. De Santo, *Experimental and Clinical Transplantation*, 2023, Suppl 2, 28–32 (malaria as a Papal disease)
B. K. Lombe, T. Zhou, L. Caputi, K. Ploss and S. E. O'Connor, *Angew. Chem. Int. Ed.*, 2025, 64, e202418306 (origin of the methoxy group in quinine and related alkaloids)

35. SCOPOLAMINE (AKA HYOSCINE)

A. Ladenburg, *Annalen*, 1880, 206, 274–307, esp. 299 (isol.)

E. Schmidt, *Archiv der Pharmazie*, 1890, 228, 435–441 (on the discovery of scopolamine)
G. Sharp, *BMJ*, 1895, 2, 1547–1549 (early use of scopolamine)
G. Geis, *J. Crim. L. & Criminology*, 1959–1960, 50, 347–357 (scopolamine as a truth drug)
T. K. Hardy and D. Wakely, *Anesthesia*, 1962, 17, 331–336 (the amnesic properties of scopolamine)
A. Ardila and C. Moreno, *Brain and Cognition*, 1991, 15, 236–245 (burundanga and scopolamine)
R. Glaser, D. Shiftan and M. Drouin, *Can. J. Chem.*, 2000, 78, 212–223 (cryst. struct.)
J. Timbrell, *The Poison Paradox*, Oxford, OUP, 2005, pp. 147–148, 213–214 (scopolamine and mandrake)
M. G. Uribe-Granja, C. L. Moreno-López, A. Zamora-Suárez and P. J. Acosta, *Acta Neurol. Colomb.*, 2005, 21, 197–201 (ER Admissions in Colombia from burundanga intoxication)
A. Winter, *Bulletin of the History of Medicine*, 2005, 79, 500–533 (truth serum history)
M. Wink and B. E. van Wijk, *Mind-Altering and Poisonous Plants of the World*, Portland, Timber Press, 2008 (plants containing scopolamine)
J. Ziegler and P. J. Facchini, *Annu. Rev. Plant Biol.*, 2008, 59, 735–769 (biosynthesis of scopolamine)
J. Emsley, *Molecules of Murder*, Cambridge, Royal Society of Chemistry, 2008, pp. 23–45 (the Crippen case)
B. Madea and F. Musshoff, *Dtsch. Arztebl. Int.*, 2009, 106, 341–347 (knockout drugs reviewed)
D. R. Foran, B. E. Wills, B. M. Kiley, C. B. Jackson and J. H. Trestrail III, *J. Forensic Sci.*, 2011, 56, 233–240 (claims for contradictory DNA evidence in Crippen case)
F. Khakpai, M. Nasehi, A. Haeri-Rohani, A. Eidi and M. R. Zarrindast, *Behavioural Brain Research*, 2012, 231, 1–10 (scopolamine-induced memory impairment, possible mechanisms)
E. Saussereau, M. Guerbet, J.-P. Anger and J.-P. Goullé, 'Memory impairment after drug-related crimes', in: P. Kintz (ed.), *Toxicological Aspects of Drug-Facilitated Crimes*, Amsterdam, Elsevier, 2014, pp. 121–138 (memory impairment after drug-facilitated crimes)
E. Le Garff, Y. Delannoy, V. Mesli, V. Hédouin and G. Tournel, *Forensic Science International*, 2016, 261, e17–e21 (fatal datura poisoning during a robbery)
K. J. Lusthof, I. J. Bosman, B. Kubat and M. J. Vincenten-van Maanen, *Forensic Science International*, 2017, 274, 79–82 (scopolamine-facilitated robberies)
S. Reichert, C. Lin, W. Ong, C. C. Him and S. Hameed, *Canadian Family Physician*, 2017, 63, 369–370 (crime committed during involuntary scopolamine intoxication)
A. M. Lakstygal et al., *ACS Chem. Neurosci.*, 2019, 10, 2144–2159 (DARK Classics in Chemical Neuroscience: Atropine and Scopolamine)

36. 'SPICE'

J. W. Huffman, D. Dai, B. R. Martin and D. R. Compton, *Bioorganic & Medicinal Chemistry Letters*, 1994, 4, 563–566 (cannabimimetic 1-alkyl-2-methyl-3–(1-naphthoyl)indoles, including JWH-018)
V. Auwärter, S. Dresen, W. Weinmann, M. Müller, M. Pütz and N. Ferreirós, *J. Mass. Spectrom.*, 2009, 44, 832–837; N. Uchiyama, R. Kikura-Hanajiri, N. Kawahara and Y. Goda, *Forensic Toxicol.*, 2009, 27, 61–66 (identification of JWH-018 in herbal products)
I. P. Buchler, M. J. Hayes, S. G. Hedge, S. L. Hockerman, D. E. Jones, S. W. Kortum, J. G. Rico, R. E. Tenbrink and K. K. Wu, *Indazole derivatives as CB1 receptor modulators and their preparation and use in the treatment of CB1-mediated diseases*, Patent WO 2009/106982 (2009)
K. A. Seely, J. Lapoint, J. H. Moran and L. Fattore, *Prog. Neuropsychopharmacol. Biol. Psychiatry*, 2012, 39, 234–243 (pharmacology and toxicology of synthetic cannabinoids)

C. Lovett, D. M. Wood and P. I. Dargan, *Réanimation*, 2015, 24, 527–541 (review)

S. D. Banister, M. Longworth, R. Kevin, S. Sachdev, M. Santiago, J. Stuart, J. B. C. Mack, M. Glass, I. S. McGregor, M. Connor and M. Kassiou, *ACS Chem. Neurosci.*, 2016, 7, 1241–1254 (pharmacology of AMB-FUBINACA and related drugs)

A. J. Adams, S. D. Banister, L. Irizarry, J. Trecki, M. Schwartz and R. Gerona, *N. Engl. Med. J.*, 2017, 376, 235–242 ('zombie' deaths associated with AMB-FUBINACA in 2016 in New York)

P. Adamowicz, E. Meissner and M. Maślanka, *Clinical Toxicology*, 2019, 57, 1103–1108 (fatality)

O. Corazza, S. Coloccini, S. Marrinan, M. Vigar, C. Watkins, C. Zene, A. Negri, A. Aresti, S. Darke, R. Rinaldi, A. Metastasio and G. Bersani, *Front. Psychiatry*, 2020, 11, 450 ('Spice' users in gaols)

P. L. Morrow, S. Stables, K. Kesha, R. Tse, D. Kappatos, R. Pandey, S. Russell, O. Linsell, M. J. McCarthy, A. Spark, D. Vertes, Y. Triggs, S. McCarthy, N. Cuthers and R. Massey, *EClinicalMedicine*, 2020, 25, 100460 (deaths associated with AMB-FUBINACA in Auckland NZ in 2017–2019)

J. A. Hvozdovich, C. W. Chronister, B. K. Logan and B. A. Goldberger, *Journal of Analytical Toxicology,* 2020, 44, 298–300 (synthetic cannabinoids including AMB-FUBINACA and deaths in Florida prisons)

C. Norman, G. Walker, B. McKirdy, C. McDonald, D. Fletcher, L. H. Antonides, O. B. Sutcliffe, N. Nic Daéid and C. McKenzie, *Drug Test Anal.*, 2020, 12, 538–554 (detection of spice in paper in prisons)

O. Apirakkan, A. Frinculescu, H. Denton, T. Shine, D. Cowan, V. Abbate and N. Frascione, *Forensic Chemistry*, 2020, 18, 100227 (detection and identification of AMB-FUBINACA in paper)

C. Norman, S. Halter, B. Haschimi, D. Acreman, J. Smith, A. J. Krotulski, A. L. A. Mohr, B. K. Logan, N. Nic Daéid, V. Auwärter and C. McKenzie, *Drug Testing and Analysis*, 2021, 13, 841–852 (synthetic cannabinoids in prison vs. the general population)

R. Roque-Bravo, R. S. Silva, R. F. Malheiro, H. Carmo, F. Carvalho, D. D. da Silva and J. P. Silva, *Annu. Rev. Pharmacol. Toxicol.*, 2023, 63, 187–209 (synthetic cannabinoids reviewed)

37. STREPTOMYCIN

W. H. Feldman and H. C. Hinshaw, *Proc. Staff Meet., Mayo Clinic.*, 1944, 19, 593–599 (effect on guinea pigs)

A. Schatz, E. Bugie and S. A. Waksman, *Proc. Soc. Exp. Biol. Med.*, 1944, 55: 66–69; D. Jones, H. J. Metzger, A. Schatz and S. A. Waksman, *Science*, 1944, 100, 103–105 (discovery of properties of streptomycin)

H. C. Hinshaw and W. H. Feldman, *Proc. Staff Meet., Mayo Clinic.*, 1945, 20, 313–318; H. C. Hinshaw, W. H. Feldman and K. H. Pfeutze, *J. Am. Med. Assoc.*, 1946, 132, 778–782; W. McDermott, C. Muschenheim, S. J. Hadley, P. A. Bunn and R. V. Gorman, *Ann. Intern. Med.*, 1947, 27, 769–822 (streptomycin and TB in humans)

F. A. Kuehl, R. L. Peck, C. E. Hoffhine and K. Folkers, *J. Am. Chem. Soc.*, 1948, 70, 2325–2330 (structure of streptomycin)

D. D. Martin, F. S. Spring, T. G. Dempsey, C. L. Goodacre and D. E. W. Seymour, *Nature*, 1948, 161, 435 (p-aminosalicylic acid in the treatment of tuberculosis)

J. H. Comroe, *Am. Rev. of Resp. Dis.*, 1978, 117, 773–781 and 957–968 (story of streptomycin)

M. Wainwright, *Miracle Cure: Story of Antibiotics*, Wiley-Blackwell, 1990

F. Ryan, *Tuberculosis: The Greatest Story Never Told – The Search for the Cure and the New Global Threat*, Swift Publishers, 1992

A. Schatz, *Actinomycetes*, 1993, 4, 27–39 (Schatz's account of the development of streptomycin)
R. B. Mason, *Can. Med. Ass. J.*, 1996, 154, 921–922 (TB and opera)
M. Ormrod and P. Lindley, eds., *The Black Death in England*, Donington, Shaun Tyas, 1996
T. Dormandy, *The White Death: A History of Tuberculosis*, New York Hambledon Continuum, 1998
C. K. Stover et al., *Nature*, 2000, 405, 962–966 (PA 824)
K. Andries et al., *Science*, 2005, 307, 223–227 (discovery of bedaquiline)
M. Protopopova, E. Bogatcheva, B. Nikonenko, S. Hundert, L. Einck and C. A. Nacy, *Med. Chem.*, 2007, 3, 301–316 (TB cures, reviews)
A. H. Diacon et al., *N. Engl. J. Med.*, 2009, 360, 2397–2405 (TMC207, now Bedaquiline, for multidrug-resistant tuberculosis)
Q. Huang, J. Mao, B. Wan, Y. Wang, R. Brun, S. G. Franzblau and A. P. Kozikowski, *J. Med. Chem.*, 2009, 52, 6757–6767 (2-Methylbenzothiazoles as treatments for TB)
A. M. Ginsberg, M. W. Laurenzi, D. J. Rouse, K. D. Whitney and M. K. Spigelman, *Antimicrob. Agents Chemother.*, 2009, 53, 3720–3725 (PA 824)
G. A. Marriner et al., *Top. Med. Chem.*, 2011, 7, 47–124 (major review of chemotherapies for TB)
M. T. Gler et al., *N. Engl. J. Med.*, 2012, 366, 2151–2160 (Delamanid)
H. Bynum, *Spitting Blood: The History of Tuberculosis*, Oxford, OUP, 2012
J. Murphy and A. Blank, *Invincible Microbe. Tuberculosis and the Never-Ending Search for a Cure*, Boston, Clarion Books, 2012
P. Pringle, *Experiment Eleven: Deceit and Betrayal in the Discovery of the Cure for Tuberculosis*, London, Bloomsbury, 2012 (Schatz's notebooks)
T. Paulson, *Nature*, 2013, 502, S2–S3 (worldwide distribution of TB)
A. Maxmen, *Nature*, 2013, 502, S4–S6 (combinational therapy for TB)
H. Demirci, F. Murphy, E. Murphy, S. T. Gregory, A. E. Dahlberg and G. Jogl, *Nature Communications*, 2013, 4, 1355 (how streptomycin works)
G. J. Fox and D. Menzies, *Infect. Dis. Ther.*, 2013, 2, 123–144 (bedaquiline treating multi-drug resistant tuberculosis)
V. Skripconoka et al., *Eur. Respir. J.*, 2013, 41, 1393–1400 (Delamanid)
J. Avorn, *JAMA*, 2013, 309, 1349–1350 (FDA approval for bedaquiline)
M. Fiorillo, C. Scatena, A. G. Naccarato, F. Sotgia and M. P. Lisanti, *Cell Death & Differentiation*, 2021, 28, 2797–2817 (how bedaquiline works)
S. Khoshnood, M. Goudarzi, E. Taki, A. Darbandi, E. Kouhsari, M. Heidary, M. Motahar, M. Moradi and H. Bazyar, *Journal of Global Antimicrobial Resistance*, 2021, 25, 48–59 (bedaquiline review)
P. Sarris, *Past and Present*, 2022, 254, 315–346 (plagues, notably in England)
C. Lange, A. Vasiliu and A. M. Mandalakas, *Lancet Microbe*, 2023, 4, E964–965; B. Derendinger et al., *Lancet Microbe*, 2023, 4, E972–982 (emerging bedaquiline-resistant tuberculosis)
N. A. Khosravi, M. Sirous, A. D. Khosravi and M. Saki, *Journal of Clinical Laboratory Analysis*, 2024, 38, e25091 (review of Bedaquiline and Delamanid)
J. Green, *Everything is Tuberculosis*, London, Ebury Press, 2025 (TB in the Third World)

38. SULFANILAMIDE AND OTHER SULFA DRUGS

P. Gelmo, *J. Prakt. Chem.*, 1908, 77, 369–382 (synth. of sulfanilamide)
G. Domagk, *Deutsche Medizinische Wochenschaft*, 1935, 61, 250–253 (first publication)
G. Domagk, *Klinische Wochenschrift* (*Journal of Molecular Medicine*), 1936, 15, 1585–1590 (Hildegarde Domagk case)
L. Colebrook et al., *The Lancet*, 1936, 227, 1279–1281; 228, 1319–1322 (Prontosil and puerperal sepsis)

E. Fourneau, J. Trefouel, F. Nitti and D. Bovet, *Compt. Rend. Soc. De Biol.*, 1936, 122, 652–654 (treatment of streptococcal infections by sulfanilamide)
L. Colebrook et al., *The Lancet*, 1937, 230, 1291–1294 (sulfanilamide and puerperal sepsis)
A. G. N., *Can. Med. Assoc. J.*, 1937, 37, 590 (elixir of sulfanilamide)
L. E. H. Whitby, *The Lancet*, 1938, 231, 1210–1212 (M+B 693 and pneumonia)
G. M. Evans and W. F. Gaisford, *The Lancet*, 1938, 232, 14–18 (M+B 693 and pneumonia)
E. M. Geiling and P. R. Cannon, *J. Amer. Med. Assoc.*, 1938, 111, 919–926 (pathologic effects of elixir of sulfanilamide poisoning)
J. H. Weatherby and G. Z. Williams, *J. Amer. Pharm. Assoc.*, 1939, 28, 12–17 (studies on elixir of sulfanilamide)
D. D. Woods, *Brit. J. Exp. Pathol.*, 1940, 21, 74–90 (PABA and sulfanilamide)
I. Galdston, *Behind the Sulfa Drugs*, New York, Appleton-Century Co., 1943
A. I. Vogel, *A Textbook of Practical Organic Chemistry*, London, Longmans, 3rd edition, 1959, pp. 1005–1009; F. G. Mann and B. C. Saunders, *Practical Organic Chemistry*, London, Longmans, 4th edition, 1960, pp. 179–180 (syntheses)
G. H. Hitchings, *J. Infect. Dis.*, 1973, 128, Suppl 3, S433–S436 (Trimethoprim-Sulfamethoxazole)
T. Triglia, J. G. T. Menting, C. Wilson and A. F. Cowman, *Proc. Natl. Acad. Sci. USA*, 1997, 94, 13944–13949 (mechanism of sulfonamide resistance)
J. E. Lesch, in: G. J. Higby and E. C. Stroud (eds.), *The Inside Story of Medicines: A Symposium*, Madison, American Institute of the History of Pharmacy, 1997, pp. 101–119 (M+B 693 and pneumonia)
C. M. De Costa, *Medical Journal of Australia*, 2002, 177, 668–671 (rev. of puerperal sepsis)
P. A. Masters, T. A. O'Bryan, J. Zurlo, D. Q. Miller and N. Joshi, *Arch. Intern. Med.*, 2003, 163, 402–410 (Trimethoprim-Sulfamethoxazole)
J. Bankston, *Gerhard Domagk and the Discovery of Sulfa*, Bear, Mitchell Lane, 2003
R. J. Franklin, *Medic! How I Fought World War II with Morphine, Sulfa and Iodine Swabs*, Lincoln, University of Nebraska Press, 2006
J. E. Lesch, *The First Miracle Drugs: How the Sulfa Drugs Transformed Medicine*, Oxford, OUP, 2006
T. Hager, *The Demon Under the Microscope: From Battlefield Hospitals to Nazi Labs, One Doctor's Heroic Search for the World's First Miracle Drug*, New York, Harmony, 2006
T. Leslie, M. I. Mayan, M. A. Hasan, M. H. Safi, E. Klinkenberg, C. J. M. Whitty and M. Rowland, *J. American Medical Association*, 2007, 297, 2201–2209 (antimalarials)
G. L. Patrick, *An Introduction to Medicinal Chemistry*, Oxford, Oxford University Press, 4th edition, 2009, pp. 426–432

39. TAXOL

J. Caesar, *De Bello Gallico*, Book 6, section XXXI; vide: R. Carruthers (ed.), *Caesar's Conquest of Gaul*, Barnsley, Pen and Sword, 2012, p. 173
M. C. Wani, H. L. Taylor, M. E. Wall, P. Coggon and A. T. McPhail, *J. Am. Chem. Soc.*, 1971, 93, 2325–2327 (isolation and structure of taxol)
P. B. Schiff, J. Fant and S. B. Horwitz, *Nature*, 1979, 277, 665–667; P. B. Schiff and S. B. Horwitz, *Proc. Natl. Acad. Sci. USA*, 1980, 77, 1561–1565 (taxol's mechanism of action)
J. N. Denis, A. E. Greene, D. Guenard, F. Gueritte-Voegelein, L. Mangatal and P. Potier, *J. Am. Chem. Soc.*, 1988, 110, 5917–5919 (semi-synthesis of taxol)
D. Guenard, F. Gueritte-Voegelein and P. Potier, *Acc. Chem. Res.*, 1993, 26, 160–167 (taxol and taxotere review)
K. C. Nicolaou, R. K. Guy and P. Potier, *Sci. Am.*, June 1996, 84–88 (review)
W. M. Daniewski, M. Gumulka, W. Anczewski, M. Masnyk, E. Bloszyk and K. K. Gupta, *Phytochem.*, 1998, 38, 168–171 (insect resistance of yew and taxanes)

J. Mann, *The Elusive Magic Bullet*, Oxford, OUP, 1999, pp. 166–168
L. A. Amos and J. Lowe, *Chem. Biol.*, 1999, 6, R65–R69 (taxol and microtubule structure)
J. Goodman and V. Walsh, *The Story of Taxol: Nature and Politics in the Pursuit of an Anti-Cancer Drug*, Cambridge, Cambridge University Press, 2001
H. Tabata, *Adv. Biochem. Engin. Biotechnol.*, 2004, 7, 1–23 (paclitaxel production by plant-cell-culture technology.)
M. D. Chau, S. Jennewein, K. Walker and R. Croteau, *Chem. Biol.*, 2004, 11, 663–672 (biosynthesis of taxol)
J. Hui et al., *l'actualité chimique*, mai 2006, 297, 4–8; François Chast, *Nature*, 2006, 440, 291 (obituary and bio. of Pierre Potier)
D. Frense, *Appl. Microbiol. Biotechnol.*, 2007, 73, 1233–1240 (taxol production in yew cell cultures)
F. Hageneder, *Yew: A History*, Stroud, The History Press, 2007
O. Exposito, M. Bonfill, E. Moyano, M. Onrubia, M. H. Mirjalili, R. M. Cusido and J. Palazon, *Anti-Cancer Agents in Med. Chem.*, 2009, 9, 109–121 (taxol production)
Y. Fu, S. Li, Y. Zu, G. Yang, Z. Yang, M. Luo, S. Jiang, M. Wink and T. Efferth, *Curr. Med. Chem.*, 2009, 16, 3966–3985 (review)
P. K. Ajikumar, W.-H. Xiao, K. E. J. Tyo, Y. Wang, F. Simeon, E. Leonard, O. Mucha, T. H. Phon, B. Pfeifer and G. Stephanopoulos, *Science*, 2010, 330, 70–74 (*E. coli* bacteria engineered to produce taxadiene)
M. Köksal, Y. Jin, R. M. Coates, R. Croteau and D. W. Christianson, *Nature*, 2011, 469, 116–122 (taxadiene synthase structure)
T. Grobosch, B. Schwarze, D. Stoecklein and T. Binscheck, *J. Anal. Toxicol.*, 2012, 36, 36–43 (fatal poisoning with *Taxus baccata*)
O. Piskač, J. Stříbrný, H. Rakovcová and M. Malý, *Cor et Vasa*, 2015, 57, e234–e238 (cardiotoxicity of yew)
X. Xiong et al., *Nat. Plants*, 2021, 7, 1026–1036 (the *Taxus* genome)
C. Li, X. Yin, S. Wang, S. Sui, J. Liu, X. Sun, J. Di, R. Chen, D. Chen, Y. Han, K. Xie and J. Dai, *Angew. Chem. Int. Ed.*, 2024, 63, e202407070 (a cytochrome P450 enzyme catalyses oxetane ring formation in paclitaxel biosynthesis)
C. Yang, Y. Wang, Z. Su, L. Xiong, P. Wang, W. Lei, X. Yan, D. Ma, G. Zhao and Z. Zhou, *Nature Communications*, 2024, 15, 2339 (biosynthesis of the tetracyclic core skeleton of taxol)
Y. Zhao, F. Liang, Y. Xie, Y.-T. Duan, A. Andeadelli, I. Pateraki, A. M. Makris, T. G. Pomorski, D. Staerk and S. C. Kampranis, *J. Am. Chem. Soc.*, 2024, 146, 801–810 (oxetane ring formation in taxol biosynthesis catalysed by a bifunctional cytochrome P450 enzyme)
Y. Yin, M. Lai, X.-Y. Yu, D.-D. Su, X.-Y. Xiong and Y.-L. Li, *Horticulture Research*, 2025, 12, uhae346 (comprehensive strategies for paclitaxel production)
C. Li, X. Yin and J. Dai, *ChemBioChem*, 2025, 26, e202400947 (review, oxetane ring formation in paclitaxel biosynthesis)
C. J. McClune, J. C.-T. Liu, C. Wick, R. De La Peña, B. M. Lange, P. M. Fordyce and E. S. Sattely, *Nature*, 2025, 643, 582–592 (biosynthesis of baccatin III)
F. Liang, Y. Xie, C. Zhang, Y. Zhao, M. S. Motawia and S. C. Kampranis, *Nature Synthesis*, 2025, 4, 1212–1222 (final steps in taxol biosynthesis)

40. TERYLENE

J. R. Whinfield, *Textile Research Journal*, 1953, 23, 289–293 ('the development of Terylene')
W. T. Astbury and C. J. Brown, *Nature*, 1946, 158, 871; R. de P. Daubeny, C. W. Bunn and C. J. Brown, *Proc. Roy. Soc. A*, 1954, 226, 531–542 (the crystal structure of polyethylene terephthalate)

S. Fenichell, *Plastic. The Making of a Synthetic Century*, London, HarperCollins, 1996, esp. pp. 228–230
J. Emsley, *Molecules at an Exhibition*, Oxford, OUP, 1998, pp. 228–231 (PET bottles)
P. Black and C. Driscoll, *Australasian Journal of Popular Culture*, 2012, 1(1), 67–76 (Emma Peel's catsuit)
B. Geyer, G. Lorenz and A. Kandelbauer, *Express Polym. Lett.*, 2016, 10, 559–586 (chemical recycling of PET reviewed)
A. T. Archibald and D. D. Parrish, *J. Indust. Environ. Chem.*, 2021, 5(4), 2 pp (brief review)
E. Barnard, J. J. R. Arias and W. Thielemans, *Green Chem.*, 2021, 23, 3765–3789 (depolymerisation of PET reviewed)
R. L. Smith, S. Takkellapati and R. C. Riegerix, *ACS Sustain. Chem. Eng.*, 2022, 10, 2084–2096 (recycling Polyethylene Terephthalate (PET) in the USA)
K. Ghosal and C. Nayak, *Mater. Adv.*, 2022, 3, 1974–1992 (chemical recycling of PET waste into value added products)
A. M. Smelik, *Fashion Practice*, 2023, 15, 279–299 (Polyester: a cultural history)
W. L. Soon, H. Q. Chong, J. L. Foo and M. W. Chang, *Nature Chemistry*, 2025, 17, 982–983;
N. W. Johnson, M. Valenzuela-Ortega, T. W. Thorpe, Y. Era, A. Kjeldsen, K. Mulholland and S. Wallace, *Nature Chemistry*, 2025, 17, 1020–1026 (biocompatible conversion of PET into paracetamol)
L. Undagas, R. Orozco, K. Heeley, W. Thom and B. Al-Duri, *Polymer Degradation and Stability*, 2025, 234, 111246 (recycling PET by hydrolysis)
E. Francis, G. B. Kurup and R. Mucherla, ACS Appl. Polym. Mater., 2025, 7, 17074–17086 (PET recycling by methanolysis)

41. TETRACYCLINE

F. A. Hochstein, C. R. Stephens, L. H. Conover, P. P. Regna, R. Pasternack, P. Gordon, F. J. Pilgrim, K. J. Brunings and R. B. Woodward, *J. Am. Chem. Soc.*, 1953, 75, 5455–5475 (structure of terramycin, oxytetracycline)
E. J. Bassett, M. S. Keith, G. J. Armelagos, D. L. Martin and A. R. Villanueva, *Science*, 1980, 209, 1532–1534 (tetracycline in Nubian bone from AD 350)
D. E. Brodersen, W. M. Clemons, A. P. Carter, R. J. Morgan-Warren, B. T. Wimberly and V. Ramakrishnan, Cell, 2000, 103, 1143–1154 (structural basis for the action of the antibiotics Tetracycline, Pactamycin, and Hygromycin B on the 30S Ribosomal Subunit)
M. Pioletti, F. Schlünzen, J. Harms, R. Zarivach, M. Glühmann, H. Avila, A. Bashan, H. Bartels, T. Auerbach, C. Jacobi, T. Hartsch, A. Yonath and F. Franceschi, *EMBO J.*, 2001, 20, 1829–1839 (cryst. struct. of tetracycline complex of the small ribosomal subunit)
L. R. Peterson, *International Journal of Antimicrobial Agents*, 2008, 32, S215–S222 (tigecycline reviewed)
L. Principe, S. D'Arezzo, A. Capone, N. Petrosillo and P. Visca, *Ann. Clin. Microbiol. Antimicrob.*, 2009, 8, 18–27 (tigecycline in combinational therapy)
N. E. Holmes and P. G. P. Charles, *Clinical Medicine: Therapeutics*, 2009, 1, 471–482 (doxycycline safety and efficacy review)
L. B. Pickens and Y. Tang, *Metab. Eng.*, 2009, 11, 69–75 (biosynth. rev.)
M. L. Nelson, A. Dinardo, J. Hochberg and G. J. Armelagos, *Amer. J. Phys. Anthropol.*, 2010, 143, 151–154 (mass spec. characterization of tetracycline in Nubian bones)
H. Petković, T. Lukežič and J. Šušković, *Food Technol. Biotechnol.*, 2017, 55, 3–13 (biosynth. of oxytetracycline)
R. Ramachanderan and B. Schaefer, *ChemTexts*, 2021, 7, 18 (tetracycline antibiotics reviewed)
E. M. Graber, *Dermatological Reviews*, 2021, 2, 321–330 (treating acne with tetracyclines)

H. Wang, L. Wang, K. Fan and G. Pan, *Chinese Journal of Natural Medicines,* 2022, 20, 773–794 (discovery and biosynth. of tetracycline)
D. Kounatidis, M. Dalamaga, E. Grivakou, I. Karampela, P. Koufopoulos, V. Dalopoulos, N. Adamidis, E. Mylona, A. Kaziani and N. G. Vallianou, *Biomolecules*, 2024, 14, 783 (third-generation tetracyclines)
A. Załęski, M. Sapuła, A. Lembas and A. Wiercińska-Drapało, *Venereology*, 2024, 3, 1–14 (doxycycline in STI Prophylaxis reviewed)
I. Szondy, F. A. Meznerics, K. Lõrincz, L. V. Kemény, A. Walter, A. A. Mohammed, P. Hegyi, N. Kiss and A. Bánvölgyi, *International Journal of Infectious Diseases*, 2024, 147, 107186 (doxycycline and STIs)
J. C. Pearson, E. Gillett, N. D. Gadri and B. Dionne, *CMI Communications*, 2025, 2, 105059 (tetracyclines, the old and the new)
K. S. Blake, Y.-P. Xue, V. J. Gillespie, S. R. S. Fishbein, N. H. Tolia, T. A. Wencewicz and G. Dantas, *Nature Communications*, 2025, 16, 1452 (tetracyclines and resistance mechanisms)

42. TETRAETHYL LEAD

C. Löwig, *J. Prakt. Chem.,* 1853, 60, 304–310 (first synthesis of tetraethyllead)
G. B. Buckton, *Proc. R. Soc. London,* 1859, 9, 309, 685–690; *Ann.*, 1859, 109, 218–227
M. Murozumi, T. J. Chow and C. Patterson, *Geochim. Cosmochim. Acta*, 1969, 33, 1247–1294 (lead in annual ice layers from northern Greenland and Antarctica)
A. D. Beattie, M. R. Moore and A. Goldberg, *Lancet*, July 1, 1972, 12–15 (four cases of tetra-ethyl lead poisoning)
S. Hong, J.-P. Candelone, C. C. Patterson and C. F. Boutron, *Science,* 1994, 265, 1841–1843 (lead in Greenland ice)
H. L. Needleman, J. A. Riess, M. J. Tobin, G. E. Biesecker and J. B. Greenhouse, *JAMA*, 1996, 275, 363–369 (bone lead levels and delinquent behaviour)
H. L. Needleman, *Environmental Research Section A*, 1998, 78, 79–85 ('Clair Patterson and Robert Kehoe: Two Views of Lead Toxicity')
W. Shotyk, D. Weiss, P. G. Appleby, A. K. Cheburkin, R. Frei, M. Gloor, J. D. Kramers, S. Reese and W. O. Van Der Knaap, *Science*, 1998, 281, 1635–1640 (measurements of atmospheric lead deposition for 12,370 years from a peat bog in Switzerland)
C. Warren, *Brush with Death. A Social History of Lead Poisoning*, Baltimore, MD, John Hopkins University Press, 2000
R. Nevin, *Environmental Research Section A*, 2000, 83, 1–22; *Environmental Research*, 2007, 104, 315–336 (crime, unmarried pregnancy, IQ and lead exposure)
T. Midgley IV, *From the Periodic Table to Production: The Life of Thomas Midgley, Jr., the Inventor of Ethyl Gasoline and Freon Refrigerants*, Corona, CA, Stargazer Publ. Co., 2001
G. Markowitz and D. Rosner, *Deceit and Denial, the Deadly Politics of Industrial Pollution*, Berkeley, University of California Press, 2002
D. Seyferth, *Organometallics*, 2003, 22, 2346–2357; 5154–5178 (rise and fall of tetraethyllead)
W. Kovarik, *International Journal of Occupational and Environmental Health*, 2005, 11, 384–397 (issues with tetraethyl lead in petrol)
J. W. Reyes, *J. Economic Analysis Policy.*, 2007, 7(1), 1–51 (childhood lead exposure and crime)
L. Denworth, *Toxic Truth: A Scientist, a Doctor, and the Battle over Lead*, Boston, Beacon Press, 2009
B. K. Wills, J. Christensen, J. Mazzoncini and M. Miller, *J. Med. Toxicol.*, 2010, 6, 31–34 (severe neurotoxicity following ingestion of tetraethyllead)
G. Oudijk, *Environmental Forensics*, 2010, 11, 17–49 ('The Rise and Fall of Organometallic Additives in Automotive Gasoline')
L. Wolf, *Chem. Eng. News*, 2014, 92(5), 27–28 ('The crimes of lead')

J. L. Epstein, *Bull. Hist. Chem.*, 2016, 41(1/2), 38–43 ('The Legacy of Tetraethyl Lead')
M. P. Taylor, M. K. Forbes, B. Opeskin, N. Parr and B. P. Lanphear, *Environmental Health*, 2016, 15, 23 (atmospheric lead emissions and aggressive crime in Australia)
J. Watts, *The Many Lives of James Lovelock: Science, Secrets and Gaia Theory*, Edinburgh, Canongate, 2024, pp. 123–128 (tetraethyllead levels inside motor cars)

43. TETRODOTOXIN

A. Furusaki, Y. Tomiie and I. Niita, *Bull. Chem. Soc. Japan*, 1970, 43, 3322–3341 (crystal structure of tetrodotoxin hydrobromide)
C. Turkington, *The Poisons and Antidotes Sourcebook*, New York, Checkmark Books, 1999, pp. 280–282, 331–332
J. Timbrell, *The Poison Paradox*, Oxford, Oxford University Press, 2005, pp. 161, 252–254
E. D. Brodie III, C. R. Feldman, C. T. Hanifin, J. E. Motychak, D. G. Mulcahy, B. L. Williams and E. D. Brodie Jr., *Journal of Chemical Ecology*, 2005, 31, 343–356 (TTX arms races)
B. Venkatesh, S. Q. Lu, N. Dandona, S. L. See, S. Brenner and T. W. Soong, *Current Biology*, 2005, 15, 2069–2072 (reason for pufferfish's immunity)
R. Ritson-Williams, M. Yotsu-Yamashita and V. J. Paul, *PNAS*, 2006, 103, 3176–3179 (tetrodotoxin in a deadly flatworm)
M. Wheatley, in: R. H. Waring, G. B. Steventon and S. C. Mitchell (eds.), 'Wheatley - "Tetrodotoxin"', *Molecules of Death*, London, Imperial College Press, 2007, pp. 387–414 (review)
S. Jal and S. S. Khora, *Journal of Applied Microbiology*, 2015, 119, 907–916 (origin and production of tetrodotoxin)
K. Murakami, T. Toma, T. Fukuyama and S. Yokoshima, *Angew. Chem. Int. Ed.*, 2020, 59, 6253–6257 (total synthesis of tetrodotoxin)
R. González-Cano, M. C. Ruiz-Cantero, M. Santos-Caballero, C. Gómez-Navas, M. Á. Tejada and F. R. Nieto, *Toxins*, 2021, 13, 483 (review: tetrodotoxin, drug for pain relief)
S. A. Goldlust, M. Kavoosi, J. Nezzer, M. Kavoosi, W. Korz and K. Deck, *Toxins*, 2021, 13, 235 (tetrodotoxin pain killer trials)
G. M. Bucciarelli, M. Lechner, A. Fontes, L. B. Kats, H. L. Eisthen and H. B. Shaffer, *Toxins*, 2021, 13, 517 (evolution of tetrodotoxin and its potential as a therapeutic)
P. Katikou, C. Gokbulut, A. R. Kosker, M. Campàs and F. Ozogul, *Mar. Drugs*, 2022, 20, 47 (tetrodotoxin reviewed)
M. Á. Huerta, J. de la Nava, A. Artacho-Cordón and F. R. Nieto, *Marine Drugs*, 2023, 21, 316 (tetrodotoxin in treating cancer-related pain)
W. Mi and S. Liu, *Front. Microbiol.*, 2024, 15, 1413741 (tetrodotoxin reviewed)
P. Chen, J. Wang, S. Zhang, Y. Wang, Y. Sun, S. Bai, Q. Wu, X. Cheng, P. Cao and X. Qi, *Nature Communications*, 2024, 15, 679 (total syntheses of tetrodotoxin and 9-epitetrodotoxin)
B. Nißl, M. Mülbaier, F. Grisoni, D. Trauner, M. Bermúdez, C. Dialer and D. B. Konrad, *Angew. Chem. Int. Ed.*, 2025, 64, e202502404 (chemistry and biology of tetrodotoxin)
W.-S. Chung, N. D. Kurniawan, N. J. Marshall and F. Cortesi, *Current Biology*, 2025, 35(5P), R169–170 (blue-lined octopus males envenomate females to facilitate copulation)
C. Varini, M. Manganelli, S. Scardala, P. Antonelli, C. Losasso and E. Testai, *Toxins*, 2025, 17, 76 (tetrodotoxin toxicity in contaminated seafood in Europe)
Y. Cai, Q. Li, A. K. Banga, U. Wesselmann and C. Zhao, *Advanced Healthcare Materials*, 2025, 14, 2401549 (pen device to deliver tetrodotoxin painkiller)

44. THALIDOMIDE

W. G. McBride, *Lancet*, 1961, 2, 1358 (thalidomide and congenital abnormalities)
W. Lenz, *Lancet*, 1962, 1, 271–272 (thalidomide and congenital abnormalities)

J. Sheskin, *Clin. Pharm. Ther.*, 1965, 6, 303–306 (thalidomide and erythema nodosum leprosum (ENL))

R. A. Fine, *The Great Drug Deception. The Shocking Story of Mer/29 and the Folks Who Gave You Thalidomide*, New York, Stein and Day, 1972

Sunday Times Insight Team (P. Knightley, H. Evans, E. Potter and M. Wallace), *Suffer the Children: The Story of Thalidomide*, London, Andre Deutsch, 1979

M. Wallace and M. Robson, *On Giants' Shoulders: The Terry Wiles Story*, London, Times Books, 1976

D. A. Blake, G. B. Gordon and S. P. Spielberg, *Teratology*, 1982, 25, 28A–29A

W. Lenz, *Teratology*, 1992, 46, 417–418 (a personal perspective on the thalidomide tragedy)

M. T. Miller and K. Strömland, *Teratology*, 1999, 60, 306–321 (review)

T. D. Stephens and R. Brynner, *Dark Remedy: The Impact of Thalidomide and Its Revival as a Vital Medicine*, New York, Perseus Publishing, 2001

A. Raza, *The Biochemist*, February 2002, 21–23 (thalidomide as a drug)

R. von Moos, R. Stolz, T. Cerny and S. Gillessen, *Swiss Medical Weekly*, 2003, 133, 77–87 (review: thalidomide as a drug)

J. N. Gordon and P. M. Goggin, *Postgraduate Medical Journal*, 2003, 79, 127–132 (uses: thalidomide emerging from the wilderness)

J. Timbrell, *The Poison Paradox*, Oxford, OUP, 2005, pp. 56–59

T. Paravar and D. J. Lee, *International Review of Immunology*, 2008, 27, 111–135 (mechanisms of action)

S. W. Smith, *Toxicological Sciences*, 2009, 110, 4–30 (chiral toxicology and thalidomide)

T. Stephens, *Birth Defects*, 2009, A 85, 725–731 (thalidomide and chicken embryos)

C. Therapontos, L. Erskine, E. R. Gardner, W. D. Figg and N. Vargesson, *Proceedings of National Academy of Sciences*, 2009, 106, 8573–8578 (thalidomide and limb defects)

J. Avorn, *N. Engl. J. Med.*, 2011, 365, 2151–2153 (thalidomide and drug safety)

T. Ito and H. Handa, *Congenital Anomalies*, 2012, 52, 1–7 (deciphering the mystery of thalidomide teratogenicity)

N. Vargesson, *ISRN Developmental Biology*, 2013, Article ID 241016 (thalidomide embryopathy: an enigmatic challenge)

C. Mahony, L. Erskine, J. Niven, N. H. Greig, W. D. Figg and N. Vargesson, *PNAS*, 2013, 110, 12703–12708 (Pomalidomide is nonteratogenic in chicken and zebrafish embryos)

J. Krönke et al., *Science*, 2014, 343, 301–305 (Lenalidomide)

G. Lu et al., *Science*, 2014, 343, 305–309 (Lenalidomide)

A. K. Stewart, *Science,* 2014, 343, 256–257 (how Thalidomide works against cancer)

45. VANILLIN

N.-T. Gobley, *J. de Pharmacie Chimie*, 1859, 34, 401 (original isolation)

D. A. Krueger and H. W. Krueger, *J. Agric. Food Chem.*, 1985, 33, 323–325 (fraudulent C-13 labelling)

R. I. Aylott, A. H. Clyne, A. P. Fox and D. A. Walker, *Analyst*, 1994, 119, 1741–1746 (authenticating whisky)

A. S. Ranadive, in: G. Charalambous (ed.), 'Vanillin - cultivation, curing, chemistry', *Spices, Herbs and Edible Fungi (Developments in Food Science)*, Amsterdam, Elsevier, 1994, pp. 517–577

M. B. Hocking, *J. Chem. Ed.*, 1997, 74, 1055–1059 (review)

R. Hoffmann, 'Fraudulent Molecules', *American Scientist*, 1997, 85, 314–317 (detecting genuine vanilla extract)

G. S. Remaud, Y.-L. Martin, G. G. Martin and G. J. Martin, *J. Agric. Food Chem.*, 1997, 45, 859–866 (SNIF-NMR)

H. Priefert, J. Rabenhorst and A. Steinbüchel, *Appl. Microbiol. Biotechnol.*, 2001, 56, 296–314 (review, synthesis through microorganisms)
N. J. Walton, M. J. Mayer and A. Narbad, 'Vanillin', *Phytochemistry*, 2003, 63, 505–515 (rev, esp. biosynth)
C. Clarke, *The Science of Ice Cream*, Cambridge, Royal Society of Chemistry, 2004, pp. 54–55
T. Ecott, *Vanilla: Travels in Search of the Ice Cream Orchid*, London, Grove/Atlantic, 2004
P. Rain, *Vanilla: The Cultural History of the World's Favorite Flavor and Fragrance*, New York, Tarcher, 2004
E. J. Tenailleau, P. Lancelin, R. J. Robins and S. Akoka, *J. Agric. Food Chem.*, 2004, 52, 7782–7787 (authenticating vanillin using natural abundance ^{13}C NMR)
C. Sell, ed., *The Chemistry of Fragrances: From Perfumer to Consumer*, Cambridge, RSC, 2nd edition, 2006, pp. 44–45, 116–118 (in perfumes)
A. Pérez-Silva, E. Odoux, P. Brat, F. Ribeyre, G. Rodriguez-Jimenes, V. Robles-Olvera, M. A. García-Alvarado and Z. Günata, *Food Chemistry*, 2006, 99, 728–735 (volatiles in *Vanilla planifolia* from Tuxtepec region)
D. F. Taber, S. Patel, T. M. Hambleton and E. E. Winkel, *J. Chem. Educ.*, 2007, 84, 1158 (lab. synth)
M. Takahashi, Y. Inai, N. Miyazawa, Y. Kurobayashi and A. Fujita, *Biosci. Biotechnol. Biochem.*, 2013, 77(3), 601–605 (key odorants in *Vanilla tahitensis*)
S. Guyader, F. Thomas, E. Jamin, M. Grand, S. Akoka, V. Silvestre and G. S. Remaud, *Flavour Fragr J.*, 2019, 34, 133–144 (^{13}C and ^{2}H SNIF-NMR isotopic fingerprints of vanillin)
R. Abreu-Runkel, *Vanilla: A Global History*, London, Reaktion Books, 2020
C. Pironti, M. Ricciardi, O. Motta, F. Camin, L. Bontempo and A. Proto, *Foods*, 2021, 10, 2635 (authenticating vanillin using natural abundance ^{13}C NMR)
Y. Liu, L. Sun, Y.-X. Huo and S. Guo, *Microbial Cell Factories*, 2023, 22, 147 (review: improving the production of bio-based vanillin)
L. Xu, F. Liaqat, J. Sun, M. I. Khazi, R. Xie and D. Zhu, *Renewable and Sustainable Energy Reviews*, 2024, 189, 113905 (vanillin synthesis and biotransformation reviewed)
A. Perez-Silva, E. Pena-Mojica, A. Ortega-Galeana, J. I. Lopez-Cruz, C. A. Ledesma-Escobar, M. Rivera-Rivera and E. Paz-Gamboa, *Plants*, 2025, 14, 300 (*Vanilla cribbiana* Soto Arenas)
R. Hernandez-Peña, J. L. Lorenzo-Manzanarez, L. A. Cruz-Ramirez, D. Reyes-Lopez, C. Hernandez-Dominguez, F. Pascual-Ramirez and J. J. Ordaz-Ortiz, *J. Agric. Food Chem.*, 2025, 73, 19094–19106 (vanillin biosynthesis)

46. VASKA'S COMPOUND

M. Angoletta, *Gazz. Chim. Ital.*, 1959, 89, 2359–2370 (iridium carbonyl phosphine complexes)
L. Vaska and J. W. DiLuzio, *J. Am. Chem. Soc.*, 1961, 83, 2784–2785 (synthesis)
L. Vaska and J. W. DiLuzio, *J. Am. Chem. Soc.*, 1962, 84, 679–680 (H_2 adduct)
L. Vaska, *Science*, 1963, 140, 809–810 (synthesis of O_2 adduct)
S. J. Laplaca and J. A. Ibers, *J. Am. Chem. Soc.*, 1965, 87, 2581–2586 (structure of O_2 adduct)
L. Vaska, *Science*, 1966, 152, 769–771 (synthesis of CO adduct)
P. B. Chock and J. Halpern, *J. Am. Chem. Soc.*, 1966, 88, 3511–3514 (CH_3I adduct)
J. D. Collman and W. R. Roper, *J. Amer. Chem. Soc.*, 1966, 88, 3505–3508; U. Bellucco, U. Croatto, P. Uguagliati and R. Pietropaolo, *Inorg. Chem.*, 1967, 6, 718–721; H. C. Clark, P. W. R. Corfield, K. R. Dixon and J. A. Ibers, *J. Amer. Chem. Soc.*, 1967, 89, 1360–1361 (early references to 'Vaska's compound')
P. D. Brotherton, C. L. Raston, A. H. White and S. B. Wild, *J. Chem. Soc., Dalton Trans.*, 1976, 1799–1802 ($HgCl_2$ adduct)
W. H. Thompson and C. T. Sears, *Inorg. Chem.*, 1977, 16, 769–774 (CH_3I adduct mechanism)

F. DeMartin and N. Mascioccini, *Acta Cryst.*, 1983, C39, 1225–1226 (structure of $[IrCl_3(CO)(PPh_3)_2]$)
M. R. Churchill, J. C. Fettinger, L. A. Buttrey, M. D. Barkan and J. S. Thompson, *J. Organomet. Chem.*, 1988, 340, 257–266 (cryst. struct. of $[IrCl(CO)(PPh_3)_2]$)
A. J. Blake, E. A. V. Ebsworth, H. M. Murdoch and L. J. Yellowlees, *Acta Cryst.*, 1991, C47, 657–659 (cryst. struct. of $[IrCl(CO)(PPh_3)_2]$)
A. J. Blake, E. A. V. Ebsworth, S. G. D. Henderson, H. M. Murdoch and L. J. Yellowlees, *Z. Krist.*, 1992, 199, 290–292 (SO_2 adduct)
M. Rahim and K. J. Ahmed, *Inorg. Chem.*, 1994, 33, 3003–3004 (synthesis)
F. Abu Hasanayn, T. Emge, J. Maguire, K. Krogh-Jespersen and A. Goldman, *Organometallics*, 1994, 13, 5177 (structure of CO adduct)
J. P. H. Charmant, N. C. Norman, A. G. Orpen and G. R. Whittell, *Acta Cryst.*, 2004, E60, m162-m163 (structure of $[IrHCl_2(CO)(PPh_3)_2]$
H. Lebel, C. Ladjel, F. Bélanger-Gariépy and F. Schaper, *J. Organomet. Chem.*, 2008, 693, 2645–2648 (revised structure of O_2 adduct)
R. U. Kirss, *Bull. Hist. Chem.*, 2013, 38, 52–60 (review)
R. R. Contreras, E. C. Villalba and B. Fontal, *Novasinergia*, 2020, 3, 96–110 (review)
P. Friederich, G. dos Passos Gomes, R. De Bin, A. Aspuru-Guzik and D. Balcells, *Chem. Sci.*, 2020, 11, 4584–4601 (modelling H_2 activation by Vaska's compound)

47. VX AND THE EARLIER NERVE AGENTS

W. Lange and K. von Krueger, *Berichte der Deutschen Chemischen Gesellschaft*, 1932, 65, 1598–1601 (pioneering work)
B. C. Saunders, *Some Aspects of the Chemistry and Toxic Action of Organic Compounds Containing Phosphorus and Fluorine*, Cambridge, Cambridge University Press, 1957 (DFP)
R. Harris and J. Paxman, *A Higher Form of Killing: The Secret Story of Chemical and Biological Warfare*, New York, Hill and Wang, 1982, pp. 53–67, 138–139
D. E. C. Corbridge, Phosphorus: *An Outline of Its Chemistry, Biochemistry, and Technology*, Amsterdam, Elsevier, 5th edition, 1995, pp. 574–598
P. Zurer, *Chem. Eng. News*, 1998, 76(35), 7 ('Japanese cult used VX to slay member')
E. O'Connor, N. Fryer and T. P. Kee, *Education in Chemistry*, 1998, 123–125 (sarin and VX)
S. W. Wiener and R. S. Hoffman, *Journal of Intensive Care Medicine*, 2004, 19(1), 22–37 ('Nerve Agents: A Comprehensive Review')
J. Smythies and B. Golomb, *J. R. Soc. Med.*, 2004, 97, 32 (nerve gas antidotes)
J. B. Tucker, *War of Nerves; Chemical Warfare from World War I to Al-Qaeda*, New York, Pantheon Books, 2006
J. Burdon, in: R. W. Waring, G. B. Steventon and S. C. Mitchell (eds.), 'Nerve Gases', *Molecules of Death,* London, Imperial College Press, 2nd edition, 2007, pp. 209–231
C. D. Fleming, C. C. Edwards, S. D. Kirby, D. M. Maxwell, P. M. Potter, D. M. Cerasoli and M. R. Redinbo, *Biochemistry*, 2007, 46, 5063–5071 (hCE1 complexes with soman and tabun)
B. Sanson, F. Nachon, J.-P. Colletier, M.-T. Froment, L. Toker, H. M. Greenblatt, J. L. Sussman, Y. Ashani, P. Masson, I. Silman and M. Weik, *Journal of Medicinal Chemistry*, 2009, 52, 7593–7603 (structures of AChE with soman before and after aging)
A. C. Hemmert, T. C. Otto, M. Wierdl, C. C. Edwards, C. D. Fleming, M. MacDonald, J. R. Cashman, P. M. Potter, D. M. Cerasoli and M. R. Redinbo, *Molecular Pharmacology*, 2010, 77, 508–516 (hCE1 complexes of cyclosarin and sarin)
D. D. Haines and S. C. Fox, *Forensic Science Review*, 2014, 26, 98–114 (chemical weapons in the Iran-Iraq war)
H. John, M. J. van der Schans, M. Koller, H. E. T. Spruit, F. Worek, H. Thiermann and D. Noort, *Forensic Toxicol.*, 2018, 36, 61–71 (forensics of fatal sarin poisoning in Syria 2013)

E. J. Hulse, J. D. Haslam, S. R. Emmett and T. Woolley, *British Journal of Anaesthesia*, 2019, 123, 457–463 (treatment of organophosphorus nerve agent poisoning)
D. Kaszeta, *Toxic*, London, Hurst & Company, 2020 (nerve agents)
A. T. Tu, *Global Security: Health, Science and Policy*, 2020, 5(1), 48–56 (use of VX by Aum Shinrikyo in Japan and the death of Kim Jong-Nam in Malaysia)

48. WATER

F. Franks, *Polywater*, Cambridge, MA, MIT Press, 1981
F. Franks, *Water*, London, Royal Society of Chemistry, 1983
R. J. Gillespie and I. Hargittai, *The VSEPR Model of Molecular Geometry*, Boston, Allyn and Bacon, 1991, p. 89
N. N. Greenwood and A. Earnshaw, *Chemistry of the Elements*, Oxford, Butterworth-Heinemann, 2nd edition, 1997, pp. 620–632
P. Ball, *H_2O. A Biography of Water*, London, Weidenfeld & Nicolson, 1999
W. Gratzer, *The Undergrowth of Science*, Oxford, Oxford University Press, 2000, pp. 83–95, (polywater)
R. Rosenberg, *Physics Today*, December 2005, 58, pp. 50–55 (why ice is slippery)
R. Kumar, J. R. Schmidt and J. L. Skinner, *J. Chem. Phys.*, 2007, 126, 204107 (hydrogen bonds in water)
J. Grant, *Corrupted Science*, Wisley, Facts, Figures & Fun, 2007, pp. 92–94 (polywater)
R. Wright, *Take Me to the Source. In Search of Water*, London, Harvill Secker, 2008
G. Malenkov, *J. Phys. Condens. Matter*, 2009, 21, 283101 (review of properties of liquid water and ice)
S. Solomon, *Water*, New York, Harper Collins, 2010
B. Fagan, *Elixir: A History of Water and Humankind*, London, Bloomsbury Press, 2011
J. Salzman, *Drinking Water*, New York, Overlook Duckworth, 2012
A. Jha, *The Water Book*, London, Headline, 2015
C. J. Vörösmarty, A. Y. Hoekstra, S. E. Bunn, D. Conway and J. Gupta, *Science*, 2015, 349, 478–479; J. G. Hering, D. L. Sedlak, C. Tortajada, A. K. Biswas, C. Niwagaba and T. Breu, *Science*, 2015, 349, 479–480 (water management)
L. Piani, Y. Marrocchi, T. Rigaudier, L. G. Vacher, D. Thomassin and B. Marty, *Science*, 2020, 369, 1110–1113 (origin of Earth's water)

49. WILKINSON'S CATALYST

J. A. Osborn, G. Wilkinson and J. F. Young, *Chem. Commun.*, 1965, 17 ($RhCl_3(PPh_3)_3$)
J. F. Young, J. A. Osborn, F. H. Jardine and G. Wilkinson, *Chem. Comm.*, 1965, 131–132; J. A. Osborn, F. H. Jardine, J. F. Young and G. Wilkinson, *J. Chem. Soc. (A)*, 1966, 1711–1732; F. H. Jardine, J. A. Osborn and G. Wilkinson, *J. Chem. Soc. (A)*, 1967, 1574–1578 (synth. and catalysis)
M. A. Bennett and P. A. Longstaff, *Chem. Ind. (London)*, 1965, 846 (synth.)
R. S. Coffey (ICI), British Patent 1,121,642, 1965 (catalysis)
J. A. Osborn and G. Wilkinson, *Inorganic Syntheses*, 1967, 10, p. 67 (synth)
J. P. Candlin and A. R. Oldham, *Discuss. Faraday Soc.*, 1968, 46, 60–71 (hydrogenation rates with various substrates)
M. J. Bennett and P. B. Donaldson, *Inorg. Chem.*, 1977, 16, 655–660 (cryst. struct. of allotropes of $RhCl(PPh_3)_3$
J. C. Chabala, H. Mrozik, R. L. Tolman, P. Eskola, A. Lusi, L. H. Peterson, M. F. Woods and M. H. Fisher, *J. Med. Chem.*, 1980, 23, 1134–1136 (use in synth. of Ivermectin)
F. H. Jardine, *Prog. Inorg. Chem.*, 1981, 28, 63–202 (comprehensive review)

J. Halpern, *Inorg. Chim. Acta*, 1981, 50, 11–19 (mechanisms of catalytic hydrogenation)

J. M. Brown, P. L. Evans and A. R. Lucy, *J. Chem. Soc. Perkin Trans II*, 1987, 1589–1596 (hydrogenation intermediates)

S. B. Duckett, C. L. Newell and R. Eisenberg, *J. Am. Chem. Soc.*, 1994, 116, 10548–10556 (hydrogenation intermediates)

F. H. Jardine, *Rhodium Express*, 1996, 16, 4–8 ('The Contributions of Sir Geoffrey Wilkinson, F. R. S. (1921–1996) to Rhodium Chemistry')

S. A. Cotton, *Chemistry of Precious Metals*, London, Blackie, 1997, pp. 89–96, 106–107, 124–132 ($RhCl(PPh_3)_3$ and other rhodium phosphine complexes)

M. L. H. Green and W. P. Griffith, *Platinum Met. Rev.*, 1998, 42, 168–173 (Geoffrey Wilkinson and platinum metals chemistry)

50. ZINGERONE

W. J. Sanders and J. L. Seidel, *J. Agric. Food Chem.*, 1992, 40, 2, 263–265 (synth. of zingerone)

G. Charalambous, ed., *Spices, Herbs and Edible Fungi*, Amsterdam, Elsevier, 1994 (esp. pp. 579–694 for ginger)

A. Dalby, *Dangerous Tastes. The Story of Spices*, Berkeley, University of California Press, 2000, esp. pp. 21–26

K.-h. Tan and R. Nishida, *J. Chem. Ecol.*, 2000, 26, 533–546 (*Bulbophyllum patens* attracts male *Bactrocera* fruit flies using zingerone)

J. Turner, *Spice*, London, Harper Collins, 2005

T. J. Zachariah, 'Ginger', in: V. A. Parthasarathy, B. Chempakam and T. J. Zachariah (eds.), *Chemistry of Spices*, Wallingford, CAB International, 2008, pp. 70–96

B. Ahmad, M. U. Rehman, I. Amin, A. Arif, S. Rasool, S. A. Bhat, I. Afzal, I. Hussain, S. Bilal sand M. R. Mir, *The Scientific World Journal*, 2015, Article ID 816364 (review of pharmacological properties of zingerone)

Y.-S. Kim, C. Hong, S. W. Lee, J. H. Nam and B. J. Kim, *International Journal of Molecular Medicine*, 2016, 38, 1905–1914 (effects of ginger and its constituents on TRP channels)

B. A. Mir et al., *Pharmacognosy Magazine*, 2018, 14, S434–439 (zingerone counteracts side-effects of cyclophosphamide-induced hepatic toxicity in rats)

B. L. Hanssen, S. J. Park, J. E. Royer, J. F. Jamie, P. W. Taylor and I. M. Jamie, *Scientific Reports*, 2019, 9, 19332 (zingerone as a lure for Jarvis's fruit fly, *Bactrocera Jarvisi*)

Y. Yin, Y. Dong, S. Vu, F. Yang, V. Yarov-Yarovoy, Y. Tian and J. Zheng, *Br. J. Pharmacol.*, 2019, 176, 3364–3377 (activation of TRPV1 channels by pungent compounds in ginger)

A. F. Wali, M. U. Rehman, M. Raish, M. Kazi, P. G. M. Rao, O. Alnemer, P. Ahmad and A. Ahmad, *Molecules*, 2020, 25, 5127 (zingerone protects rats from sepsis associated multi-organ damage)

Y. Zhu, C. Wang, J. Luo, S. Hua, D. Li, L. Peng, H. Liu and L. Song, *Food Funct.*, 2021, 12, 3120–3131 (zingerone has potential to relieve asthma)

M. K. Samota, M. Rawat, M. Kaur and D. Garg, *Sustainable Food Technol.*, 2024, 2, 1652–1669 (gingerol and health)

B. De Vos, A. E. Kasonga, A. M. Joubert and T. T. Nyakudya, *Metabolites*, 2024, 14, 693 (zingerone and bone health)

Y. Wang, L. Dong, S. Han, Y. You, M. Zhang, B. Sun, H. Ni, R. Ge, J. Liu and J. Yu, *Front. Pharmacol.*, 2025, 16, 1543594 (zingerone alleviates inflammatory pain in mice)

Index

D

G

H

I

N

Q

R